APPLIED BUSINESS STATISTICS

Paul Marston
Senior Lecturer in Quantitative Methods, Preston Polytechnic

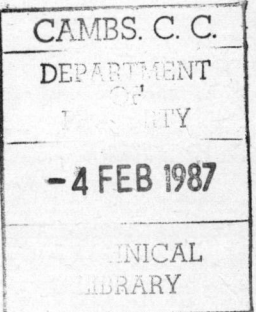

HOLT, RINEHART AND WINSTON
London·New York·Sydney·Toronto

Holt, Rinehart and Winston Ltd: 1 St Anne's Road,
Eastbourne, East Sussex BN21 3UN

British Library Cataloguing in Publication Data

Marston, Paul
   Applied business statistics.
   1. Business mathematics   2. Mathematical
   statistics
   I. Title
   519.5'024658   HF5691

ISBN 0 03 910366 8

Typesetting and artwork by Oxford Print Associates

Printed in Great Britain by Mackays of Chatham Ltd, Chatham, Kent

Copyright © 1982 Holt, Rinehart and Winston Ltd. *All rights reserved.* It is illegal to reproduce any part of this publication except by special permission of the copyright owners. Reproduction of this material by any duplication process whatsoever without authorization is a violation of copyright.

Last digit is print number: 9 8 7 6 5 4 3 2 1

Contents

PREFACE　　　　　　　　　　　　　　　　　　　　　　　vii

SYMBOLS USED　　　　　　　　　　　　　　　　　　　　　ix

THE NATURE AND ROLE OF STATISTICS

1 Statistics and Statistical Method　　　　　　　　　　　1

　1.1 *The purpose of statistics* 1; 1.2 *The tasks of statistical method* 3

2 Handling Numbers　　　　　　　　　　　　　　　　　4

　2.1 *Tools for statistics* 4; 2.2 *Kinds of numbers* 6; 2.3 *Rounding and truncating* 7; 2.4 *Accuracy and errors* 9

SOURCES OF DATA

3 Secondary Data　　　　　　　　　　　　　　　　　　18

　3.1 *The nature and use of secondary data* 18; 3.2 *Sources of published statistics* 20

4 Data Collection and Surveys　　　　　　　　　　　　27

　4.1 *Basic tasks and aims* 27; 4.2 *Kinds of sampling* 30; 4.3 *Sources of error and bias* 35; 4.4 *Conducting a survey* 43

PRESENTING DATA

5 Tabulation 47

5.1 The need for and nature of tabulation 47; 5.2 Guidelines for tabulation 49; 5.3 Time series 52; 5.4 Measurable variables – ungrouped frequencies 54; 5.5 Measurable variables – grouped frequencies 55

6 Charts and Diagrams 64

6.1 Purpose and principles of diagrammatic representation 64; 6.2 Pictograms 66; 6.3 Bar charts 68; 6.4 Pie charts 71

7 Simple Graphs 77

7.1 Coordinate axes and scattergraphs 77; 7.2 Simple time graphs 80; 7.3 Band curve charts 83; 7.4 Logarithmic graphs 84; 7.5 Z-charts 90

8 Diagrams for Grouped Frequencies 94

8.1 The histogram 94; 8.2 Frequency polygons 103; 8.3 Cumulative frequency curves 104; 8.4 Lorenz curves 106

SUMMARY FIGURES

9 Measures of Average 111

9.1 The arithmetic mean 111; 9.2 The median 116; 9.3 The mode 120; *9.4 Other averages 123; 9.5 Distribution shapes and averages 124; 9.6 Choosing the right 'average' 127; 9.7 Summary 130

10 Measures of Dispersion 134

10.1 The concept of dispersion 134; 10.2 The range 134; *10.3 The mean deviation 135; 10.4 Variance and standard deviation 136; *10.5 Alternative formulae and large numbers 141; 10.6 Dispersion and quartiles 144; 10.7 Summary 145

11 Index Numbers 148

11.1 Price indexes 148; 11.2 Linking and chaining 157; 11.3 Practical price indexes 160; 11.4 Quantity indexes 165;

11.5 *Practical quantity indexes* 169; 11.6 *Index number references* 170; 11.7 *Summary* 172

PROBABILITY AND DISTRIBUTIONS

12 Elementary Probability — 178

12.1 *The importance and meaning of probability* 178; 12.2 *The basic features of probability* 180; *12.3 *The addition law* 183; *12.4 *The multiplication law* 185; *12.5 *Using the laws* 188; *12.6 *Probability distributions and expectancy* 188; 12.7 *Summary* 190

*13 Binomial and Poisson — 197

13.1 *The binomial distribution* 197; 13.2 *The Poisson distribution* 203; 13.3 *Summary* 204

14 The Normal Distribution — 208

14.1 *Continuous probability distributions* 208; 14.2 *The normal distribution* 212; 14.3 *Using normal tables* 220; 14.4 *Summary* 227

15 The Distribution of the Sampling Mean — 232

15.1 *The concept of the sampling mean distribution* 232; 15.2 *The central limit theorem* 236; 15.3 *The standard error of the mean* 236; 15.4 *Applications* 238; 15.5 *Summary* 241

SIGNIFICANCE AND CONFIDENCE

16 Significance Testing — 244

16.1 *The significance test concept* 244; 16.2 *Formalizing testing* 248; 16.3 *Levels, errors and conclusions* 251; 16.4 *One tailed and two tailed tests* 253; 16.5 *Summary* 257

17 Estimation — 262

17.1 *Bias, error and precision* 262; 17.2 *Confidence intervals* 265; 17.3 *Summary* 268

vi–*Applied business statistics*

*18   Further Significance Testing   272

18.1 *Basic elements of significance tests* 272; 18.2 *Tests of population means – unknown standard deviation* 273; 18.3 *Difference between two means* 276; 18.4 *Tests on proportions* 277; 18.5 *Chi-square contingency tests* 280

REGRESSION AND CORRELATION

19   Regression   285

19.1 *The need for regression* 285; *19.2 *A simple three-point method* 287; 19.3 *Least-squares regression* 289; 19.4 *Summary* 295

20   Correlation and Causality   297

20.1 *The correlation coefficient* 297; 20.2 *Significance tests on r* 301; 20.3 *Linearity and assumptions* 303; 20.4 *Correlation and causation* 305; *20.5 *Rank correlation* 306; 20.6 *Summary* 307

TIME SERIES

21   Time Series   313

21.1 *Time series and basic models* 313; 21.2 *Moving averages* 315; 21.3 *Forecasting* 320; 21.4 *De-seasonalizing* 321; 21.5 *Quarterly figures* 321; 21.6 *Summary* 324

FINANCIAL MATHEMATICS

22   Series and Finance   329

22.1 *Basic concepts* 329; 22.2 *Financial agreements* 334; 22.3 *Summary* 336

23   Investment Appraisal   341

23.1 *Business objectives in agreements* 341; 23.2 *General business objectives* 345

Appendix A   Statistical Tables   349

Appendix B   List of Formulae   364

Index   371

Preface

The traditional statistical calculation skills are taught in this book, but its priority is to communicate the concepts behind them. To this end unnecessary statistical jargon is avoided (or put in brackets as an optional extra), thus, where practicable, saving the reader from having to learn a new 'language' as he or she masters statistical ideas.

It is assumed that all students will have the use of an electronic calculator with at least the four basic functions and a square root key. The view is taken that, on the whole, time can be better spent than in teaching the 'short cut' methods which were useful before calculators were generally available, and this aspect is minimized. As calculators inevitably become more powerful, the sections on the use of 'assumed means' and so on may be omitted entirely.

At the end of each chapter there are two kinds of exercises. The first, the 'Recall and Exercise Questions', should be answerable by anyone who has carefully read the text and followed the methods taught. The second, the 'Problems', involve either data searches through published sources, or a much deeper level of analysis and usually presentation in the form of a report. Some of these 'Problems' may take students some time to do (depending on level and ability), and may best be done with help and supervision where appropriate. In the 'Problems' there may also be elements of earlier work. The aim of the 'Problems' is to provide a vehicle to help students to integrate their newly learned skills in the context of a problem. In all examples, exercises and problems, realistic data and situations have been striven for. There are too many books supposedly about 'business statistics' which seem to be mainly concerned with billiard balls, examination results, and mysterious variables X and Y which appear to have no units or dimensions. In this book published data or data based on privately obtained figures is generally used both to develop ideas and to set examples.

In courses in which examinations are set externally and call for little beyond calculation skills, the 'Problems' could be omitted from the course. But it is the author's hope that, especially as we see the spread of minicomputers and their software, more courses will move away from a purely skill-based to a problem-orientated approach.

The material in this book has been developed over more than ten years of teaching in technical college and polytechnic, on a variety of courses with different levels and abilities of students. I have designed syllabuses, taught, examined and moderated on BEC National and Higher courses, Professional Accounting body levels 1 and 2, IM, RVA, DMS, HNC/D BEC/TEC Computer Studies, and Business and Accounting degrees. The Book can be used, then, in a variety of contexts and levels, for although the foundation material is common, the text contains a built-in option on depth of treatment: starred sections and more difficult problems can be omitted without losing continuity. Thus a BEC National course, for example, would omit most starred sections, while for a higher-level course the book in its entirety might be used on a wider quantitative syllabus.

The primary function of the book is likely to be as a course text, with students receiving guidance and assistance from teaching staff, and discussion on some of the problems. Nevertheless, the text endeavours to be self-explanatory, and the book could be recommended by staff who prefer to teach from their own notes but find a 'back-up text' useful for students.

Paul Marston

Symbols used

The following symbols will be assumed to be known.

$+ \; - \; \div \; \times$	plus, minus, divide, multiply
$\leqslant$	less than or equal to (e.g. $4 \leqslant 6$)
$\geqslant$	greater than or equal to (e.g. $7 \geqslant 7$)
$<$	less than (e.g. $34 < 78$)
$>$	greater than (e.g. $67 > 21$)
x^2	x squared $= x \times x$ (e.g. $6^2 = 6 \times 6 = 36$)
x^3	x cubed $= x \times x \times x$ (e.g. $5^3 = 5 \times 5 \times 5 = 125$)
x^n	x raised to the power n (x multiplied by itself n times).
	Note: This must not be confused with x_n which means the nth observation of the value of x in a set of such values.
%	per cent from a Latin phrase meaning 'as a proportion of 100' (e.g. one quarter or 0.25 is 25%)
$\sqrt{x}$	square root: the number which when multiplied by itself gives x (e.g. $\sqrt{81} = 9$)

Some other symbols will be introduced in the text, and these are listed in Appendix B. Appendix B also contains a list of formulae, including some forms in use in other circles but not used in this book.

1
Statistics and Statistical Method

1.1 The purpose of statistics

The word 'statistics' can simply mean a set of numbers. Statisticians, however, often refer to such numbers as 'data', and by the word 'statistics' may mean the *methodology* developed to gather, handle and analyse those numbers.

Sometimes data are collected and analysed simply out of curiosity, but this is very unusual. More usually the intention is that, ultimately at least, the numerical information should form a basis for decision and action. The purpose of a study of statistical method could then be stated thus: 'Statistical methodology allows the effective collection, analysis, and presentation of numerical information, on the basis of which optimal decisions can be made.' It will be useful always to keep this basic purpose in mind, for it is surely true that occasionally people become engrossed in producing useless statistics, or statistics in unnecessary detail, just for the sake of it.

It might be useful here to mention two opposite but equally misguided approaches to statistics. The first is a naive credulity that anything with numbers quoted in it 'must be right', especially if produced by a computer! This is foolish, for statistics can certainly be used to mislead, for instance:

(a) by presenting analysis of opinion based on a questionnaire in which questions are worded to 'lead' the person being interviewed into giving particular responses (see Chapter 4);

(b) by fiddling the scales on diagrams (see section 7.2); or

(c) by ascribing changes to one particular factor when in reality they are a product of many factors.

The reader may be able to think of other ways in which statistics can be used to mislead.

This has sometimes led others into the opposite approach – an equally misguided refusal to accept *any* kind of proof based on numbers, since 'you can prove anything with statistics'. While misuse of statistics may fool some people, a person with any knowledge of statistical method will not be fooled. He or she will be able to spot the misuse, either pointing to the correct conclusion or stating where the information provided is inadequate to reach any proper conclusion. This is why some knowledge of statistical method is so very useful.

Computers. Vast amounts of numerical data are now stored on computers, and computers are very good at doing statistical analysis quickly. Does this mean, then, that we can leave all decisions to computers? Not really: *people* make decisions, while computers do as they are told! When a computer produces an analysis, or issues an instruction, it is because a person has programmed it to do so. People still have to decide what analysis is appropriate, interpret the results, decide on criteria for decision making, and so on. They also have to monitor computer systems and decide whether modifications are needed. In the foreseeable future this is likely to necessitate some knowledge of statistical method in order for a meaningful dialogue with a computer to take place. At present many tasks which could conceivably be computerized are still being done by manpower.

Areas of application. There are so many diverse areas in which statistical understanding is essential that only a few can be mentioned here:

1. For government and public administration it is useful to know about numbers and types of businesses, numbers and types of households, income distributions, costs, prices, wages and many other numerical facts about society.
2. Companies make constant use of statistics in monitoring production, quality, stock levels, and so on.
3. Marketing and market research sections need to know trends of sales and what kinds and numbers of products are likely to be sold.
4. Accountants use statistical methodology in areas like budgetary control and auditing based on samples.
5. Banks and other financial institutions need to understand financial trends, as well as using statistics in their day-to-day business.

1.2 The tasks of statistical method

The basic purpose of statistics has been noted. But it may be useful to note that there are several distinguishable aspects or tasks relating to this which may form part of any particular statistical technique. Of the five categories listed below, this book describes tasks which fall into all except the last.

Data collection. As we shall see, data may come from various sources internal or external to an organization, and there are techniques for collecting it correctly in the various circumstances.

Presentation. Some statistical techniques, particularly the clear design of tables and diagrams, are intended simply to present a particular numerical pattern clearly to the mind. A mass of information may be available, but it is also necessary for a decision-maker to 'see' the overall situation clearly.

Summary. Another way in which someone can be helped to 'see' an overall situation is for a set of figures to be summarized into a single figure, for example an average or a total. Statistical method helps us to decide what 'summary figures' are appropriate and how to calculate them.

Analysis of causes. As well as summarizing, we may wish to calculate figures which will tell us whether it is likely that a particular factor *caused* some observed effect. Statistical techniques like correlation and significance testing help us to do this.

Optimization. Finally, usually after some or all of the other functions have been performed, we may wish to build a 'decision model'. This will involve a number of decision criteria: what we are trying to achieve, maximize (for example, profit) or minimize (for example, costs). Any human decisions, of course, involve value judgements, even if these only involve the acceptance of criteria laid down by others. Neither a computer, nor statistical methodology, nor anything else other than a human being can make these value judgements. What the methodology can do, or try to do, is to offer a course of action which promises to offer the best chance of the objectives given to it being achieved.

2
Handling Numbers

2.1 Tools for statistics

Because arithmetic is tedious and slips are easy to make, people have long sought tools to make it easier. Around the mid-1960s by far the most common tools were the 'four-figure tables' (including logarithms and square roots) and the sliderule. Computers had become fast and reliable, but were too expensive for everyday arithmetic. Desk-top calculators (manual or electric) were available, but were bulky and inefficient. So the norm for private use and for examinations was four-figure tables or a sliderule.

Since that time microchip technology has brought a revolutionary change. The electronic pocket calculator has become increasingly powerful in the things it can be programmed to do, and has become cheaper and cheaper. A basic small calculator now costs less than many scientific textbooks. Virtually all statistics courses (degree, DMS, BEC, professional-body, etc.) now allow such calculators to be used in examinations. A calculator is now, in short, a part of the basic equipment for anyone wishing to study statistics, and the assumption made in this book is that readers will have the use of a calculator with at least the essential features. Both sliderules and four-figure tables are obsolete and situations in which they rather than calculators are in use will become increasingly rare. Time spent in mastering them would be better spent in improving understanding of statistical concepts, so neither logarithmic tables nor sliderules are referred to in this textbook. *Statistical* tables are, however, included, for at the time of writing they are still in common use.

Calculator features. As technology stands at the time of writing, the following is a guide as to what is likely to be found on calculators:

1. Essential features: add (+), subtract (−), divide (÷), multiply (×), square root (√).

2. Useful general features: internal store (M, M+, RM), change sign (+/−), reciprocal (1/x), natural logs and e^x *or* logs and antilogs, powers (y^x), roots ($x\sqrt{y}$).
3. Useful statistical functions: means ($\bar{x}$), standard deviation (s or σ), correlation coefficient (r), regression, normal distribution, binomial distribution, Poisson distribution, random number, chi-squared (χ^2), t-distribution, F-distribution.
4. Scientific functions not useful for statistics: sin and $\sin^{-1}$, cos and $\cos^{-1}$, tan and $\tan^{-1}$, π (pi).

Look out for these features if you are shopping around for a calculator. And one other word of warning: some machines (for reasons best known to their designers) are programmed to do things like *automatically* adding the result of any calculation to the internal store. This is a nuisance in statistics, so do not buy one. There is also a difference between an LED and an LCD machine. The latter is preferable, since the batteries will last very much longer.

When you have bought or borrowed a machine, familiarize yourself with it by doing a lot of sums for which you know the answers. This textbook will assume that readers are familiar with addition, subtraction, multiplication, division, decimals, positive and negative numbers, the meaning of a square and square root, and the meaning of percentages.

Accuracy. Many current calculators can display only eight digits. This means that if a recurring or long decimal or a very large number results from a calculation, then it cannot be shown or stored completely on most calculators. The implications of this are considered in section 2.4.

Calculators and the future. At the time of writing there are statistical pocket calculators available (for less than the price of most portable typewriters) which will do nearly all the calculation procedures in this book automatically: one just enters the numbers and presses the right buttons. Others, more expensive, are actually fully programmable pocket computers, into which a recorded program can be input. In view of this, the question has to be asked whether, if one's primary object is to understand and be able to use statistics, the actual calculations still need to be mastered at all. I would answer 'Yes' for three reasons. First, while a mere rote-learning of calculation procedures would be unhelpful to the understanding, this book does not adopt a rote-learning approach. Rather, it tries wherever possible to link calculations with the concepts behind them. Thus calculation is a valuable aid to understanding the concepts themselves. Secondly, at present it is reasonable to assume universal availability of basic calculators, but not of those with greater statistical facility. Thirdly, the average calculator may be unable to handle very large numbers without suitable adjustments to procedure.

The first of these reasons seems to me to be likely always to remain with us. The other two may well not. It must, moreover, be added that at present there is a very rapid development in desk-top computers. These are not much dearer than advanced calculators, but have greater storage and computing facilities, and can handle numbers of any size likely to be met in practice. As these and the advanced calculators become more common, we may well need to make further reconsideration of the most appropriate approach to the study of statistical methods.

2.2 Kinds of numbers

Statistics obviously concerns numbers, but it will be useful to think a little further about different kinds of numbers, and to introduce the various relevant technical terms.

Measurements are made of particular kinds of characteristics (measurable characteristics!) of individuals or groups, for example: height (in centimetres); income (in pounds); age (in years); turnover (in thousands of pounds).

Attributes, in contrast, are not measurable, but are sets of characteristics which groups or individuals either do or do not possess, for example: marital status (single, married, divorced, separated, widowed, etc.), sex (male, female), social class (AB, C1, C2, DE).

Counts are the number in a group or population which possess a particular attribute, or are of a particular size. Such counts are often called *frequencies*, for example: in an office of 57 people, the frequency of married people might be 25, of unmarried 30, and of legally separated 2.

Used in this technical sense, the word 'frequency' does not mean 'how often', but is just a count, of whatever size. Thus in our example the number 2 is the 'frequency' of legally separated persons in the group, although they are few.

Variable is a term used for a measurement which varies from one individual or observation to another. In some situations the measurement could be a kind of frequency such as the number of unemployed, which varies between regions. We are usually, in such instances, dealing with non-measurable categories (such as 'unemployed'). Where we are dealing with measurable variables such as heights then the 'size of variable' will usually be distinct from the 'frequency' with which that particular size occurs. I shall refer to an individual variable as an 'observation' or a 'value' of the variable, and shall refer to a 'set of observations' or a 'set of values'. These last two phrases mean the same – and 'set of values' in this context has nothing to do with value judgements. Examples of 'values' are numbers unemployed, heights, numbers of people in households, weights, lengths, household incomes, and so on.

Discrete and continuous

There are two basic kinds of variable, discrete and continuous, defined according to the potential values which the variable may take.

For some variables, the possible values occur at set intervals, with a gap between any given possible value and the next one below or above it. Thus, for example, the number of cars in a car park might be 120 or 121, but could not be, say, 120.6 cars. The variable 'number of cars' has a potential value of 120, but the next potential value above this comes at 121, with a gap of one unit between them into which its value cannot fall. Variables with gaps like this between the values it is possible for them to take are called *discrete* variables.

Other variables have no gaps between one potential value and the next. Thus for example a steel rod might be exactly 30.5 cm in length. There is no gap between 30.5 and the next potential value above it – a value could be infinitesimally greater. Thus between 30.5 cm and 31 cm there are a very large number (technically an infinite number) of potential values the variable could take. Such a variable is called *continuous*.

We shall return to look in more detail at this distinction between discrete and continuous, and at its relevance, in section 5.5.

2.3 Rounding and truncating

Suppose that in an agricultural trial the wheat grown on a particular acre sprayed with 'losmothrin' is carefully harvested and weighed. The total weight comes to 2104.538 grams.

It would really be quite unnecessary to give the figure this accurately. Suppose, then, that we wished to give the figure to the nearest gram. There are 2104 complete grams, and the remainder is 538/1000 of a gram. Since this is *greater than half*, it is nearer to one gram than to zero grams, so we take $2104 + 1 = 2105$ grams as the figure to the nearest gram.

In general, we first decide on the smallest units we wish to record (grams, tenths of a gram, tens of grams etc. – in this instance it is grams). We take the complete number of these units (in this case 2104). We then see if the remainder is greater than half a unit (in this case greater than half a gram). If it is, then we add one on to the total of complete units. This is called *rounding*.

It is also possible to ask for figures to be rounded to a certain number of *significant figures*. A significant figure is any digit from 1 to 9 inclusive, or a 0 if it occurs between two non-zero digits. The 'most significant' figures are the digits to the left hand end of any number, that is those signifying the largest units. Here, 2105 is 2104.538 to FOUR significant figures.

Lastly, if digits after the decimal point are to be retained, a figure may be said to be rounded to a certain number of *decimal places*: for example, rounded to two decimal places, or 'correct to two decimal places'. If as here we round to the nearest gram, this is not applicable.

Further examples

1. 2104.54 is 2104.538 rounded: 'to the nearest one hundredth of a gram'
   'to two decimal places'
   'to six significant figures'

2. 2104.5 is 2104.538 rounded: 'to the nearest one tenth of a gram'
   'to one decimal place'
   'to five significant figures'

3. 2105 is 2104.538 rounded: 'to the nearest gram'
   'to whole numbers'
   'to four significant figures'

4. The third digit of 2104.538 is a zero, so this means that we have a special case and

   2100 is 2104.538 rounded: 'to the nearest ten grams' *or* 'to the nearest 100 grams'
   'to three significant figures' *or* 'to two significant figures'

5. 2000 is 2104.538 rounded: 'to the nearest thousand grams' (to the nearest kilogram)
   'to one significant figure'

Truncating. Some calculators truncate rather than round. This means that the least significant decimal places are simply chopped off. Thus:

2104.53 is 2104.538 truncated to two decimal places
2104.5 is 2104.538 truncated to one decimal place
2104 is 2104.538 truncated to whole numbers

Needless to say, if we are consciously and purposely cutting down the number of decimal places given in a number, we would prefer to round rather than to truncate as, on average, rounding gives a figure nearer to the actual one.

2.4 Accuracy and errors

In statistics, the word 'error' does not necessarily mean a mistake. In a situation where we have a 'stated value' which differs from the actual value then we define:

$$\text{Absolute error} = \text{Stated value} - \text{Actual value}$$

$$\text{Relative error} = \frac{\text{Absolute error}}{\text{Actual value}} \times 100\%$$

This means that the absolute error is in the same units (grams, acres, years, etc.) as the original, while the relative error is a pure percentage without units.

Example

If an actual figure of 2104.538 is rounded to the nearest gram:

$$\text{Absolute error} = (2105 - 2104.538) = +0.462 \text{ grams}$$

$$\text{Relative error} = \frac{+0.462}{2104.538} = 0.02195\%$$

Errors (in this technical sense) can arise in a number of ways:

1. *Mistakes* such as transcription errors. These can be eliminated with care and with the use of checking procedures, and are comparatively unimportant if handled correctly.

2. *Incomplete returns.* A failure to obtain results on all members of a group (for whatever reason) can cause errors: for instance differences between actual averages and calculated averages.

10–*Applied business statistics*

3. *Wrong survey procedure.* Errors due to use of 'leading questions', to wrong sample selection, and so on, are considered in Chapter 4.
4. *Measurement errors.* Limitations in measuring instruments (clocks, callipers, scales, etc.) can cause these. Usually this is really a form of 'rounding' (see below).
5. *Calculator rounding or truncating.* These errors occur automatically in the course of calculations (see below).
6. *Purposeful rounding.* This is deliberate - a decision to round to a certain number of figures.

Laws of errors

When we add, subtract, multiply or divide numbers containing errors there are laws which give the errors in the calculated results in terms of the errors in the original numbers. These laws will now be stated and illustrated, although learning them by rote would be rather a waste of time. The real point of stating them is so that the reader may work through examples and become aware of their implications.

Briefly, there are four laws; if we have two numbers A and B which contain errors:

1. *Addition* (this law holds exactly)
   Absolute error $(A + B)$ = Absolute error (A) + Absolute error (B)
2. *Subtraction* (this law holds exactly)
   Absolute error $(A - B)$ = Absolute error (A) − Absolute error (B)
3. *Multiplication* (this law is approximate)
   Relative error $(A \times B)$ = Relative error (A) + Relative error (B)
4. *Division* (this law is approximate)
   Relative error $(A \div B)$ = Relative error (A) − Relative error (B)

Example

Suppose that field X (an acre treated with losmothrin) yields 2104.538 grams of wheat and field Y (an untreated acre) yields only 1863.876 grams. A farmer with 236.73 acres very similar to those used in the trials would like to know the relationship between yields on treated and untreated acres, and how much he could expect if he treats all his land. Suppose that yields are rounded to the nearest hundred grams, and the farmer's acreage is rounded to the nearest ten acres:

Yield for field X: Absolute error = 2100 − 2104.538 = −4.538 g

$$\text{Relative error} = \frac{-4.538}{2104.538} \times 100 = -0.2156\%$$

Yield for field Y: Absolute error = 1900−1863.876 = +36.124 g

Relative error = $\dfrac{+36.124}{1863.876} \times 100 = 1.938\%$

Acreage: Absolute error = 240−236.73 = 3.27 acres

Relative error = $\dfrac{+3.27}{236.73} \times 100 = 1.381\%$

The laws may be illustrated thus (the reader may wish to check in each case):

1. Addition: The sum of yields from field X and field Y:

   Actual yield: 2104.538+1863.876 = 3968.414 grams
   Stated yield: 2100+1900 = 4000 grams
   (Absolute error: (−4.538)+(+36.124) = 31.586 grams

2. Subtraction: The difference between the yields from field X and field Y:

   Actual difference: 2104.538−1863.876 = 240.662 grams
   Stated difference: 2100−1900 = 200 grams
   (Absolute) error: (−4.538)−(+36.124) = −40.662 grams

3. Multiplication: The expected yield on the whole acreage if all the land is sprayed will be:

   Actual total: 2104.538×236.73 = 498 207.2807 grams
   Stated total: 2100×240 = 504 000 grams
   (Relative) error: (−0.2156%)+(+1.381%) = 1.1654% (say 1.17%
   since this is approximate)

4. Division. The ratio of the treated to untreated yields is:

   Actual ratio: $\dfrac{2104.538}{1863.876} = 1.129\ 119$

   Stated ratio: $\dfrac{2100}{1900} = 1.105\ 263$

   (Relative) error: (−0.2156%)−(+1.938%) = −2.1536% (say −2.15%
   since this is approximate)

It will be noted that in these various laws the *signs* of the errors are important. For example in the addition example shown, one number has a positive error (an overestimate of the actual), and the other a negative error (an underestimate). Add the two numbers together and the errors tend to *compensate* for each other.

In general we call errors *compensating* if they tend to cancel out and *cumulative* if they add up to an overall error greater than either individually. This concept can be especially important in situations where we are doing a lot of calculations or summing a number of figures: Table 2.1 shows the effects of rounding and truncating yield figures for six different acres which are being added.

Table 2.1 *Effects of rounding and truncating yields*

	Accurate yield figure	Rounded to nearest gram	Truncated to nearest gram
Field 1	2 104.538	2 105	2 104
Field 2	2 212.739	2 213	2 212
Field 3	2 119.392	2 119	2 119
Field 4	2 196.219	2 196	2 196
Field 5	2 158.971	2 159	2 158
Field 6	2 129.853	2 130	2 129
Total	12 921.712	12 922	12 918
Error	0	0.288	−3.71

Rounding means that sometimes the stated value is an overestimate and sometimes an underestimate; that is, sometimes the error is positive and sometimes negative. When the figures are added, therefore, the positive errors tend to cancel out the negative ones, and the error in the total is quite small. The *truncated* figures, however, always underestimate (the error is always negative in sign). When the truncated figures are added, therefore, the effect is cumulative – and the overall error is much higher.

Procedures which have a definite tendency to push errors in a particular direction are said to cause *bias*. In this instance the error has been introduced by a calculation procedure, but biased error can also arise from features of the method used to collect data. We will return to this important subject of bias again, seeing, for instance, that the use of 'leading questions' in questionnaires can cause bias by consistently influencing answers in a particular direction (see Chapter 4).

Accuracy

Some kinds of error, like 'leading questions' in questionnaires, are always undesirable, and it is both possible and inexpensive to eliminate them. But in general total accuracy may be impossible or, given the expense of achieving it, undesirable, so in deciding on the level of accuracy to expect or aim for we need to ask:

1. For what purpose is the information likely to be used?
2. What degrees of accuracy are possible at all in practice?
3. What would these different degrees of accuracy cost in terms of time and money?

For example, survey organizations like Gallup recognize that greater accuracy could usually be achieved using random samples and elaborate checks than using the more common quota samples (for meanings of this see Chapter 4). However, the improvement would be so marginal in terms of the use to which the figures are put that it is not worth while.

A second question concerns the amount of rounding to do during processing or presentation of the figures. The following are guidelines:

1. Avoid misleading *spurious accuracy*. For example, in the wheat example introduced above it is plainly impossible to ensure that every grain of wheat is harvested, so to give a figure accurate to several decimal places would be misleading, even if the scales used were that accurate. If the errors arising in ways already mentioned restrict the figure's accuracy to, say, the nearest gram, then it is better to round the stated figure so that this degree of accuracy is clearly recognized (so round to the nearest gram).
2. When rounding figures before or during calculations, be careful that the effect is not cumulative, or that what appears to be the introduction of a small relative error does not cause a much larger one in a later calculation made on the data. This can happen in the measurement of dispersion for example (see Chaper 10).
3. When rounding final figures for presentation, ask what the purpose is of the information presented. What degree of accuracy does this purpose require? Rounding (for example, to the nearest million) may help someone to get a clearer mental picture of the situation as long as a more accurate figure is not required, perhaps for detailed planning.

Calculator accuracy. In section 2.1 we noted briefly that many calculators cannot display or deal with numbers with more than eight digits. We need now to consider the implications of this. Let us consider two calculations:

$$45.125 \times 63.257 = 2854.472\ 125$$
$$45\ 125 \times 63\ 257 = 2\ 854\ 472\ 125$$

14—Applied business statistics

The first of these results in a number with ten digits. Calculators which display only eight digits will display and store this as 2854.4721, either rounding or truncating depending on the calculator. The least significant decimal places will be dropped.

In the second calculation the result is also a ten digit number. A calculator limited to eight digits may do any of several things:
1. Print or display rubbish - numbers unconnected with the calculation.
2. Display E or ERROR, and not do the calculation.
3. Display 28545 05, 2.8545 09 or 2854472 03.

The first two are self-explanatory. The third shows the number in a form using an *exponent* (the number 05, 03, or 09 at the right). This exponent tells us to move the decimal point so many places to the right if positive as here, or to the left if negative. Thus 28545 05 and 2.8545 09 mean 2854500000, and 2854472 03 means 2854472000. The answers, then, are calculated, but are rounded or truncated to five or to seven significant digits.

Usually the pocket calculator is adequate to manipulate numbers of the size met with in practice, especially if suitable calculating procedures are adopted. Care must be taken with large numbers, however, for in some calculations a loss of accuracy due to rounding done automatically by the machine would be crucial.

Limits of accuracy. In some instances we do not know the original or true and accurate figures, but we do know the interval within which they must lie. For example, suppose we are given a figure for wheat yield of 2105 grams. If we know that this is a figure rounded to the nearest gram from an original accurate figure, we know that this original must lie within the range 2104.5 to 2105.5 for it to have been rounded to 2105. This range is sometimes written

$$2105 \pm 0.5 \quad \text{or} \quad 2105 \pm 0.023\ 75\%$$

where the symbol ± (pronounced 'plus or minus') indicates a range from minus the figure following it to plus this figure.

Sometimes it is assumed that the degree of accuracy can be told from the figure stated with no further information. Thus, for example, it is presumed that 5100 has been rounded to two significant figures and really means 5100 ± 50 (i.e. 5050 to 5150). If we have a whole set of figures then this may be valid, but if we have a single figure it could be misleading, since 5100 could be accurate to the nearest one unit.

Suppose that we *add* two values for which we have such limits, for example:

$$(2105 \pm 0.5) + (1863 \pm 0.5)$$

If both are really + 0.5 the upper limit of the result will be:

$$(2105 + 0.5) + (1863 + 0.5) = 3968 + (0.5 + 0.5)$$

If both are really −0.5 the lower limit of the result will be:
$$(2105 - 0.5) + (1863 - 0.5) = 3968 - (0.5 + 0.5)$$
Thus the range within which the sum must lie is:
$$(2105 + 1863) \pm (0.5 + 0.5)$$
$$= \quad 3968 \quad \pm \quad 1.0$$

What we are saying here is that if we know the signs and magnitudes of the errors then we know whether they compensate or cumulate. If we do not know their signs, then in order to be sure to include the true value in our stated range we must assume the worst possible situation (cumulating errors) and *add* the two individual error ranges to get the overall range.

This principle applies to the absolute errors in addition and subtraction and to the relative errors in multiplication and division. If we are working in *possible* error ranges, then we always add individual ranges to get an overall one.

One final point: suppose that we add up six figures all rounded to the nearest gram (as in Table 1.1). If we do not know the original weight each one must be given a range of ± 0.5:

$$(2105 \pm 0.5) + (2212 \pm 0.5) + (2119 \pm 0.5) + (2196 \pm 0.5)$$
$$+ (2159 \pm 0.5) + (2130 \pm 0.5) = 12\,922 \pm 3.0$$

The range ± 3.0 tells us the range within which, assuming the figures to be correctly rounded from accurate originals, the true total must lie. But it is not generally a very realistic figure to work on. For the total to really be as high as $12\,922 + 3.0$ every one of the six figures would have had to be on the highest point of its range. For this to arise by chance would be very unlikely. In some circumstances, therefore, it is useful to have an idea not only of the range within which the actual value *must* lie, but the range within which it *probably* lies. The general theory of such ideas is beyond our present scope, though some special cases (based on sampling rather than rounding errors) are dealt with in sections 6.1 and 6.2.

Recall and exercise questions

1. Define the terms: (a) variable, (b) frequency, (c) attribute.
2. List the ways in which errors in data may arise.
3. Define 'absolute error' and 'relative error'.
4. Explain the meaning of the symbol $\pm$.
5. In a particular year, the turnovers (recorded in *Business Monitor*) for the ten male clothing companies with the largest turnovers were as shown in Table 2.2.

16–*Applied business statistics*

Table 2.2 *Turnovers of the ten largest male clothing companies*

Name of company	Turnover (£000s)
The Burton Group Ltd	145 763
William Baird Ltd	81 716
Raybeck Ltd	49 557
Selincourt Ltd	48 227
Foster Brothers Clothing Ltd	44 066
UDS Tailoring Ltd	44 061
Austin Reed Group Ltd	31 283
Lee Cooper Group Ltd	28 700
J. Hepworth and Son Ltd	28 554
Thomas Marshall Investments Ltd	22 165
Total	524 092

(a) (i) Present a table showing these figures rounded to the nearest 100 (i.e. to the nearest £100 000 since the figures are given in £000s), and showing the total of the rounded figures.

   (ii) List the absolute error for each figure, and show that they add to the absolute error in the total.

   (iii) On the whole did the errors tend to compensate or cumulate?

(b) Repeat the steps in (a) but: (i) round to the nearest 1000;
   (ii) truncate to the nearest 100;
   and compare the results.

(c) (i) Find the percentage of the total turnover for the top ten which went to each company. List these percentages in a table showing them correct to one decimal place, and check that they add up to 100 per cent.

   (ii) List the percentages to the nearest whole number, and find the error in their total when shown in this form.

6. A County Council lists its main areas of expenditure as: Education £402m; Social Services £28m; Police £63m; Highways and Public Transport £56m; and Other Services £69m. Assuming that these figures are accurately rounded to the nearest million:

   (a) For each figure list:
      (i) the range within which it must lie;
      (ii) the maximum relative error which might have been introduced by the rounding.

   (b) For the total give:
      (i) the range within which it must lie;
      (ii) the maximum relative error which might have been introduced by the rounding.

7. A company has a section producing whatsits. Employed in that section at any time there are three grades of people: grades A, B and C. The numbers of each grade employed vary with time, but next month are expected to be (to the nearest 10): 120 grade A, 150 grade B and 200 grade C. The average labour cost for the month is estimated (to the nearest £10) to be about £550 per grade A, £450 per grade B, and £410 per grade C employee. The overheads for the month are estimated to be about £120 000 (to the nearest £1000 and it is estimated that raw materials will cost £150 000 (plus or minus 5 per cent). About 15 200 whatsits (to the nearest 100) are likely to be made.

(a) Estimate the total labour costs for each of the three grades, giving limits for the relative errors in each case, and hence deriving limits for absolute errors.
(b) Estimate the total costs of labour, overheads and raw materials for the section, giving limits of absolute error.
(c) Estimate the likely cost per whatsit produced, giving the limits for relative error, and hence deriving the limits for absolute error.
(d) Comment briefly on the likelihood in practice that the cost per whatsit would be on the upper or lower limit from (c).

Problem

The Times 1000 list of the largest UK industrial companies lists the profits (net before interest and tax) for the ten with largest turnover as follows: British Petroleum, 4 823 200; Shell Transport and Trading, 2 662 400; BAT Industries, 528 000; ICI 721 000; Unilever Ltd, 335 200; Imperial Group, 177 543; Ford Motor Company, 425 000; Esso Petroleum Ltd, 639 057; Shell UK, 621 416; BL, 40 200. (All figures are in £000s.)

Write a short report on the apparent rounding implicit in these figures, indicating

(a) the probable limits of accuracy for each;
(b) the possible maximum relative error due to rounding in each case;
(c) why you think that the compilers may have used figures rounded by different amounts.

3
Secondary Data

2.1 The nature and use of secondary data

In statistics a basic distinction is made between 'primary data' and 'secondary data'. The terms do not concern the data itself, but the use to which it is being put. When data is being used for a purpose for which it was expressly collected then it is called 'primary'; when it is being used for a different purpose it is called 'secondary'.

Other things being equal, one would obviously prefer to use primary data. Data specially collected for the purpose in hand is always likely to be better. But often this may be impossible, for example if information about a whole nation or a whole industry is required, or else prohibitively expensive. In such instances secondary data may be available from a number of sources. It could be data collected for a different purpose by another department (or even the same department) of the company. It may be data published in a commercially available market research survey (organizations like Social Services (Gallup Poll) Ltd do surveys on a wide variety of business topics). It may be data which is published by the government.

Here are a few examples of possible uses of secondary data (remember that it is the *use* rather than the data as such which makes it secondary):

1. The public relations department of a company might use a survey conducted by the market research department to find areas in which they could improve.
2. An accountant concerned with budgetary control might use information obtained by the buying department giving quotations on machine costs and delivery times in order to budget for the future.
3. New piecework rates might be calculated on the basis of information originally compiled for job costing.

4. The research and development department of a company might use records kept by their service department on claims under guarantee.
5. A commercially available survey on the credit card market might be used by a new chain of motels in deciding which schemes to offer.
6. A commercially produced readership survey might be used to decide on features of a new magazine.
7. Statistics on the toy industry published by the Government might be used by someone setting up a toyshop to discover the types of goods sold and the seasonal pattern of sales.
8. Published regional statistics on wage rates might be used by a company wishing to site a new factory.
9. Government statistics on population and population trends might be used by a supplier of childrens' clothing to estimate future trends.
10. Government statistics in the *Business Monitors* (see below) may be used by companies to estimate changes in their market share and compare their performance with their competitors'.
11. The Government itself uses secondary data, for example when compiling import-export statistics from returns made for customs purposes.

Problems of secondary data

There are questions to be asked and problems to be borne in mind when we use secondary data.

Quality of data. Since the data collection has been done by another agency, we have no control over its quality. The sample size may have been too small, the sample poorly selected or the questionnaire poorly worded. In fact any of the aspects of survey design described later in this chapter may have been done poorly. In many cases, moreover, we may not even know how well the survey was done, since the publication may not give the relevant details.

Date. The data used may not be recent enough, and changes in consumer habit, law, taxes and so on, may have altered the situation since the survey took place.

Coverage. The geographical area of a survey may differ from what we want. One must be careful, for example, not to use 'UK' and 'England and Wales' figures together for comparative purposes. The coverage may exclude certain groups of people we should wish to see included. For example, a register of unemployed will exclude groups of the unemployed important for social services purposes, and a study of registered drug addicts will again exclude categories important for health and social services purposes.

Definition. Sometimes the definitions used for compiling the original information may be unknown. Does 'clothing', for example, include footwear and handbags? A different instance of this comes in the use for other purposes of some financial accounts statistics on costs or values. Depending on the system used, the 'book value' of stock or machinery may be very different from the real value, and different again from the replacement value.

In some instances information may be available to help us decide whether any of the above will produce problems, or we may be able to get in touch with the agency which did the original work. In some cases we just have to do the best we can, but at least we can be aware of the problems and try to minimize them by suitable adjustments where appropriate.

3.2 Sources of published statistics

Most of the chapters in this textbook are intended to develop skills, understanding and ability to analyse and present numerical information clearly. Any issues of simple recall of information are minimal and incidental. The topic of published sources is different because the central point does not concern skills, understanding, analytical ability or presentation, but rather a knowledge of the *kinds* of information available, its usefulness, and how to go about locating it.

One approach to this would be to learn lists of contents of various publications, perhaps aiming at examination questions like: 'Describe the contents of the *Monthly Digest of Statistics*'. This, however, seems to be a poor use of time. It is far better to have some idea of the range of statistics available and to have guidelines as to the main publications to look at, and then to undertake projects involving library searches to prepare information on a given topic or topics.

This chapter has been written to aid in developing this approach, but three basic points should be noted. First, the kind of project undertaken will depend on access to publications. The library of any college mounting business courses should be encouraged to obtain at least one copy of each of the publications listed below, but they are expensive and many colleges will be unable to buy all the up-to-date issues.

Second, this chapter lists a number of books about sources and books containing sources. The information is correct at the time of writing, but changes do occur, and details may have changed by the time it comes to be read. The Government publishes up-to-date information on official sources, which should be obtained and used in preference to the information listed below.

Finally, in a practical situation one way to locate available information is to contact the relevant agencies. One thing which must be remembered in this connection is that each Government Department collects and publishes its own statistics. The Departments may be contacted, and are usually very helpful to inquirers, but I know from personal experience that one Department does not necessarily know what information another Department has. In theory the Central Statistical Office should reveal all. In practice ... let us say that it is advisable to contact individually any Department which looks remotely relevant. The most useful general contact is probably the *Statistics and Market Intelligence Library* (Export House, 50 Ludgate Hill, London EC4M 7HU; 01 248 5757).

Government Departments

The addresses at the time of writing of the most important Government Departments are listed below; other Departments' addresses are contained in *Government Statistics: A Brief Guide to Sources.*

Central Statistical Office (CSO), Press and Information Office
  Great George Street, London SW1P 3AQ (01 233 7317)
Department of Industry, Business Statistics Office
  Cardiff Road, Newport, Gwent NP1 1XG (Newport 56111)
Office of Population Censuses and Surveys
  St Catherine's House, Kingsway, London WC2B 6JP (01 242 0262)
Department of Employment
  Orphanage Road, Watford, Herts WD1 1PJ (Watford 28500)
Department of Health and Social Security
  14 Russell Square, London WC1B 5EP (01 636 6811)

Uses of sources

At the start of this chapter we looked at a few examples of secondary statistics in use. The following suggest possible areas of usefulness in more detail:

1. *Marketing.* Sales trends in an industry will generally appear in the appropriate *Business Monitor.* Trends in consumer expenditure appear in the *Family Expenditure Survey* and *National Food Survey.* Details of potential

markets classified by area and type of person may be assessed from the *Census of Population, Regional Trends, Social Trends*, and others. The import figures available from the *Customs and Excise Bill of Entry Service* can help to assess foreign competition. *Eurostats* publications can give valuable information on European markets, and the numerous market research studies, commercially produced and in journals, can be relevant to individual companies.

2. *Contracts.* The *Monthly Bulletin of Construction Indices* and various other price indexes may be useful if contracts have escalation clauses.

3. *Accounting.* Price index numbers used in cost accounting are published in *Price Index Numbers for Current Cost Accounting* and monthly in *Business Monitor MM17* (see also Chapter 11).

4. *Buying.* Sales trends and wholesale prices of particular inputs can be followed through *Business Monitors* and the appropriate wholesale price indexes.

5. *Personnel.* The *Department of Employment Gazette* and other publications contain figures on earnings and wages trends, overtime, unemployment, vacancies and work stoppages due to industrial disputes on a regional and industrial breakdown. They also publish the Index of Retail Prices, and the Taxes and Prices Index, which are central to many wage claims.

6. *Management efficiency and finance.* Costs and operating ratios can be compared with those in an industry generally by consulting appropriate *Business Monitors* and the *Census of Production*. The financial indicators against which financial decisions must be made are collected in the monthly *Economic Trends*. The monthly *Financial Statistics* may also be relevant to investments. Newspaper and journal articles often contain useful statistics or statistical analyses in this area.

There are, of course, vast numbers of more specialized statistics available both from Government agencies and from journals and other non-government sources. For details of these, and for up-to-date details on all sources, the reader will have to make his or her own search using the various statistical guides available.

Secondary data—23

Books about sources

Unless they are published annually or frequently revised, such books may soon come to have out-of-date information. Regular publications are listed first below:

1. *Government Statistics: A Brief Guide to Sources* (Annual – free) Central Statistical Office
2. *Guide to Official Statistics* (Annual) Central Statistical Office
3. *Guide to Public Sector Financial Information* (Annual) Central Statistical Office
    A brief, a comprehensive, and a more specialized official guide.
4. *Research Index* (twice monthly) Business Surveys Ltd
5. *Research*
5. *Reports Index* (every two months) Business Surveys Ltd
    Descriptions of articles (*Research Index*) and 'one-off' surveys (*Reports Index*) in business fields.
6. *BIM Reading List* (Occasional) British Institute of Management
    A library list of additions, including statistical sources etc. Other libraries such as London Business School issue similar lists.
7. *Statistical News: Developments in British Official Statistics* (Quarterly) HMSO
8. *Eurostat News* (Quarterly) Statistical Office of the European Community
    Give news of changes in official publications for the UK and Europe.
9. *Sources of Statistics* J. M. Harvey (1971) Bingley
10. *Sources of Economic and Business Statistics* B. Edwards (1972) Heinemann
11. *Sources of Official Data* K. Pickett (1974) Longman
12. *Open University: Statistical Sources Course Team D291, Units 1–16* (1975)
13. *Recommended Basic Statistical Sources: International* (1975) The Library Association
14. *Recommended Basic United Kingdom Statistical Sources for Community Use* (1975) The Library Association
15. *Statistics Europe: Sources for Social, Economic and Market Research* J. M. Harvey (1976) CBD Research
16. *Subject Index to Sources of Comparative International Statistics* F. C. Pieper (1978) CBD Research

24—Applied business statistics

17. *Reviews of UK Statistical Sources* (Ed.) W. F. Maunder (Volumes issued 1974 onwards) Pergamon
18. *Studies in Official Statistics* (Various dates) HMSO
    On various topics (such as the Index of Industrial Production) giving details of how particular statistics are compiled. Periodically updated.

Specifically with a marketing orientation there are:

19. *Principal Sources of Marketing Information* C. Hull (Annual) Times
20. *IPC: Consumer Marketing Manual of the UK* (Annual) IPC
21. *IPC: Industrial Marketing Manual of the UK* (Annual) IPC
22. *European Directory of Market Research Surveys* T. Landeau (1974) Gower
23. *Sources of UK Marketing Information* G. Wills and E. Tupper (1975) Benn
24. *International Directory of Published Market Research,* 4th Edition (1980) British Overseas Trade Board/Arlington

Important sources

At the time of writing the following are some of the most important (mainly official) sources. Please note that this list is not intended to be comprehensive, and other areas of particular interest might be considered on a statistics course:

1. *General digests.* The *Monthly Digest of Statistics* and the more comprehensive *Annual Abstract of Statistics* contain the main series of all Government Departments – on population, social, industrial, finance, wages, and so on. *Social Trends* (annual) concentrates more on the social than the industrial statistics. *Regional Trends* (annual) gives general information broken down into the UK standard regions.

2. *General economic. Economic Trends* (monthly) provides a background to the UK economy. The *National Income and Expenditure 'Blue Book'* presents the national accounts, including consumer expenditure. *Price Index Numbers for Current Cost Accounting* (annual) and *Business Monitor MM17* (monthly) have already been mentioned. The Treasury's free monthly *Economic Progress Report* contains main series and some comment. The Department of Industry publication *British Business* (weekly) contains information on sales, capital expenditure, stocks, hire purchase, production, company acquisitions, liquidity and insolvencies and home and EEC economic indicators.

3. *Financial.* *Financial Statistics* (monthly) is the key general publication. *Inland Revenue Statistics* (annual), on taxation, includes information on personal wealth.

4. *Manpower, earnings and prices.* The *Department of Employment Gazette* (monthly) includes articles and tables on manpower, employment, hours worked, wage rates, earnings, labour costs, retail prices and stoppages. It also contains occasional articles explaining or analysing the statistics it contains. *British Labour Statistics: Year Book* (annual) has longer coverage and notes on its series. *New Earnings Survey* (annual) relates to earnings from employment by industry, occupation and region. *Time Rates of Wages and Hours of Work* (annual) covers what its title says.

5. *Production and distribution.* Statistics on these areas are published in the *Business Monitor* series. For production the annual *Census of Production* obtains data on total purchases, total sales and work done, stocks, work in progress, capital expenditure, transport costs, employment, and wages and salaries. This is issued separately for different categories under the *Standard Industrial Classification* (SIC) in the *Business Monitor PA* series. Quarterly (*PQ* series) and monthly (*PM* series) *Business Monitors* cover value and quantities of sales and some other aspects of turnover, stocks, orders in hand, exports. For distribution there are separate *Business Monitor* series, in particular the annual (*SDA25*) and monthly (*SDM28*) publications covering retail trades.

6. *Population and household expenditure.* Various publications are issued by the Office of Population Censuses and Surveys, and the Registrars General for Scotland and Northern Ireland. These include a decennial census (last taken in 1981), and *OPCS Monitors*. On consumer expenditure the *National Food Survey* (annual), the *Family Expenditure Survey* (annual - see section 4.15 below), and the rather wider scoped *General Household Survey* (annual) are available at the time of writing. It is possible that some streamlining in this area may take place, however.

7. *Non-government sources.* In view of the great number of occasional or one-off studies, I shall here mention only a few of the most important regular publications. *Financial Times* (daily) and *Economist* (weekly) contain useful articles and reviews. Various banks and groups of banks publish reviews. Jordans Ltd publish studies of particular industries, drawing together various published sources.

When government figures are shown by industrial breakdown, the 'Standard Industrial Classification' (SIC) is used. The 1968 SIC is being replaced by a new one for series based on 1980 (from 1983). The SIC is an important concept, and details of the revised code are available in *Statistical News*, November 1979 p. 47, 1-2, and in *Standard Industrial Classification 1980 Revised* (HMSO). Basically, the code breaks industry down into major categories, with sub-categories within each of the major ones.

Some further comment on sources, in particular on official published index numbers, is made in section 4.18.

Recall questions

The best way to achieve familiarity with sources is to use them, but, although a couple of recall questions are given below, the answers to problems about sources really depend on ideas of presentation developed later in the book. Most of the problems concerning source searches are therefore to be found in chapters later in this textbook.

1. What is the definition of 'secondary data'?
2. What problems may arise in the use of secondary data?
3. To which government publication or publications would you turn first for data on each of the following:
    (a) Employment
    (b) Prices of retail goods
    (c) Patterns of consumer expenditure
    (d) Sales in an industry
    (e) Wages and earnings
    (f) Retail trade
    (g) Production

4
Data Collection and Surveys

4.1 Basic tasks and aims

To gather primary data there are two basic methods: direct observation or some kind of questionnaire. Direct observation might be used, for example:

(a) to survey traffic prior to design of a new one way system,
(b) to measure television viewing using machines attached to TV sets,
(c) to analyse consumer behaviour in walking around a supermarket, or,
(d) for organization-and-methods or job-costing time estimates.

Commonly, however, data is collected by some kind of questionnaire. The basic decisions to be made in conducting such a survey are as follows:

1. *Objectives*. What is the information for, what information is wanted, who are the 'target group' from whom it is required?
2. *Respondents*. Are all of the target group to be approached, or a sample? If a sample, how is it to be chosen and contacted?
3. *How to contact*. The three basic alternatives are telephone, personal interview or postal questionnaire; or some combination may be used.
4. *Questionnaire and approach*. The questionnaire has to be designed, and the approach by letter or personal introduction and manner chosen.

Reliability and validity

There are, of course, situations in which an individual decision maker may wish simply to get a subjective 'feel' of a mood or situation, and no numerical data will be involved. For any situation which is objectively analysable, however, data collected will have some numerical basis either in measurement or in numbers in various categories. Implicitly in such situations there is usually some 'true' value for a figure, which one is trying to estimate accurately by observation or survey. This relationship between our estimate and the 'true' figure will occupy the next few paragraphs.

A first obvious question one could ask is whether the same estimating procedure, if carried out again by a different researcher, would give the same kind of result. Any estimate based on a procedure which could be repeated with the same results is called a 'reliable' estimate. In the technical sense the word 'reliable' does not necessarily mean that the estimate is accurate in reflecting the true value - simply that the same procedure would always give about the same result. If, however, a survey depends so much on subjective viewpoint that it would not do so, it is 'unreliable'. For example, in some government surveys 'colour of respondent' is assessed by interviewers, so an estimate of the proportions of people of different colour in an area might depend on the interviewer - it would not be very 'reliable'.

Even if a procedure gives consistent results, they could be consistently wrong! A first and obvious question is whether or not the measure used actually measures the concept in which we are interested. For example, if one were interested in total unemployment in an area, would this be measured by the *registered* unemployed? How much *disguised* unemployment would there be among those married women who would like to be employed but do not register because they are ineligible for unemployment benefit? Again, if one wished to measure the *aptitude* of applicants for a particular post, would IQ tests be a valid measure of that aptitude?

Obviously it is much more difficult to assess 'validity' in some instances than in others. A person's aptitude is much harder to measure than (say) their age in years. But it is no use trying to analyse a business situation using concepts for which no valid measures can be devised. For example, it might be useful to know how many more people would have bought a particular product *if it had* been cheaper - but how could one estimate this? One could conduct a survey and ask people. This might be 'reliable', but people are notoriously bad at predicting their own behaviour in hypothetical situations and it is to be doubted whether their answers would really validly reflect reality.

Bias, error and precision

When one has selected - or ingeniously devised - valid and reliable ways to estimate the values one wishes, one still has to ask how accurate these measures will be. In Chapter 2 we defined:

$$\text{Absolute error} = \text{Stated value} - \text{Actual value}$$

Any individual estimate, then, has an 'error', the difference between it and the true value.

In most practical surveys, and any which depend upon estimates made from samples, if we repeat the same procedure over and over again we will get a slightly different estimate each time. The error of the estimate will therefore vary from one estimate to the next, and will depend on chance. Now we are usually not in a position to know the error for any given estimate, since to know this we should need to know the true value. But we may know something about the *pattern* of errors we would expect if the procedure for obtaining an estimate were repeated over and over again. Two important questions can be asked about this pattern.

Bias. In some instances the pattern of errors shows a tendency to consistently over- or underestimate. For example, suppose we conducted a survey to estimate the number of English electors intending to vote Conservative by ringing names selected haphazardly from a telephone book. This method would tend consistently to overestimate the proportion since more Conservative voters than Labour voters are likely to be on the telephone (both telephone ownership and voting habit correlate with class). Due to chance, any particular estimate might happen to *underestimate* the proportion, but on average the survey will lead to an overestimate, that is it has an 'upward bias'. If, of course, we knew from past surveys the actual amount by which a method overestimates it would be possible to allow for it and remove the bias; but the more common situation is for us to realize that a method is biased but not know by how much. In any event one of the objectives in survey design must be to avoid bias, other things being equal.

Precision. We have seen that if the pattern of errors shows a consistent tendency to over- or underestimate, then this is bias, but we might also ask within what limits we could normally expect the error to lie. This is referred to as the precision of an estimate. Note that it refers not to one single estimate but to the pattern of such estimates which would be obtained by repeating the same procedure over and over again.

Example

A council has twenty thousand ratepayers. It wishes to estimate the proportion of these who paid their rates by the required date. Each individual record contains the information accurately for that ratepayer, but they do not wish to go through them all. From the twenty thousand, 100 are selected in such a way that each ratepayer has an equal chance of being included. The proportion of the 100 who paid up on time is found to be 0.45 (45 per cent).

In this instance, the value of 45 per cent is the best estimate for the whole group. We would not, of course, have any way of knowing its actual error (how much it differs from the figure for the whole twenty thousand), but we could say that it is *unbiased*: in repeated use of this method of estimating we should overestimate as often as we underestimate.

But we could go further: using methods dealt with in section 17.2, we could say that in 95 per cent of cases we should expect this method to get an answer within about 10 per cent of the true one, so the true value is likely to lie within the range 35 to 55 per cent (that is, within 10 percentage points of 45 per cent) (45 ± 10 per cent).

As one might expect, a larger sample would improve this precision. For example, if the estimate of 45 per cent was derived from a sample of 500, then we could be 95 per cent sure that this was within about 4.5 percentage points of the true figure (i.e. 45 ± 4.5 per cent).

A higher precision, then, means that in a long run of repeating the survey the error would on average be smaller.

Our aim is to find a survey design which gives us an unbiased estimate with high precision. In this way we will maximize our chances of getting an estimate with a low error.

4.2 Kinds of sampling

For most business purposes it is neither necessary nor desirable to survey the whole of a target population, and we base our estimated values on a sample. Before looking at the implications for bias and precision which this entails, we need to describe the basic forms of sampling. These fall into two fundamental categories: random and non-random sampling.

A random sample is one in which each member of the target population has a known chance of falling in the sample. Various variations of it are used, but in each instance there has to be an actual or potential list of all the members of the target population from which to select. This list is called a *sampling frame*. Examples of sampling frames are: the Electoral Register, the rating records, the General Practitioner list, a local education authority list of schools, and a list of book club subscribers. In random sampling the proportion of the total target population which is included in the sample is called the *sampling fraction*. While this is of interest, it is found in practice that the actual size of sample is more important in determining accuracy than the proportion of the population it constitutes.

In a non-random sample it is not possible to calculate any figure for the chance of a particular member of a target population being included in the sample. In the following sections giving types of sample, the first four are random and the last two non-random.

Simple random sampling. In this it is as though all the names of the target population were written on cards, mixed together well in a drum, and the required number taken out. Thus any given sub-set is as likely to occur as any other. In practice, of course, the use of a drum would be crude, and it is likely to be done by numbering the target population and using random number tables to choose the required numbers.

Stratified sampling. Sometimes a sampling frame may give certain basic facts about the members listed such as sex, age group or type of dwelling. A stratified sample ensures that the proportions of such groups in the sample match those of the population. For example a list of ratepayers might list 3000 as 'industrial', 2000 as 'private rural', and 15 000 as 'private urban'. These would constitute three *strata* for this population. A stratified sample which took a sampling fraction of 1:100 from each would require 30 'industrial', 20 'private rural' and 150 'private urban'. This would give a sample of 200, in which the proportions in the three listed strata matched those of the population. Since, however, the sampling fraction is less important than the actual numbers, researchers sometimes deliberately sample more of the smaller groups, readjusting their final answers to allow for this.

Multi-stage random sampling. This may be used to survey a widespread population when time and money are limited. For example, if we wished to contact a sample of ratepayers in the United Kingdom we might first select randomly 15 counties, stratifying for different kinds of county if desired. From within each county we might randomly select 30 per cent of its rating authorities (borough councils). Then from within each rating authority we might take, say, 0.1 per cent of its ratepayers. This would give each ratepayer an equal chance of being included, but it would mean that the final sample we had to contact were concentrated into certain areas.

Panels. Respondents (usually random samples) are asked to keep a diary over a period of time in which transactions or reactions are noted. If this is done on a semi-permanent basis this constitutes a panel. There are obvious problems in this in that only particular kinds of people may be prepared to do this, and also in that by thinking about the subject over a period their behaviour itself may change. To lessen this problem panel members are changed after a certain period in appointment.

Simple non-random sampling. There are many forms of this. For example a newspaper might invite its readership to write in and give opinions on some issue. If this is then used to draw inferences about the whole readership it will be biased, for it will include only those who feel motivated to write. A company might include a form with its products asking opinions of its brand image, but this will suffer from a similar problem. An interviewer simply told to go and interview 100 people may tend to choose them all from one particular social group because that group are easiest to reach. Again, this may systematically bias the results in favour of the views of that group.

Quota sampling. This is a form of non-random sampling which is widely used. Basically the interviewer is told to obtain views from ten people who between them cover various cross categories. The most common would be age (given in four or five age ranges), sex and 'social class of chief wage earner'. There are a number of ways used to define 'social class' (every Government department uses a different one!). That most commonly used in market research comprises six socioeconomic groups defined by the Institute of Practitioners of Advertising:

A   (top 2 per cent: A successful business or professional person, such as a top civil servant, physician or person of private means.

B   (about 10 per cent): Senior but not top professional people, such as a university lecturer, pharmacist, matron or employed accountant.

C1   (about 24 per cent): Tradespeople and non-manual workers in administrative or supervisory posts, such as teachers, bank clerks, insurance agents or clerical officers.

C2   (about 32 per cent): Skilled manual workers, such as a bricklayer, plumber, fitter, welder or head cowman.

D   (about 22 per cent): Semi-skilled and unskilled, such as a fisherman, ticket collector, traffic warden, shop assistant or army private.

E   (about 9 per cent): Casual labourers, old age pensioners, widows on state pension, and any who are little above subsistence level incomes.

These are sometimes telescoped together into four: AB; C1; C2; DE. Thus an interviewer might be given a quota sheet like Fig. 4.1. As marked, this quota sheet shows that so far the interviewer has interviewed one man aged 25-34, working and in class C1, and one woman aged 45-64, not working and in class DE. Each person interviewed fills one quota in each category.

    Quota samplers are allowed to fill their quotas however they wish, either by street interviews or by calling on houses. They are encouraged to try to fill 'difficult' categories first, because otherwise they might be left to find,

```
┌─────────────────────────────────────────────────────────┐
│                    QUOTA SHEET                          │
│                                                         │
│  As each contact is interviewed, cross through one      │
│  circle (i.e. O) in each quota category:                │
│                                                         │
│  Quota Category:              Men:       Women:         │
│  (Total)                      (5)        (5)            │
│                                                         │
│  Age:       16-24              O          O             │
│             25-34              ⌀          -             │
│             35-44              -          O             │
│             45-64              OO         ⌀O            │
│             65+                O          O             │
│                                                         │
│  Employment:  Working         ⌀OOO       OOOO           │
│               Not Working      O          ⌀             │
│                                                         │
│  Social Class:  AB             -          O             │
│                 C1             ⌀          -             │
│                 C2             OO         OO            │
│                 DE             OO         ⌀O            │
└─────────────────────────────────────────────────────────┘
```

Figure 4.1.

say, a 16-24-year-old, non-working AB male – which could prove difficult. Quotas could obviously include many other things, such as car ownership (or ownership of a particular make) and smokers and non-smokers.

Any random sample is, in theory, unbiased. This does *not* mean (as some books assume) that random sampling *always* produces a representative sample and accurate estimates. What it means is that there is no *consistent* over- or underestimation. A second point is that for any random sample, in theory, we can assign accurate precision limits. That is, we can make a statement like: 'We are 95 per cent certain that the true percentage of the population who have seen the Bloggo advert is between 27 and 33 per cent'.

These are important theoretical advantages of a random sample over a non-random one. A random sample gives an unbiased estimate to which we can find precision limits. But we need to remember one important point. This assumes that the sampling frame is complete and perfect, and that every one of the sample selected was accurately surveyed. This is often far from the case. Suppose, for example, a random survey is done to find how many people watched a particular television programme. It is quite possible that, in spite of repeated calls at their houses, only 80 per cent are actually contacted. What then? It could well be that the other 20 per cent were difficult to contact because they are busy people. But busy people may have less time, on average, to watch the television. This could mean that our estimate is systematically biased in an upwards direction. Using this procedure we will tend to interview a higher proportion who watched it than is contained in the target population.

A simple non-random sampling approach is usually used only by those who know nothing about sampling. It is really just too prone to bias, which in many cases can be very large indeed. Quota sampling is, however, widely used as an alternative to random sampling. Obviously, notwithstanding the quotas, those interviewed will tend to be atypical in the sense that they were easily accessible while many others were not. The important question is whether that atypicality extends to the subject under study. If it does then the results will be biased: a repetition of similar methods will tend to systematically over- or underestimate. The obvious problem is that one is almost never in a position to know for certain whether or not the particular estimate *is* significantly biased. It has been found by market researchers not to be in most cases, but because of the uncertainty random sampling would be preferred, other things being equal.

But in practice all other things are not equal. In almost all commercial surveys one is trying to maximize accuracy for given time and money. Random surveys only give their promised benefits if a high proportion of the sample is actually contacted and interviewed. To achieve this some respondents may need to be called upon three or more times, which is expensive and time-consuming. Moreover, a simple random sample may give low precision unless it is large. Stratification would usually improve the precision, but useful characteristics for stratification may not be available from the sampling frame, in which case stratified sampling is impossible. Quota sampling does at least ensure that *some* characteristics of the sample are representative of the population as a whole. Whether these are in fact the key characteristics can never be certain.

I have said that, strictly speaking, precision limits can be derived only for a random sample. In practice market research organizations have found that limits can generally also be obtained for quota samples.

Example

Suppose that in a simple random sample of 500 adults (all of whom were contacted) 45 per cent said that all other things being equal they would prefer to own a foreign car. Using methods similar to those in section 17.2 it could be calculated that we could be 95 per cent certain that the true population figure was somewhere between 41.5 and 49.5 per cent.

Suppose that instead we had a quota sample of 500, based on the principles applied by Gallup Poll. A figure of 45 per cent would in their experience enable them to say with 95 per cent certainty that the true population figure was somewhere between 39 and 51 per cent. Precision from a quota sample of 900 would match the precision of the simple random sample of 500.

Which is best: the random or the quota sample? The answer is that it depends on whether a good sampling frame is available *and* a high proportion of the chosen sample can be interviewed or surveyed by post successfully, with speed and economy. If so, then a random sample will usually be better. It will be unbiased and enable us to find valid precision limits. But if this is not the case, then a quota sample may give more value for money.

4.3 Sources of error and bias

In sample surveys, no methodology can *guarantee* a representative sample and accurate answers. Some error is inevitable, though we try to minimize it, and luck may determine how large it is. The most important factor determining the chances of low or high error is the size of the sample: a larger sample is generally more accurate, so we may be balancing cost and accuracy.

To minimize bias we should be aware of ways in which it can arise. In general there are three main sources. The first is selection bias, meaning that the sample actually interviewed is non-representative. The second is bias introduced by the structure and wording of questions asked. The third is bias introduced by an interviewer, either stimulated in the respondent or introduced into the recording of answers. Each of these will be dealt with in the sections below.

Questions of selection

In sampling we aim to choose and interview a representative sample. However, no procedure can guarantee this in every instance, and we shall expect an error, the size of which will to some extent be decided by chance. Selection *bias* occurs when the procedures adopted mean that the questionnaires actually completed overrepresent on average particular groups of the target population. The main sources of selection bias are non-random sampling, sampling frame deficiencies and non-response from a section of the population.

Non-random sampling. This has already been mentioned. The very fact that those interviewed for a quota sample were easily accessible may make them, on average, atypical in certain respects. This could lead to bias – but the question is whether it leads to a significant bias in respect of the matter under study. All that one can do, however, is avoid any means of obtaining a sample which would give obvious sources of bias, and try to include in the quota specifications any obviously relevant factors.

Sampling frame deficiencies. A good sampling frame lists all those in the target population once and once only. Often, however, the frame actually available will be deficient in some way, and many of these deficiencies can cause bias; it can be:

1. *Incomplete or inadequate.* Incomplete means that some members of the target population are omitted from the frame, but that there is no particular pattern to this; inadequate means that some classes of members are omitted. Incompleteness would not cause bias, but inadequacy may do. For example, it is estimated that around 15 per cent of electors in a district will be recent arrivals and not listed on the Register of Electors at the end of that Register's life. If these differ from the average because of their apparently greater mobility or for some other reason, then this will cause bias.

    It should also be noted that, for example, a telephone directory is a good sampling frame for those telephone subscribers who are not ex-directory. As a frame for the general population, however, it is not satisfactory because it is 'inadequate': certain classes of people are excluded. Since the class of non-subscribers will very likely differ in many respects from the class of subscribers (speaking in general), the use of a telephone directory to sample the general population will lead to bias.

2. *Inaccurate.* This means that some information the sampling frame contains is wrong. If this is purely haphazard then it will not lead to bias, but if the errors are in any way systematic then it could.

3. *With duplicates.* This could cause overrepresentation of certain classes, which would cause bias.

4. *With blanks.* That is, the sampling frame lists some who are not in our target population. This may not mean that they are wrongly on the list: often we may have to use a list of a more general character. For example, to interview a random sample of smokers we might use the Register of Electors and simply discard the non-smokers. The presence of blanks will only cause bias if they are not detected as blanks and removed.

Non-response. This is where we fail to obtain some or all of the required answers from particular potential respondents. The very fact that we failed with these respondents while we succeeded with others marks them out as in some sense atypical. It may well be that they have other characteristics which arise more often than among the whole target population. If any of these relate to the subject under survey then bias will be caused. Of the four main categories listed below, the first three can arise with either quota or random sampling, but the last applies only to random sampling.

1. *Refusals.* Some people refuse to answer a personal interviewer or simply fail to return a mail questionnaire. They must, obviously, be different kinds of people from those who do respond. If they are different on average in other respects related to the survey, this will cause bias.

   The personal interviewer can reduce the number of refusals, usually down to about 3 to 7 per cent, by dressing neatly but neutrally, approaching in a pleasant but businesslike manner, and conveying that he or she is not selling anything. Unfortunately, in recent years we have been plagued by salespeople for encyclopaedias or double-glazing posing as researchers, and the general public are more suspicious. Researchers working for commercial groups such as Gallup Poll seek to reassure by beginning the interview by announcing the name of their organization first and showing a card. The residual refusal rate, however, is unavoidable, and it has been shown to be correlated with age, sex and social class. The quota sampler automatically replaces a refusal by someone with roughly similar characteristics in these respects, but it would be as well to check the random sample for these aspects when completed, as no replacement occurs. In any case, it is impossible to be certain that refusal is not associated with some other characteristic which will therefore be underrepresented in the final sample.

   In a mail questionnaire the response rate (i.e. percentage returned) is crucial. On a subject about which respondents feel strongly and see a gain in expressing opinion there is little problem – for example a survey of top industrialists for views of the Government's economic policy, or of pharmacists on the problems of prescription payments.

   A high response rate, say 70 per cent, is difficult to achieve otherwise, but the rate is increased by the use of free prize draws for respondents; these are commonly used by commercial organizations. The response rate is also increased by sending reminders. Classically, the sequence may be questionnaire, reminder letter, second reminder with duplicate questionnaire, with suitable times between postings. This does of course depend on the time available.

2. *Unsuitable for interview.* Both random and quota sample interviewers are allowed to abandon interviews with anyone who proves to be too deaf,

senile or mentally subnormal to understand, or who does not speak English These groups will not be represented in surveys, which could cause some bias.

3. *Partial non-response.* This means that on particular questions the respondent leaves the questionnaire blank, or the interviewer fails to get a proper answer. Interviewers are instructed how to deal with this (see below) to lessen the problem, but with mail questionnaires it would generally be too expensive to recontact respondents. If it happens often, it could cause bias: for example it might be that only those with a naive view of a situation respond. It can be lessened by careful wording of questions.

4. *Failure to contact.* This obviously only applies to random samples: it is in a sense already implicit in quota sampling because by definition the respondents are easily accessible. There are two groups: those out at the time of call, and those absent during duration of survey. There is often little that can practicably be done about the latter group (which includes those permanently moved away). For the former group, the problem can be lessened by calling back. On a general population survey, one might expect to contact about 40 per cent of those finally contacted on the first call, 80 per cent by the second call, and 99 per cent by the fourth call, so more than four calls is uneconomic. It is not, of course, a solution to substitute an alternative respondent, for the fact that the substitute is available and the original is not already marks them off as different. Substitution restores sample size but does not solve problems of bias.

Typical response rates for commercial surveys vary between types of survey. Postal surveys can vary from a 35 per cent response up to 60 to 65 per cent for a motivated group. On a door-to-door random survey, typical response rates would be about 65 per cent for named individuals and 70 per cent for households.

Questionnaire design

Some differences exist between postal questionnaires and questionnaire recording schedules for use in personal interviews. The former have to be totally self-explanatory and simple enough for anyone to fill in, whereas where a trained interviewer is to fill the record in one can assume more understanding. Also, in a postal questionnaire one cannot assume that answers to earlier questions are unaffected by later ones. But many of the points to bear in mind are similar.

Overall questionnaire format. It is important, especially with postal questionnaires, for the form to be pleasing to the eye and to *look* easy to fill in; so avoid masses of small print. It is also an advantage to keep the questionnaire as short as possible. Other things being equal, the longer it is, the lower the response rate is likely to be, and the more the bias caused by non-response.

The questionnaire should follow a logical order and personal details – occupation, age and perhaps income range – should be left until the end in a random survey. However, in a quota survey these details will be better near the beginning so that no time is wasted interviewing people not required for the quotas. In order to reassure the respondent, such questions may be introduced with brief explanation: 'In order that my company can check on the representativeness of the sample, could you please tell me your occupation . . .'

Question content and format. Information required can be crudely divided into two categories: fact and opinion. Facts could be name, age, sex, occupation, income, recent attendance at a leisure centre, possession of a particular make of car, brand of washing-up liquid currently in use, and so on. Opinion might relate to agreement or disagreement with a particular viewpoint, image or brand-names. There is, in fact, a third kind of category, that of 'hypothetical action': 'What *would you do* in such and such a situation?'. This sounds as if it is factual, but in practice people are very bad at predicting their own actions. One may ask such questions, but the answers must be taken with a pinch of salt.

Some points should be noted about the kind of information asked for. For either fact or opinion it is not good to rely on respondents' memories too much, so questions relating to times too far past are best avoided. Factual questions which require too much calculation should also be avoided; for example 'About how much did you spend on public transport last year?'. If this kind of information is really needed, then cooperation in keeping a diary should be sought. Where questions do ask for estimated numbers, it is useful to give some idea of the required accuracy (to the nearest week, nearest £100, etc.).

For either fact or opinion there are two basic question formats – open-ended and pre-coded. Examples of open-ended questions are:

What is your occupation?
What factors are most important to you when choosing a brand of toothpaste?
What is your opinion of the Employment Protection Act?

Examples of pre-coded questions are:
1. How strongly would you agree or disagree with the following statement?
   'The Government should hold at least a 51 per cent shareholding in all companies extracting North Sea oil.'
   Strongly agree ☐ Agree ☐ Neither ☐ Disagree ☐ Strongly disagree ☐

2. Please place a cross in the box under the company which, in your opinion:

	BP	Esso	Mobil	Shell	Don't know
(a) makes the best products	☐	☐	☐	☐	☐
(b) is most concerned to meet customer needs	☐	☐	☐	☐	☐
(c) is leading new developments	☐	☐	☐	☐	☐

For a personal interview two other variations are available. One is that the question is asked as an open question, but the interviewer has certain pre-coded answers. For example:

To which religious denomination, if any, would you say that you belong?
Church of England ☐ Roman Catholic ☐ Methodist ☐ United Reformed ☐
Baptist ☐ Jehovah's Witness ☐ Pentecostal ☐ None ☐ Don't know ☐
Other: please specify ....

This enables a quick entry to be made on the questionnaire recording schedule and enables easy coding later, but does not influence the respondent's reply by actually offering alternatives. At the other extreme, one might wish deliberately to offer alternatives. This may be done either by reading them out or by handing to the respondent a card with them written on.

There are various relative advantages and disadvantages to be considered in choosing between open and pre-coded questions. For purely factual information the choice will largely depend on the number of possible alternatives; for example, for occupation, it would be impractical to list them all, whereas one can easily list a set of age ranges. With opinion questions, if the respondent is offered a set of pre-coded answers this will often tend to channel his or her answer into one or other category when a free answer would not have fitted so easily into any of them. This can be an advantage in that it is easier to code for analysis and it is a decision of the respondent rather than of an analyst which category the answer 'really' falls into. On the other hand, if the real answer is not adequately covered by any of the alternatives offered, forcing the respondent into one of them may be a form of bias. It could also cause the researchers to miss significant details which might have come out of free answers. This problem is lessened by the inclusion of an 'Other – please specify ...' category, though many respondents will still take the easy path and pick one of those offered. It may also be possible to combine open questions, to pick up any ideas not thought of by the research designers, and pre-coded questions, to enable objective analysis.

If a survey is to be analysed by computer it is useful to have a right-hand column in which answers can be coded by numbers. This enables the cardpunch or tape-punch operators to do their work quickly without having to search through the questionnaire.

Wording of questionnaires. Correct wording of questionnaires is very important, and apparently innocuous changes may radically alter the pattern of responses – in some cases for no logical reason. The following are points to bear in mind:

1. *Simplicity.* Simple words are always preferable to complicated and unnecessarily technical ones.
2. *Brevity.* Avoid long questions, with too many 'ifs'.
3. *Ambiguity.* An example of an ambiguous question is 'Would you go on strike if your company introduced black labour?'. Does this mean non-union labour or negro workers? Such ambiguity should be avoided.
4. *Leading questions.* 'How strongly do you favour government control of imports in textiles?' is a leading question: it assumes that the respondent *does* favour controls. Such leading should be avoided. Similarly, if a selection of 'agree/disagree' categories is offered in a pre-coded multiple-choice question, it should not contain a predominance of one particular direction.
5. *Emotive wording.* Avoid words like 'big business', 'reasonable'. 'Should oil extraction be controlled by big-business interests or be under the control of the nation?' and 'Should oil extraction be controlled by private enterprise or be under the control of the Government?' mean exactly the same, but the emotional impact is different. A classic study by Rugg (in 1941) found that the questions 'Do you think that the United States should allow public speeches against democracy?' and 'Do you think that the United States should forbid public speeches against democracy?' produced quite different opinion results. In this context it is better to put both alternatives, that is 'Do you think that the United States should allow or should forbid . . .' Care must be taken, however, that the two alternatives offered really are the only alternatives.
6. *Aspersions.* Try to avoid questions which reflect on the respondent. This may not be possible, for example if asking whether he or she has any convictions for drunken driving. Various ways of trying to lessen the extent to which the respondent feels threatened have been suggested (see *Statistician*, 29 (4) (1980) p. 237. None is completely successful.
7. *Personalizing.* I have found that asking 'What do you think happens to people after death?' produces different results from 'What do you think will happen to you after you die?'. Make sure that any 'personalized'

questions are intentional; they produce a different emotional response from generalized questions.

The interviewer

An interviewer needs honesty, interest, accuracy, adaptability, a balanced temperament, and to be average in intelligence, dress and speech so as to be easy with all social classes. The main potential sources of error and bias fall into two categories: those stimulated in the respondents and those made in recording answers.

Most obviously, a respondent could react to an interviewer's appearance and dress, which should be neutral. The interviewer must put the respondent at ease – the survey is not a test or an education programme – and should not reveal his or her own attitude, for this could cause a reaction and introduce bias, as could altering the wording or order of questions.

Care must be taken in 'prompting' in dealing with:

(a) partial response (a question only partly answered),
(b) non-response (silence after a question is asked),
(c) irrelevant response,
(d) inaccurate response (where this is obvious) and
(e) spoken response problems: 'I can't really answer that because ...'

Prompting should be neutral: 'Uh-huh ...' or 'What exactly do you mean by this?' or a repetition of part of the answer quizzically. Interviewers should *never* suggest a rephrasing of an answer or summarize it. This could cause bias.

The second category concerns the way the interviewer records answers. Usually potential interviewers are tested to see that they can accurately enter replies – that carelessness will not cause random mistakes in entry. But there are still three ways in which bias could enter:

1. *Attitude expectation.* Vague answers are categorized according to previous ones, and may even be altered to make them consistent. When conducting a Free Methodist survey into religious opinion in a new town area I did some interviewing myself, and had a respondent who said that she did not believe there was a God, and then said that Jesus was 'God's Son'. Respondents may not be consistent, but their answers should not be altered.

2. *Role expectation.* Here vague answers are categorized according to roles expected from the kind of person the interviewer takes the respondent to be. This would cause bias.

3. *Probability expectation.* Here an interviewer tries to 'even out' replies according to the proportion of replies in each category which he or she expects overall. Again this could cause bias.

4.4 Conducting a survey

The following stages are typically involved in conducting a survey:

1. *Define objectives.* Who wants the information? Broadly, what information is required? What is it for?
2. *Define target population.* What class of people or objects are to be surveyed, and where are they located?
3. *Decide sampling method.* The fundamental decision is whether to use a random or a quota sample. This decision will be affected by: (a) the nature of the information required, (b) whether a suitable sampling frame exists or can be constructed, (c) how much time and money are available, and (d) the required accuracy of results. If a random sample is decided on it will further have to be decided whether to stratify or use a multi-stage selection, and whether to use personal interviews or a postal method. The latter decision will depend on how motivated it is felt potential respondents will be to reply, and how geographically scattered the target population are.
4. *Design questionnaire.* A decision must be made about what exactly to include questions on, and then great care should be taken in the construction and wording of the questions.
5. *Train interviewers.* This will obviously be necessary only if personal interviewers are to be used. Commercial survey organizations will already have trained interviewers.
6. *Pilot survey.* If time and money permit it is useful to do a pilot survey, a small-scale survey based on a few of the target population. This enables one to see any obvious problems which emerge concerning the location of respondents, the questionnaire, and so on. If necessary, suitable modifications may be made.
7. *Main survey and follow-up.* The main interviewing or postal 'shots' should be conducted, and any 'call back' or reminder letters sent out. In quota sampling a random check on the accuracy of work submitted by interviewers may be made afterwards.

44—Applied business statistics

8. *Analysis and presentation.* Questionnaire responses may be entered on a computer or otherwise analysed. The final presentation is important: it will affect how far the survey is read and understood by those for whom it was intended.

Recall and exercise questions

1. Briefly define and discuss the relative advantages and disadvantages of:
    (a) simple random sampling;
    (b) stratified sampling;
    (c) quota sampling.
2. Explain the meaning of 'bias', how it arises in surveys, and how it may be diminished.
3. List the points to bear in mind in designing a questionnaire.
4. Outline the stages involved and decisions to be made in conducting a market opinion survey on the general public.
5. 'The *Family Expenditure Survey* of households uses a stratified, four-stage sampling scheme, with a sample size of 10 000 to 11 000, and has a response rate of about 70 per cent.'
   Explain briefly the meaning of the terms used in this statement.
6. On a questionnaire about washing powder the following questions were included:
    (a) How many kilos of washing powder have you bought over the last year?
    (b) What is the average weight (in kilos) of your family wash?
    (c) Do you believe that the caring mother should select her washing powder with care?
    (d) How do you rate 'Sudso' as a washing powder?
       Excellent ☐ Very good ☐ Good ☐ Average ☐
    (e) Some clinicians have recently posited that the enzymal ablutive quality of washing powders generally exceeds tolerance limits for what is optimally functional. Do you agree with this?
   Criticize each of these questions, and where appropriate suggest an alternative wording to obtain the information.
7. A council wishes to obtain information on attitudes to a proposed health clinic scheme, based on a sample of residents. Comment briefly on each of the following suggestions as to how to obtain the sample:
    (a) Select 500 names from a telephone directory.
    (b) Interview 500 people as they visit the doctor.

(c) Insert an article in the local press inviting people to write in with comment.
(d) Set up an exhibition in the market square, and engage visitors in conversation.

Suggest the method you would recommend (one of the above or some other method) giving reasons.

Problems

1. You have been asked by your company to conduct a survey to find out how employees travel to and from work, and how long it takes. One purpose of this survey is in considering the introduction of 'flexi-time' (flexible working hours).
   Produce a report recommending a method to be used, including a draft questionnaire.

2. You have been asked by the makers of 'BO6' shampoo to conduct a market research survey in Mumbleshire, to find out opinion of and usage of their own product and the competing brands in the area. Write a document for consideration by their marketing department, setting out how you propose to conduct this survey and why you want to do it in this way.

3. A large mail-order company wishes to use a postal questionnaire to discover the kinds of lines its agents would wish to see expanded. It would also like to find any connections between agents' area, social class, age, and so on, and the kinds of goods (for instance clothing, houseware, garden products) and the price ranges in which they are most interested.
   Draft a postal questionnaire and include as a separate document a set of notes to explain why you have written the questions as you have.

4. A small company 'Screwit Joinery Ltd' makes window frames and doors. They have previously supplied only large builders, but are wishing to expand into custom-made conversion frames for modernizing older style properties in their area. They wish to obtain information on the market for such a development. Write a letter to Andy Plank (the managing director), suggesting:
   (a) a general plan for such a survey,
   (b) a suitable sampling frame, giving your reasons,
   (c) any secondary data which might be useful.

5. A council wishes to find the probable use of a new library to be built in the area — the likely demand in terms of numbers and types of books. Draft a brief report outlining how they might best do this and what difficulties they might encounter in obtaining valid information.

6. A recent article in the Yelland Graudian entitled 'Should this outrage be allowed?' raised the possibility of a pub and multi-storey car park being built on the site of a local vicarage. Following the article the chief planning officer received 213 letters, only 12 of them supporting the idea. From this he has concluded that the vast majority of local residents are against it.

You have been asked to write a brief presentation for the planning committee:

(a) commenting on the validity of his conclusion
(b) outlining a way in which a more accurate survey could be done to ascertain public opinion on the matter, explaining any problems of bias which might have to be overcome.

5
Tabulation

5.1 The need for and nature of tabulation

We have already noted that two important functions of statistics in business are the presentation and summary of data. Any survey or abstraction from internal records may leave us with a mass of data on individuals. Many individual characteristics will be irrelevant; to form an impression of overall pattern we may want them categorized according to size or attributes. The elimination, then, of irrelevant detail leaves us with a number of categories, and often sub-categories, cross-categories or both, each of which contains a percentage, number of observations or both having the category characteristics. One obvious way to present such information is in the form of a table. This may have categories, and sometimes sub-categories, listed down its left-hand side, and cross-categories, also sometimes with sub-categories, listed across the top. The main body of the table contains numbers or percentages or both. Sub-totals may be included in rows or columns where appropriate.

Three main kinds of category are distinguishable:

1. Non-measurable attributes, for instance geographical (countries or regions), sex or car-ownership.
2. Time periods, as in annual or monthly figures.
3. Measurable variables, such as age and income.

The various kinds of category, the use of sub-categories and cross-categories, and the use of sub-totals are illustrated in Tables 5.1, 5.2 and 5.3.

Various aspects of the different kinds of category in tabulation will be explored in more detail, but first it will be useful to give some general comment on design of tables.

Table 5.1 Households and dwellings by region, mid 1979. From Social Trends 1981 (Table 9.4).

	Number of households (millions)			Number of dwellings (millions)			Tenure of dwellings (percentages)		
								Rented from	
	In metropolitan counties[1]	Elsewhere	Total	In metropolitan counties[1]	Elsewhere	Total (=100%)	Owner-occupied	Local authority	Private owner[2]
Standard regions									
North	0.4	0.7	1.1	0.5	0.7	1.2	46	41	13
Yorkshire and Humberside	1.2	0.5	1.8	1.3	0.6	1.9	55	33	12
East Midlands	–	1.4	1.4	–	1.4	1.4	57	30	13
East Anglia	–	0.7	0.7	–	0.7	0.7	58	27	15
South East	2.7	3.6	6.3	2.7	3.7	6.5	56	27	17
Greater London	2.7	–	2.7	2.7	–	2.7	49	31	20
Other South East	–	3.6	3.6	–	3.7	3.7	62	24	14
South West	–	1.6	1.6	–	1.7	1.7	63	22	15
West Midlands	1.0	0.9	1.8	1.0	0.9	1.9	56	34	10
North West	1.5	0.8	2.3	1.6	0.9	2.4	59	31	10
England	6.9	10.2	17.0	7.0	10.7	17.7	57	30	13
Wales	–	1.0	1.0	–	1.1	1.1	59	29	12
England & Wales	6.9	11.2	18.0	7.0	11.7	18.7	57	30	13
Scotland	0.6	1.2	1.8	0.6	1.3	2.0	35	54	11
Great Britain	7.5	12.4	19.8	7.7	13.0	20.7	55	32	13
Northern Ireland	–	0.5	0.5	–	0.5	0.5	51	38	11
United Kingdom	7.5	12.8	20.2	7.7	13.5	21.2	55	32	13

[1] Including Greater London and Central Clydeside conurbations.
[2] Includes housing associations and other tenures.

Source: Department of the Environment; Scottish Development Department; Department of the Environment, Northern Ireland.

Table 5.2 *Quarterly UK passenger car production, by car size and destination*

			Home				Export			
			1600 cc and under		Over 1600 cc		1600 cc and under		Over 1600 cc	
Year	Quarter	Total	No.	%	No.	%	No.	%	No.	%
1978	Q4	2404	888	*36.9*	447	*18.6*	690	*28.7*	378	*15.7*
1979	Q1	3254	1413	*43.4*	651	*20.0*	849	*26.1*	340	*10.4*
	Q2	3167	1424	*45.0*	617	*19.5*	807	*25.5*	319	*10.1*
	Q3	1711	798	*46.6*	318	*18.6*	408	*23.8*	186	*10.8*
	Q4	2572	1084	*42.1*	469	*18.2*	649	*25.2*	369	*14.3*
1980	Q1	2981	1334	*44.7*	467	*15.7*	815	*27.3*	364	*12.2*
	Q2	2515	1086	*43.2*	351	*14.0*	732	*29.1*	346	*13.8*
	Q3	1816	834	*45.9*	317	*17.4*	487	*26.8*	177	*9.7*
	Q4	1926	1037	*53.8*	316	*16.4*	461	*23.9*	113	*5.9*

Numbers in hundreds.
Adapted from *Business Monitor*.

5.2 Guidelines for tabulation

There are two major aspects of designing a table: its content and its format. The ultimate aim is a high content of useful information, clearly presented. In this there can be no absolute rules, but there are a number of guidelines. It may be useful for the reader to check whether Tables 5.1, 5.2 and 5.3 *do* follow the eight guidelines given below.

The following guidelines relate to basic structure and content of tables:

1. *Basic structure.* A table can vary in complexity. It may have several cross-categories and sub-cross-categories; if so it is best to have some of these going across and others down to give an overall balance.

2. *Sub-totals.* Row and column sub-totals should be included where appropriate. If they are to be included then there should be no omissions, and both row and column sub-totals, if given, should add up to the overall total.

3. *Percentages.* It is often useful to show percentages as well as the actual figures. Where this is done it is preferable to have the percentages as close as possible to the actual figures, perhaps in italics so that they stand out clearly from them.

4. *Size of table.* If the completed draft for a table contains too much detail, it can make it difficult to pick out relevant points. Either some detail should be sacrificed, or the table should be split into two tables, each dealing with particular aspects.

Table 5.3 Deaths from accidents[1] in the home, 1978. From Social Trends 1981 (Table 8.21)

Accident type	Males aged: 0-4	5-14	15-44	45-64	65 and over	All ages	Females aged: 0-4	5-14	15-44	45-64	65 and over	All ages
Poisoning	5	1	154	97	53	310	11	4	118	134	93	360
Falls	10	2	55	173	618	858	14	0	32	112	1647	1805
Fires	41	22	53	61	155	332	28	19	39	67	248	401
Suffocation	80	14	82	45	40	261	71	6	24	49	55	205
Other[2]	29	10	37	35	89	200	30	6	16	37	190	279
All accidents	165	49	381	411	955	1961	154	35	229	399	2233	3050

[1] Excludes deaths undetermined whether accidentally or purposely inflicted.
[2] Includes accidents caused by explosive material (fireworks, blasting material, explosive gases), hot substances, corrosive liquid, electric shocks, etc.

Source: Office of Population Censuses and Surveys; General Register Office for Scotland.

The following guidelines relate to important aspects of presentation once the basic structure and content of a table are decided:

5. *Overall title.* A title should be placed either at the top or underneath the table in bold print, giving a brief but reasonably explicit description of its content and giving the date to which the data relates if this is not included elsewhere in the table.
6. *Sources.* The source of the information should be noted, again either at the head of or underneath the table.
7. *Units.* The units (for instance '000s) should be made explicit, either as part of row or column headings or, if all units are the same, at the head of the table.
8. *Row and column division.* Sometimes in more complex tables we need to distinguish categories, sub-categories, sub-sub-categories, and so on. To divide rows or columns effectively in such cases we may use single lines, then double lines, faint lines, bold lines, dotted lines or spacing. Thus we might divide main categories by double lines and sub-categories by single ones. Large blocks of numbers in unbroken rows and columns should be avoided. If there is no natural 'break' then a line may be inappropriate, but a slightly larger than usual space can be left, say, every five rows, to make it easier to locate a particular number required from the array.

It may be noted that Government sources are rather sparing in their use of dividing lines, and sometimes print large blocks of unbroken rows and columns of numbers. Usually, however, they are reasonably clear.

The first stage in tabulation is to decide on the information to be presented. On this basis a draft can be made of a blank table in the format sought, bearing in mind the points made above. It may be best to do this before the data is abstracted or grouped, so that all the right information is in fact obtained. In many cases today, the responses to a questionnaire are entered into a computer and the actual counting of each category is done by a specially written or a standard 'package' program. The computer services unit of my college wrote such a program for me. It classified responses on some 600 questionnaires on religious belief according to sub-categories of respondents in several different ways simultaneously. Classification of answers by the various different sub-categories of respondent 'by hand' would have taken a great deal of time. Required totals and percentages can also be written into the program, especially if the survey is likely to be repeated at a future date and the same program used.

We have laid out general guidelines on table design. At this point we move to consider points specific to the three basic kinds of category: non-measurable attributes, time periods and measurable variables.

52—Applied business statistics

The first of these, non-measurable attributes, can be dealt with briefly. A table presenting these must consist of various qualitatively different categories, with sub-categories and cross-categories if appropriate. Into each category there will fall a number of somethings. This number might be a number of £s of expenditure, as in Table 6.1, or a number of dwellings, as in Table 5.1. If it is a number of people, deaths, elephants, or most things other than money, we would call the number a frequency (see section 2.2). A frequency tells us how often an observed feature fell into the category concerned, for example, in Table 5.1, how often a dwelling came into the North West category. The use of the term does not, of course, imply that this was necessarily 'frequent' in the sense of being a high number.

5.3 Time series

A time series records changes over time of some particular measurement, percentage or number in a particular category. Figures may be annual, quarterly or monthly, and official sources now often give annual figures followed by monthly or quarterly figures for more recent periods. Such a series is illustrated in Table 5.4 which shows annual and some quarterly figures for United Kingdom retail sales. Some further features of the table are worth noting.

De-seasonalizing. Sales go up in the last quarter of each year (can you think why?) and if an industry or company wishes to know how well it is doing 'for the time of year', this seasonal effect has to be statistically removed. The technique for 'de-seasonalizing' is shown in section 21.4.

Inflation. The rise in total sales from £33 600 million to £58 500 million between 1976 and 1980 gives no indication of changes in the real volume of sales, since the pound was worth progressively less because of inflation. To get a real idea of the volume, we would need to ask what the value of sales in 1980 would have been had prices remained at 1976 levels, for example; that is, we should want 'Sales at 1976 prices'. The technical details of doing this will be looked at in Chapter 11. The third column of Table 5.4 gives sales at 1976 prices, with the quarterly figures de-seasonalized.

Weekly averages. To facilitate comparison of annual and quarterly figures the average sales per week of the year or quarter is more useful than total sales. This is given in the last column as a percentage of the 1976 sales, at 1976 prices and de-seasonalized.

Accuracy. We note that figures which are all 9.2 in the third column appear to be all slightly different in the fourth, ranging from 108.9 to 109.2. This is because the four digits give more accuracy than two digits. Avoiding giving too many digits, which clutters the table, and too few, which blurs differences, is a decision which relates both to the accuracy of the data itself and to the purpose of the table (see section 2.4).

Table 5.4 *UK retail sales, all retailers, 1976-1980*

Period		Actual sales £000 million	£000 million at 1976 prices, seasonally adjusted[1]	Index of weekly sales at 1976 prices, seasonally adjusted[1]
1976		33.6	33.6	100.0
1977		38.4	33.0	98.3
1978		44.0	34.9	103.8
1979		51.5	36.5	108.6
1980		58.5	36.7	109.3
1979	1st quarter	11.0	8.9	105.8
	2nd quarter	12.3	9.5	113.0
	3rd quarter	12.6	9.0	106.6
	4th quarter	15.6	9.2	109.1
1980	1st quarter	13.3	9.3	110.2
	2nd quarter	13.8	9.2	109.2
	3rd quarter	14.3	9.2	108.9
	4th quarter	17.3	9.2	109.0
1981	1st quarter	14.6	9.5	112.7
	2nd quarter	15.2	9.4	111.3

[1] The seasonal adjustment refers only to the quarterly figures.

Adapted from *Business Monitor*.

Sub-categories. The actual source from which these figures were derived, *Business Monitor*, also gives a breakdown by sub-categories of food retailers, mixed retailers and various kinds of non-food retailers, giving sales totals for each. This kind of division of time series into sub-categories is illustrated in Table 5.2, and its sub-categories of home and export models are further divided into smaller and larger engine size. In this instance there are row totals – the total production of all the sub-categories in each quarter – but no column totals, since adding up the total of nine quarterly figures would not be meaningful.

5.4 Measurable variables – ungrouped frequencies

Suppose that a mail-order company wishes to know the household characteristics of 1700 customers on its books. Among other questions it might ask is the size of household, and from this it could construct a table like Table 5.5.

Table 5.5 *Household sizes of 1700 customers of Milkem Mail Order Company*

Number of people in household (x)	Total number of households (f)	Percentage of households ($\%f$)
1	182	10.7
2	241	14.2
3	428	25.2
4	530	31.2
5	208	12.2
6	95	5.6
7	11	0.6
8	4	0.2
9	1	0.1
Total	1700	100.0

Household sizes must be whole numbers, and among our mail-order customers there are only nine different sizes. We may therefore list the numbers (f) of household of each size against that size (x).

In section 5.1 we noted that the number of observations falling into a particular category is termed the frequency for that category. In Table 5.5 the categories are sizes (in this case numbers of people per household) and the number of households having this particular feature is the *frequency*. Thus a column like the second column in Table 5.5 (given the symbol f) could simply be labelled 'Frequency'.

This particular table is an *ungrouped* frequency table because it lists all the *exact* individual values or sizes of the variable (1, 2, 3, 4, 5, 6, 7, 8 and 9) with frequencies for each. It does not group several values together. In practice, the last three sizes (7, 8 and 9) would be very likely to be grouped together into a '7 or more' category, as there would still only be 16 in that category out of the 1700 households.

Also, in practice, ungrouped frequency tables often contain sub-categories. For example, Milkem might wish to show a breakdown of the numbers of households of each size according to social class of head of household (or chief earning joint head). Such sub-categories could be listed to the right of the present f and $\%f$ columns, showing the numbers in each.

It may be noted that the sum of the frequencies in the second column (the fs) gives us the total (1700) in the group. The sum of the percentage frequencies (the $\%f$s) in the third column gives us 100.00, if we ignore small rounding errors in some cases. But what would we get if we added the numbers (the xs) in the first column? The column contains a list of all the possible different sizes of household, and the sum of one of each size is not a useful or meaningful figure to obtain. It is worth making this point, because adding up such columns is a mistake sometimes made.

The table we have been considering is an ungrouped frequency table, that is one in which each possible value of the variable is listed individually and the frequency with which it occurs is given next to it. For such a table to be possible the variable measured must not only be discrete, but must have only a small number of possible values. If it may take more than, say, 15 different values then the number of categories we should need to list would make the table unmanageable. In this case it is better to group the variables as shown in the next section.

5.5 Measurable variables – grouped frequencies

Suppose that as well as asking the 1700 for household size, Milkem also asked for the age of the customer at last birthday. This would give them 1700 integers between, say, 18 and 90. These could be formed into an ungrouped frequency table as before, but in this case it would have 73 lines which would be rather unwieldy and would make it difficult to see any overall pattern. It would probably be best, then, to group them, as in Table 5.6.

Table 5.6 *Age at last birthday of 1700 customers of Milkem Mail Order Company*

Age last birthday (x)	Number of customers (f)	Percentage of customers ($\%f$)
18–22	107	6.3
23–27	210	12.4
28–32	220	12.9
33–37	207	12.2
38–47	341	20.1
48–57	310	18.2
58–67	196	11.5
68–90	109	6.4
Total	1700	100.0

Terminology. We need to emphasize here some important terms used in statistics:

1. The *class intervals* are the intervals 18-22, 23-27, 28-32 etc. into which the variables are grouped.
2. The *frequencies* (*f*s) are the numbers of observations falling into each of the class intervals, and the percentage frequencies (%*f*s) the percentages falling into each of the class intervals.
3. The *class limits* are the boundaries of the respective classes, e.g. the first class interval has a *lower boundary* of 18 and an *upper boundary* of 22.

Discrete and continuous variables

In section 2.2 a basic distinction was made between discrete and continuous variables. As we shall see, when we decide on what class boundaries are appropriate the discrete or continuous nature of the variable concerned is relevant. We shall therefore review the distinction at this point.

For a *discrete* variable, the values it is possible for the variable to take occur at set intervals, with a 'gap' between each potential value and the next one on either side. For example:

1. *Number in household*. This must be a whole number between, say, 1 and 20; it could not be, for example, 5.3 persons.
2. *Age last birthday*. This must be a whole number; for example it could be 24 or 25, but not 24.3, which falls in the gap between them.
3. *Retail price in pence of a loaf of bread*. Prices may vary between, say, 18p and 45p in stages of ½p; in this instance, unlike the previous two, the set of possible values for the variable does not include only integers, but there are still gaps between possible values and so it is discrete.

For a *continuous* variable, there are an infinite number of gradations in the value that the variable can take, and there are no 'gaps' between possible values within the interval where these occur. For example:

1. *Length of steel rod*. A steel rod might be exactly 30.5 cm long, but there is no gap between this and the next possible value above it.
2. *Playing time of an LP side*. Between, say, 18 and 45 minutes there are an infinite number of gradations of time – if measured accurately enough – with no gaps between the possible values.
3. *Weight of grain harvested from a one-acre plot*. Between, say, 1.2 and 1.9 kg there are an infinite number of possible values for the variable.

Now it must be added that in practice the distinction between discrete and continuous variables is not clear-cut. All measuring instruments have a limited accuracy: for example our measuring device for steel rods may give an accuracy only to the nearest 0.001 cm; our stopwatch may give playing times only to the nearest 0.01 second; and our scales may only record weights of grain to the nearest gram.

This will mean that *as actually recorded* there will be gaps between potential values for the variables: the length of the steel rod may be recorded at 30.567 cm or 30.568 cm, but never at 30.5673 cm since the measure is accurate only to 0.001 cm.

Strictly speaking, then, since there is limited accuracy in all practical measurement, all *recorded* measurements will be discrete. But the point is that in the three examples cited above as continuous, the gaps between potential values will be very small indeed. The gaps will be small both in relation to the average size of the variables, and to the range of potential values. For all practical purposes, then, the variables may be regarded as continuous.

In practice *a variable can be treated as continuous if, within its range of variation, it may be recorded as any of a large number of values, the gaps between these being very small in relation to its range and its average value.*

This practical definition relates to the two very basic features of a continuous variable which are not true of a discrete one (both relate to the fact that within a given range a continuous variable has a very very large number of potential values, whereas a discrete variable does not):

1. In a set of values of a continuous variable there is a *very* low chance of any one given potential value coming up: in a set of values of a discrete variable there is a good chance.

2. In a set of values of a discrete variable we may well find numbers of duplicates: with a continuous variable any duplicates at all would be unlikely.

These two differences affect our treatment of the variables in some instances, for example in deciding on appropriate class boundaries.

Choosing class boundaries

There are certain basic guidelines in choosing what class boundaries to use:

1. The class intervals must between them cover the whole range of possible values for the variable.

2. A manageable number of class intervals is, perhaps, between 5 and 15.

3. Equal class intervals are preferable if the values are not too unevenly spread throughout the range of values; but if equal class intervals would result in one class containing many times the number in another, it may be better to have unequal intervals.
4. Each value of the variable should be assignable to one and only one class interval.

In our example, Table 5.6, the ages ranged between 18 and 90, so our intervals have covered all the possible values. We have eight intervals, which is a good number. We have unequal intervals, for there are proportionately fewer people in some age ranges than in others.

In considering condition 4 it may be noted that the intervals as given have gaps: for example the first interval has an upper boundary of 22 and the second a lower boundary of 23, leaving a gap of one between them. However, we remember that the variable is 'age last birthday' which must always be a whole number. We could never, then, get a value between 22 and 23, so any value will be uniquely assignable to one class interval, in spite of the gaps. It is worth considering this point in a little more depth. The following are alternative sets of class boundaries:

A	B	C	D	E
18-22	18-23	17.5-22.5	18 and under 23	18-
23-27	23-28	22.5-27.5	23 and under 28	23-
28-32	28-33	27.5-32.5	28 and under 33	28-
etc.	etc.	etc.	etc.	etc.

(A), as we have seen, enables each value to be assigned to one and only one class interval.

(B) would not be satisfactory for this example, for a customer aged 23 last birthday could be assigned to either the first or the second interval or both. It does not fulfil condition 4. Generalizing this: a *discrete* variable, as we have noted above, can take only a comparatively small number of possible values, and there is therefore a good chance that a set of observations will contain any given one of these values. It is, therefore, unsatisfactory to use as a mutual boundary - as the upper boundary for one class and the lower boundary for the next - a value which the variable may take itself. If we did so there would be a good chance that, in any set of values, a number would fall on these mutual boundaries, and so not be uniquely assignable.

(C) gets around the problem slightly differently by placing the mutual boundary *between* two possible values for the variable. Here an age of 23 *is* uniquely assignable. However, it does seem a little bizarre to clutter up a table of whole numbers with 0.5s, and I personally prefer (A).

(D) is often used, and its shorthand form (E) where only the lower boundary is stated has become popular in published sources. Both do fulfil condition 4, for any value will be uniquely assignable. However, I believe that (D) tends to give a misleading impression, for one tends to feel that the values in the first class must run between 18 and somewhere very near 23. In fact they go only from 18 to 22. (E) suffers from the same problem, and in addition is often tedious to read. Personally, then, I prefer to use a table in a form like (A), with suitable gaps, if the data is discrete.

A final point about class boundaries is that in some instances the first and last class intervals can be given as 'open-ended': in our example the last interval might be stated as '68 and over' rather than as '68-90'. Such an interval is generally used when observations above a certain value are becoming increasingly widespread, but could theoretically be very large indeed. This is especially useful when a future survey may use the same intervals, and might happen to throw up a freak value or two above all those in the present survey. Open-ended intervals are frequently used in published sources, and it might be argued that sometimes they begin too soon and contain too high a proportion of the observations.

Continuous variables

We have looked at a discrete variable, but what about a continuous one? Leaving gaps between the class intervals would leave us unable to assign any values which fell in those gaps.

Suppose, for example, Prunes Stretch Covers Ltd have 157 employees on a particular kind of piecework. From internal records for the last financial year, Prunes can find the average weekly earnings in that period for each of the 157 workers. These might be:

Mabel Smith £42.13
Joan Darby £35.57
Gwen Jones £50.19
Linda Loopy £41.79
etc.

If all the 157 values fall between £30 and £60, we might wish to consider one of the following groupings:

A	B	C	D	E
30-32	30-32.99	30-33	30 and under 33	30-
33-35	33-35.99	33-36	33 and under 36	33-
36-38	36-38.99	36-39	36 and under 39	36-
etc.	etc.	etc.	etc.	etc.

In this example the figures are given in pounds, correct to two decimal places. So within the effective range of £30 to £60 there are a large number of possible values (3001) the variable could take, and the gaps between them, 0.01, are small. Practically therefore, as we have noted above, it may be thought of as a continuous variable.

(A) will clearly not do in this case, for an earnings figure like Joan Darby's of £35.57 will be unassignable since if falls in a gap between two consecutive class intervals.

(B) will enable each earnings figure to be uniquely assigned since they are given only to two places of decimals and so we will never get a value like 35.993, which would fall in a gap. I find it rather cluttered, however, and if the figures had been given to three decimal places it would look even more so.

(C) would allow all values to be assigned, but a value of earnings of exactly £33.00 could be assigned to either the first, or the second or to both intervals. Strictly speaking, then, this style fails to allow unique assignment of all values, and on this basis some books simply pronounce it wrong. However, because of the large number of possible values of the variable in this example (that is, it is practically continuous) only something like one observation in 300 will be likely in practice to fall on a mutual boundary. It could certainly be argued that (C) is so much easier to scan that one could accept the arbitrary assignment of the odd one in 300, to the lower interval say, in order to retain this clarity. Practical statistics is not about pedantic 'precision' - where data is itself often only approximate - but about clarity.

(D) and (E) are again both satisfactory in allowing each value to be uniquely assigned to a class interval. In this example, unlike that of the discrete ages, they do *not* give a misleading impression, though they may be less easy to scan than (C).

For this kind of example, involving a nearly continuous variable, the safest format for examination purposes would probably be (D), since no examiner will object to it. For practical presentations I sometimes use (C) in preference because it is easier to read, provided that the chances are small that observations will fall on mutual boundaries, and a footnote on the treatment of any such rare borderline cases is included somewhere.

Keeping tallies

One last point needs to be made about the practicalities of counting how many observations fall into each class interval. A modern survey of any size is likely to have its data grouped and analysed on a computer, but for smaller examples it is still sometimes simpler to group values by hand. Suppose that we begin with a block of figures like that in Table 5.7, which shows the weekly earnings (in pounds) for each of the 157 Prunes workers on piecework, averaged over the last year.

Table 5.7 *Last year's average weekly earnings (in pounds) for 157 employees on piecework for Prunes Stretch Covers Ltd*

42.13√	35.57√	50.19√	41.79√	42.45√	50.98√	47.87√	43.34√	44.74√	46.87√
44.56√	43.78√	46.23	47.98	52.56	49.87	50.23	56.87	50.24	37.98
45.67	52.34	40.56	32.34	46.75	41.35	58.65	50.65	46.76	51.23
37.87	43.56	48.87	50.87	55.65	52.34	55.55	54.00	49.87	46.77
40.56	34.54	57.89	56.78	49.87	56.44	46.76	50.67	52.45	46.66
49.65	43.56	46.77	47.65	49.77	46.75	58.67	49.87	49.66	45.34
45.67	44.66	48.76	49.67	48.77	39.12	40.19	37.88	48.66	51.11
46.76	47.56	43.16	49.15	49.10	43.16	37.45	55.18	52.19	48.35
49.67	46.72	43.25	40.37	36.34	48.76	49.58	46.33	57.65	46.55
46.77	49.32	50.34	52.44	56.77	32.45	45.66	47.88	49.88	43.45
53.56	52.33	48.99	55.67	34.58	46.75	48.77	52.45	44.36	44.76
48.73	49.76	50.67	45.67	47.65	44.62	40.45	48.19	39.54	40.55
45.32	47.17	58.62	52.33	48.66	45.69	46.71	44.66	34.65	42.33
49.35	48.66	45.63	49.29	40.10	37.88	49.37	48.54	55.65	50.11
51.34	44.67	40.86	50.54	51.22	48.03	44.93	41.82	43.03	45.67
47.09	48.99	36.76	42.36	58.73	48.35	46.65			

First we decide on appropriate class intervals. Since the figures all fall within the range £30 to £60, intervals of three will give us the convenient number of ten intervals. I have listed them in Table 5.8 as 30–33, 33–36, 36–39 and so on, remembering that if any of the figures happen to fall exactly on the boundary I shall assign them to the lower interval.

To find the numbers of values in each interval one *could* first go right through the table counting those in the first interval, repeat this for the second interval, and so on. This, however, would not only be laborious, but miscounts would be quite likely. A better way is to go through the set of numbers just once, ticking off each in turn and keeping a running tally of how many fall into each interval. Table 5.8 shows how this might look after the first dozen had been ticked off.

Table 5.8 *Running tally for earnings of Prunes employees on piecework*

Earnings (x) Class interval	Number of employees (f)	
	Running tally	Total
30–33		
33–36	1	
36–39		
39–42	1	
42–45	︎卌 1	
45–48	11	
48–51	11	
51–54		
54–57		
57–60		

As each observation is ticked off in turn, a mark is made opposite the appropriate class interval. Every fifth entry in a row is a line going through the previous four to make a 'five-bar gate' as this is the most convenient way to keep count. When all have been ticked off and marked up, the totals in each row can be written in, and these will be the frequencies of observation in each interval.

Table 5.8, it should be noted, is a working table. It should not appear in this form in the final version of a statistical report or presentation, for it is simply a convenient way to count the numbers in each interval. If the data is analysed by computer it will not be needed at all.

Summary of steps

Given a mass of numbers to be put into a frequency table, the following would be the steps involved:

1. Find the smallest and largest numbers – the 'range' of the variable.
2. Find out how many possible values the variable can take within its range. If these are fewer than, say, 15, we may use an *ungrouped frequency table*. In this we simply list each possible value in the range, and count the frequency with which each value appears in the table of data.
3. If the variable is effectively *continuous*, or is discrete but there are more than 15 possible values, then a *grouped frequency table* will be appropriate. The total range of the variable should be divided up into between 5 and 15 *class intervals*. These should be equal if this does not make the disparity between the numbers of observations falling into each interval too large.
4. The *class interval boundaries* should be chosen so that each value of the variable is uniquely assignable to one class interval, but the boundaries should appear as uncluttered and clear as possible.
5. A running tally table should be set up, and a mark put in the appropriate class interval as each of the numbers in the data is ticked off in turn.
6. The final table should be laid out clearly from this running tally table.

Recall and exercise questions

1. What are the main points to be borne in mind in designing tables of data?
2. Briefly explain the meaning of the term 'frequency'.
3. What is the difference between a discrete and a continuous variable?

4. State which of the following variables you would class as discrete and which as continuous:
   (a) times of task completion in work study;
   (b) weights of pickle in bottles;
   (c) daily rainfall in centimetres;
   (d) premium bond purchase by different individuals during a month;
   (e) passengers carried on a bus.

5. In a particular year the General Household Survey used a total sample from Great Britain of 11 705 households. Of these, 485 had heads of household under 25 years old, of whom 2 per cent were outright owners, 32 per cent were mortgagors, 27 per cent rented from a local authority (including New Towns), and the remainder rented from private sources. There were 3953 households which had heads aged between 25 and 44, of whom 241 owned outright, 2112 were mortgagors, 1116 rented from local authorities, and the remainder rented privately. There were 3119 households with heads aged between 45 and 59, of whom 21 per cent rented from local authorities, and the rest rented from private sources. Of the heads of household aged 60 or more, 1750 owned outright, 173 were mortgagors, 1618 rented from local authorities, and 607 rented from private sources.

   Put these figures into a table, showing appropriate actual figures, sub-totals and totals, and percentages.

6. In a market survey, 50 customers were followed up, and the numbers in each of their households found to be as follows:

2	5	3	3	1	6	3	2	4	8
3	4	3	4	1	5	1	3	4	3
6	3	4	3	2	1	3	4	2	3
3	4	2	3	2	3	7	3	4	3
4	3	2	3	4	3	3	3	4	1

   Present this data as an ungrouped frequency table.

7. From Table 5.7 complete the compilation of a grouped frequency table with the class intervals shown in Table 5.8.

8. The following figures show the numbers of passengers carried on each of 50 regular flights of an aircraft seating 80 people:

   41 51 58 45 39 55 54 76 64 46 45 45 22 56 67 62 61 69
   60 12 16 48 33 23 80 56 27 31 59 58 54 52 51 42 75 34
   68 40 35 39 72 60 68 62 57 65 42 43 52 50

   Present this data in the form of a grouped frequency table with suitable class intervals.

6
Charts and Diagrams

6.1 Purpose and principles of diagrammatic representation

However well we design tables of figures, most people find it difficult to obtain a vivid idea of overall pattern from them. Tables 5.1 and 5.4, for example, show seasonal effects, but it would be much easier to see this if the figures were shown graphically. Table 5.3 shows great differences both by age and by sex in the numbers of fatal accidents in the home, but again the pattern could be seen more clearly if presented in a diagrammatic way.

In some books, diagrammatic presentation is regarded as a 'soft' topic, and tends to be skipped over before the 'real' statistics. Yet the need for good, clear presentation of statistics arises more often than the need for advanced complex analysis, and the too high incidence of poor standards of presentation belies the suggestion that people can 'just pick up' the principles of good presentation when needed. Underlying the treatment in this book is a belief that anyone undertaking an elementary or an advanced course of statistics will benefit from thinking about presentation.

A diagram is unsuitable for conveying precise numerical detail – a table does this better. Data is presented in diagram form to give a vivid visual impression of the relative sizes of comparable frequencies, amounts or measurements. A good diagram conveys quickly what features represent which variables or values, and gives a basis for an easy, and not misleading, visual comparison of their relative sizes.

In general, relative size is conveyed visually by means of either numbers of symbols, for example a number of little men or a number of dots, or sizes of symbols, for example big men and little men or tall blocks and short blocks. If we use sizes, then there are three possible bases for visual comparison: simple length, area and volume. One thing which we must be careful *not* to do is to present symbols with different *areas* or *volumes* in a diagram, and expect people to compare mentally only lengths. Since in general in our society volumes, packets of sugar or cement for instance, are more significant than heights, people tend subconsciously to compare volumes.

Our diagram will show different numbers of or sizes of things, and those things can either be actual pictures or simple dots or blocks. In general, pictures may make the diagram more vivid, but dots and blocks are easier to use since they are more flexible.

Finally, we must avoid trying to show too much on one diagram. It is better to have two clear diagrams than one comprehensive but confusing one.

In this chapter and the next two we shall be looking at various kinds of diagram, and the remainder of this chapter concerns ways of presenting data which has different frequencies or amounts in various qualitatively different categories. An example of this is shown in Table 6.1, where the qualitatively different categories are different kinds of services paid for by Mumbleshire County Council.

Table 6.1 *Expenditure (in £millions at current prices) of Mumbleshire County Council: 1981-2 and 1984-5*

Service provided	1981-2 £m	1981-2 %	1984-5 £m	1984-5 %
Education	306.3	55.4	440.6	53.8
Libraries	9.1	1.6	13.2	1.6
Social services	63.2	11.4	71.5	8.7
Police	54.8	9.9	103.6	12.6
Fire services	13.3	2.4	16.5	2.0
Transport and highways	49.7	9.0	75.6	9.2
Other	24.6	4.4	50.2	6.1
Inflation provision	32.1	5.8	48.3	5.9
Total	553.1	100.0	819.5	100.0

What can be shown

The sections which follow will explain various formats for visual presentation, bringing out different aspects of the figures given. In general, different things we might wish to portray are:

1. Comparison of monetary amounts, or percentages of the total, spent on the different services within a single year.
2. Comparison of monetary amounts spent on the same services in the two years.

3. Comparison of the percentage expenditure on the different services in the two years.

4. Comparison of expenditure *in real terms* on the different services in the two years. This would mean allowing for inflation reducing the real value of the monetary expenditure. This kind of thing is looked at in Chapter 11 but will not be considered at this stage. Here it should be remembered that we *are* dealing with monetary amounts and not real values.

One final point: as a means of illustrating the various diagrams I have chosen an example in which the 'amounts' in each category are expenditures; that is they are numbers of pounds, or rather millions of pounds. But these amounts could equally well be numbers of something else, such as numbers of vehicles, numbers of people or numbers of sheep. The basic situation we are looking at is any where there are a number of qualitatively different categories (in our example Council services) each containing a different number of somethings (in our example pounds spent).

6.2 Pictograms

One way of presenting relative sizes, amounts or frequencies is to use a diagram which involves pictures of what is represented. Thus, for example, Fig. 6.1 uses numbers of bags of money to represent expenditure. One could, of course, equally well use numbers of dots to represent amounts, but using bags is more striking. It is difficult to use such an approach to present precise figures – the difficulty of representing 0.3 of a bag is obvious – but the method is only intended to give a striking general impression.

One variation on this would be to draw different *sizes* of thing for each figure. Thus, for example, we could draw different sizes of money bag to present the data in Table 6.1. But this brings problems. What exactly would form the basis of our visual comparison? For example, a bag which was twice as big as another *in all three dimensions* would hold not twice but eight (2^3) times the amount. Capacity is proportional not to height, nor even to area, but to volume. Since in our everyday affairs volume is usually the most important feature of size, most people will subconsciously judge volume.

So what might we do? Money bags with the same proportions but of different sizes could be drawn so that volumes rather than heights or widths were proportional to amounts, but even this would make it difficult for most people to form a good mental comparison.

Another alternative would be to use pictures where the cross-sectional areas were the same and only the heights differed. This is possible with columns of coins, for example, but is unsuitable for many things, including money bags.

Charts and diagrams—67

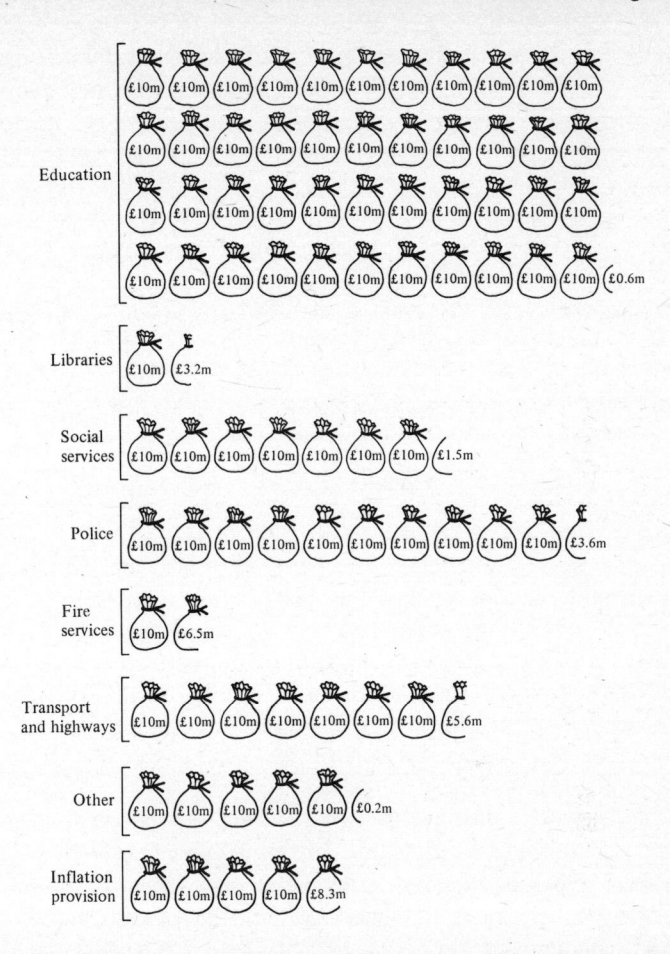

Figure 6.1 *Pictogram showing Mumbleshire County Council expenditures: 1984-85*

A third alternative would be to use only two dimensions and make *areas* proportional to amounts. This may be feasible, but there is always a danger that people will unconsciously add a third dimension to the flat diagram.

Really, it is usually best to avoid using sizes of objects to represent amounts. Yet this method continues to be used, not merely misleadingly, but without any attempt to get it right. While preparing this chapter I received a copy of a county council publication showing sources and expenditures of council money. The different amounts were written over different sizes of money chests and wallets respectively, but the sizes of these were in no way related to the amounts: for example, a chest representing £334m had, I estimate, a volume only about nine times that of a chest representing £4m.

68–*Applied business statistics*

It is best to stick to using numbers of things to represent amounts. A variation on this is shown in Fig. 12.1, where sex and marital status are shown by variations in the pictorial symbol. Another variation is shown in Fig. 11.1, where completely different pictorial symbols are used to represent different kinds of thing. This too has its pitfalls, for if the symbols are different then their relative sizes are arbitrary. A further variation involves the use of a map, on which shadings, numbers of pictorial symbols, or simply numbers, represent amounts or frequencies of variables in each area.

Diagrams where pictorial symbols are used are sometimes called *pictograms*. To sum up: in some situations they can be a vivid way of presenting data, but they lack the flexibility and accuracy of the more purely diagrammatic methods of presentation which we shall now consider.

6.3 Bar charts

If we decide against a pictogram, we may use some purely diagrammatic way to represent the different sizes of expenditures in Table 6.1.

For simplicity and clarity, such diagrams are usually made using flat, two-dimensional areas for the visual comparison. The simplest form of this uses rectangles or 'blocks' of equal width but of varying height or length. Our visual comparison will be based on area – equal to length of rectangle times width – but we may concentrate only on the one dimension of length, since the lengths will be proportional to the areas.

Such a diagram is called a 'bar chart', and Fig. 6.2 shows two such charts (one for each year) put side by side for comparison. In each bar chart the length of any block is proportional to the amount spent on the service it represents. Since all blocks are equal in width, their *areas* are also proportional to the amounts spent.

For practical purposes the blocks representing education expenditure for each year have been divided into three equal parts. This is because if the scale had been chosen so that they could be shown as single rectangles, the other amounts would have appeared as very small rectangles and comparison between them would have been difficult.

Fig. 6.2 has the blocks running horizontally, so that what they represent can be written on or alongside them. But bar-chart blocks can run vertically if preferred, as in Fig. 6.3. There are also some other differences in style sometimes seen in bar charts, such as omitting the gaps between blocks.

What do these bar charts tell us? Looking just at the bar chart for 1984–5, for example, we can see the relative sizes of expenditure on the different services. With the bar charts for the two years placed alongside one another, we may compare expenditure on the same services in those years.

Charts and diagrams—69

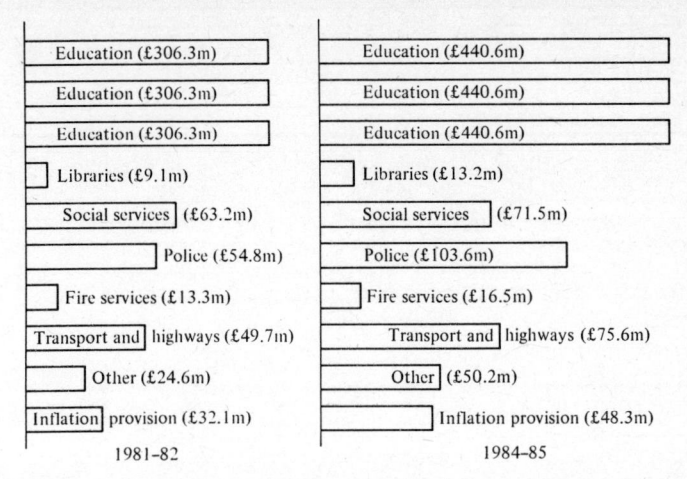

```
| Education (£306.3m)          |    | Education (£440.6m)          |
| Education (£306.3m)          |    | Education (£440.6m)          |
| Education (£306.3m)          |    | Education (£440.6m)          |
| Libraries (£9.1m)            |    | Libraries (£13.2m)           |
  Social services | (£63.2m)        Social services | (£71.5m)
              | Police (£54.8m)             Police (£103.6m)  |
  | Fire services (£13.3m)            | Fire services (£16.5m)
  Transport and | highways (£49.7m)    Transport and | highways (£75.6m)
        | Other (£24.6m)                    Other | (£50.2m)
  Inflation | provision (£32.1m)              | Inflation provision (£48.3m)
              1981–82                              1984–85
```

Figure 6.2 *Bar charts to show expenditure (at current prices) on various services by Mumbleshire County Council*

Steps in drawing

The steps in drawing a simple bar chart are set out below. It is assumed that, initially at least, the chart will be drawn on graph paper.

1. Look at the relative quantities or sizes to be portrayed. If one is *much* larger than the others then it may be best to divide its block into several parts of equal length, as for education in Fig. 6.2.

2. Find the category which will require the longest block; this may be a complete amount or part of an amount divided as in 1. Count the squares on the graph paper, and make the scale such that this longest block fills as much as possible of the available width or height of the paper.

3. Draw the blocks running horizontally or vertically as preferred.

4. Make clear which category each block represents. This can be done either with colour or shading described in a 'key', or by writing what it represents on or alongside the block itself. The latter is preferable where practicable.

5. The actual amounts represented should be conveyed somewhere. This can either be done as in Fig. 6.2, by writing them on or alongside the blocks, or by placing a labelled scale along the edge of the diagram running parallel to the blocks.

6. Ensure that the chart is clearly titled, and states the date and source of the data.

Multiple bar charts

Rather than placing two charts side by side, it is possible to 'mix' the different years on one chart. Such a multiple bar chart is shown in Fig. 6.3. The purpose of a multiple bar chart is very similar to that of drawing two simple bar charts as in Fig. 6.2. Similar techniques and comments apply. One restriction is that it becomes more cumbersome to divide a block into several parts, as for Education in Fig. 6.2.

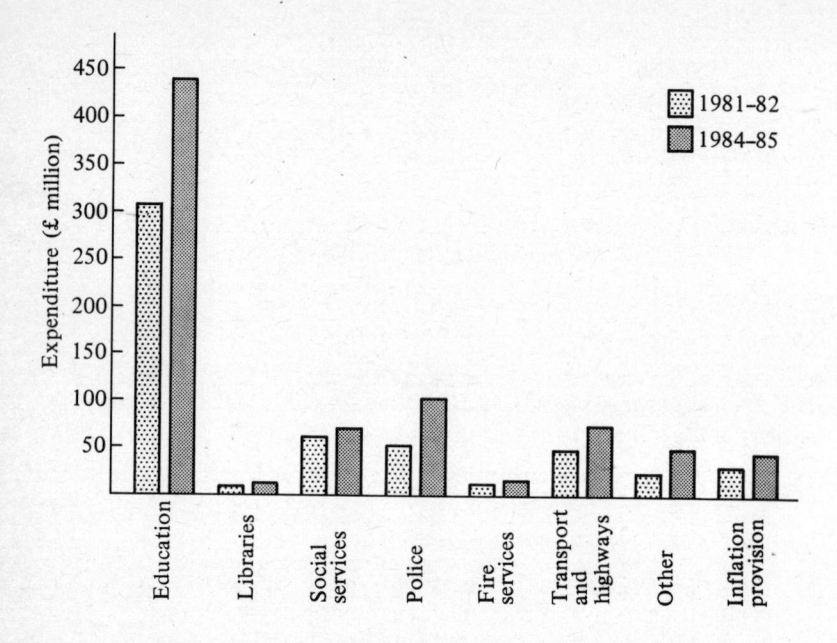

Figure 6.3 *Multiple bar chart of Mumbleshire expenditures*

Component bar charts

There are two remaining major variations on the bar chart. Both involve drawing one block or bar to represent the total for a set of figures, but dividing this into sections representing the various components of that total. In Fig. 6.4 two blocks are given, proportional to the actual total expenditures for 1981-2 and 1984-5. In Fig. 6.5 the two blocks are made the same size, so that one may concentrate simply on the *percentages* of the total expenditure going to each service in the two years without having to allow for the change in monetary total. Whether we choose an ordinary *component bar chart*, as in Fig. 6.4, or a *percentage component bar chart*, as in Fig. 6.5, will depend upon

Charts and diagrams—71

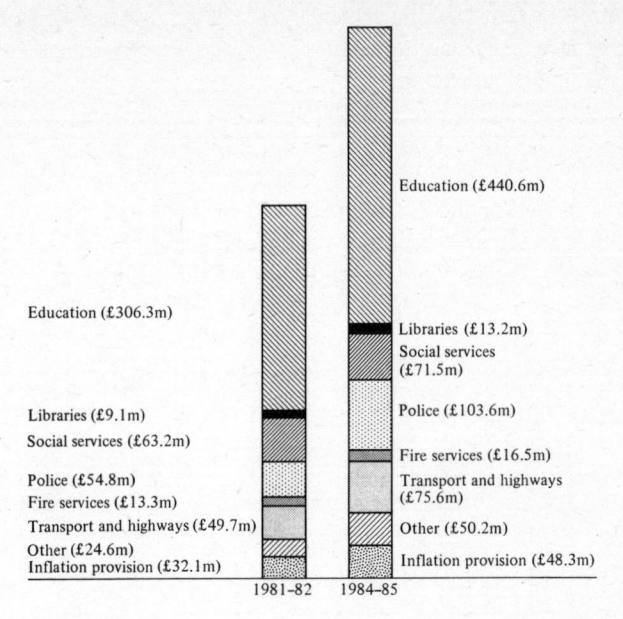

Figure 6.4 *Component bar charts showing expenditures of Mumbleshire County Council*

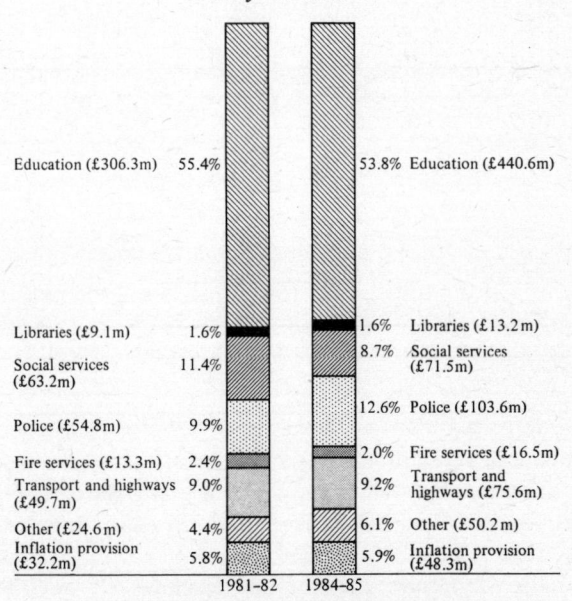

1981–82 bar:
- Education (£306.3m)
- Libraries (£9.1m)
- Social services (£63.2m)
- Police (£54.8m)
- Fire services (£13.3m)
- Transport and highways (£49.7m)
- Other (£24.6m)
- Inflation provision (£32.1m)

1984–85 bar:
- Education (£440.6m)
- Libraries (£13.2m)
- Social services (£71.5m)
- Police (£103.6m)
- Fire services (£16.5m)
- Transport and highways (£75.6m)
- Other (£50.2m)
- Inflation provision (£48.3m)

Figure 6.5 *Percentage component bar charts showing expenditures of Mumbleshire County Council*

1981–82:
- Education (£306.3m) 55.4%
- Libraries (£9.1m) 1.6%
- Social services (£63.2m) 11.4%
- Police (£54.8m) 9.9%
- Fire services (£13.3m) 2.4%
- Transport and highways (£49.7m) 9.0%
- Other (£24.6m) 4.4%
- Inflation provision (£32.2m) 5.8%

1984–85:
- Education (£440.6m) 53.8%
- Libraries (£13.2m) 1.6%
- Social services (£71.5m) 8.7%
- Police (£103.6m) 12.6%
- Fire services (£16.5m) 2.0%
- Transport and highways (£75.6m) 9.2%
- Other (£50.2m) 6.1%
- Inflation provision (£48.3m) 5.9%

72 – Applied business statistics

the feature (actual amount or percentage) which we wish to bring out. The details of how to go about drawing such charts are very similar to those given above for simple bar charts.

Further examples

Both *Social Trends* and *Economic Trends* contain various kinds of bar chart. It will be useful to give one other example here. Fig. 6.6 is a bar chart where in each category there are not expenditures but rather numbers or 'frequencies' of people – readers of Sunday newspapers by social class.

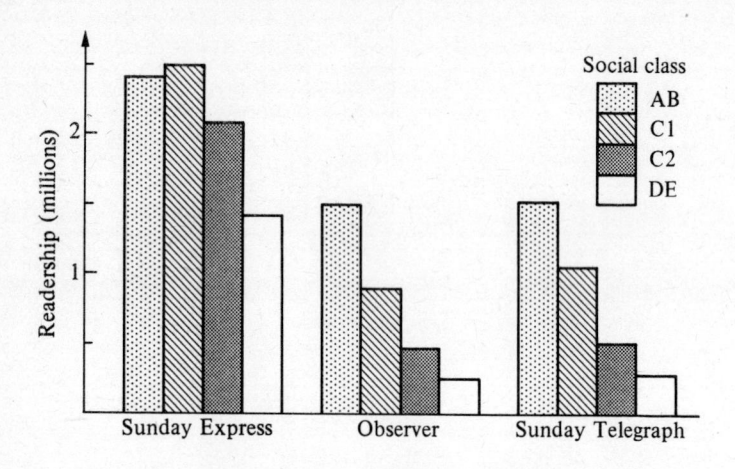

Figure 6.6 *Sunday paper readership*

6.4 Pie charts

Bar charts portrayed different amounts by areas of blocks having the same width. An alternative technique also uses area – the areas of segments of a circle. This is called a pie chart; one for each year is shown in Fig. 6.7. Looking first at the one for 1984-5, the education expenditure forms 53.8 per cent of the total. Since there are 360 degrees in a circle, we make the angle of the Education segment at the centre of the circle

$$\frac{53.8}{100} \times 360° = 193.68°$$

Similarly that for Libraries will be:

$$\frac{1.6}{100} \times 360° = 5.76°$$

and so on. The segments are then drawn with the calculated angles at their centres. Their areas will then be proportional to the amounts of expenditure they represent.

The drawing of a pie chart for each year allows a comparison between years. In Fig. 6.7 the circles for 1981-2 and 1984-5 have been drawn with the same radius. This makes these pie charts equivalent to the percentage component bar charts shown above: they enable us to visually compare percentages of expenditure between the two years, rather than the actual amounts spent on each service. However, it would be possible to draw different sized circles to represent the different total expenditures. Since the area of a circle is proportional to the radius *squared* we would have to make the radius of the circle proportional to the *square root* of the total expenditure in each year, so that the areas were proportional to the actual totals. This is possible, but is not going to be shown in detail here since it is not very satisfactory. Mentally, it is not easy to compare the areas of segments in circles of different radii. A pair of component bar charts would probably be better if we wish to compare amounts spent.

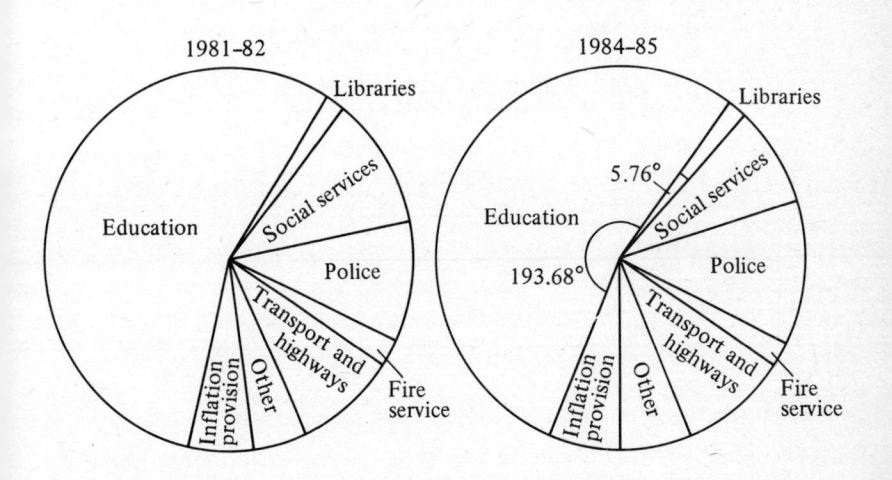

Figure 6.7 *Pie charts showing expenditures of Mumbleshire County Council*

Pie charts are frequently used both in Government published sources and in other publications. For some purposes they may be neater than bar charts, and, since we all have experience of different sizes of slices of pie, are thought by some to be more immediately understandable by the general public. They may be especially vivid in conveying the sharing out of some kind of communal 'cake' of limited size, such as the local authority expenditure budget shown here, between various worthy causes.

Further examples

Published sources contain many. In the current *Social Trends*, for example, there is one showing household types, and a set of differently sized pie charts showing reception of male remand prisoners by age. *Economic Trends* does the same for income distributions.

Recall and exercise questions

1. What is the point of using diagrams to present data?
2. Briefly comment on the general impression about social class and Sunday paper readership which is conveyed by Fig. 6.6. What other diagrams might have been used to emphasize different aspects of the data?
3. Table 6.2 shows UK Customs and Excise estimates of spirits consumption in millions of proof gallons for several years.

Table 6.2 *Estimates of spirits consumption*

	1975	1977	1979
Whisky	16.8	18.2	20.9
Gin	5.7	6.0	6.4
Vodka	2.6	3.4	4.4
Rum	3.5	3.2	3.8
Brandy	2.5	2.5	3.0
Other	1.4	1.6	2.0
Total	32.4	34.9	40.7

Compare the different years' consumption patterns using suitable
(a) pictograms
(b) bar charts (including multiple and component)
(c) pie charts

4. Use suitable diagrams to portray different market shares in the field of decorative and industrial paints for the companies shown in Table 6.3.

Table 6.3 Market shares of makers of domestic and industrial paint

Company	Decorative paint market share (%)	Industrial paint market share (%)
ICI	27	15
International Paint	4	32
BJN	17	14
Crown	17	4
Macpherson	11	2
Goodlass Wall	4	4
Blundell-Permoglaze	6	4
Ault and Wiborg	–	10
Leyland	4	–
Bestobell Paints	3	2
Other	7	13

Problems

1. The 1983 Annual Report of a large chemical company carried the figures for exports given in Table 6.4.

Table 6.4 Destination of exports

	1981 £million	1982 £million	1983 £million
EEC	318	505	532
Rest of Western Europe	110	117	125
Total Western Europe	428	622	657
Africa	100	116	141
Far East	73	87	86
North America	77	79	84
Central and South America	48	59	66
Eastern Europe and USSR	53	69	64
Middle East	48	44	55
Australasia	42	46	45
Indian subcontinent	13	13	10
Total	882	1135	1208

Prepare a brief report on the changing pattern of exports over the three years.

2. In a particular year the National Food Survey obtained the results shown in Table 6.5 for expenditure on various items by households containing two adults.

Table 6.5 Weekly expenditures (pence per head) of two-adult households

Product	\multicolumn{5}{c}{Number of children in household}
	0	1	2	3	4 or more
Butter	22.90	17.64	12.92	11.37	10.25
Margarine	10.88	8.77	7.62	9.54	7.97
Lard and compound cooking fat	3.81	3.39	1.97	2.47	2.97

Present a report showing the patterns of total and per-head expenditure on these products, and how this is affected by the size of household.

3. Obtain figures on the different type of motor vehicles currently licensed in the UK from *Transport Statistics Great Britain* or *Social Trends*. Use diagrams to present an analysis of the way in which numbers and percentages of different kinds of vehicle have changed over about the last decade.

7
Simple Graphs

7.1 Coordinate axes and scattergraphs

The previous chapter showed the use of lengths and areas to represent various magnitudes. But suppose that for each individual or entity measured we have *two* measurements of different aspects. Suppose, for example, that I find the market and the rateable values of eleven detached houses, taken at random, in Fulwood, as in Table 7.1.

Table 7.1 *Market value and rateable value of eleven houses in Fulwood*

House	Market value (y) (£thousand)	Rateable value (x) (£)
1	51.0	360
2	59.0	302
3	72.5	298
4	72.5	352
5	45.5	290
6	47.5	333
7	38.8	274
8	39.5	289
9	36.0	242
10	30.0	272
11	27.0	215

One way to show both measurements at the same time is to have a vertical length to show market value, and a horizontal one to show rateable value, or vice versa. This means using the *coordinate axes* with which most readers will be familiar. Fig. 7.1 shows such a graph of the data in Table 7.1.

78 – Applied business statistics

A vertical scale, the 'y-axis', shows market value, and a horizontal scale, the 'x-axis', shows rateable value. For each house a point on the graph, marked here by a dot, shows market value by its distance up the vertical scale and rateable value by its distance along the horizontal scale. To make this clearer, in Fig. 7.1 numbers are placed by the sides of the dots to show which point represents which of the eleven houses. House 9, for example, is 36 (thousand pounds) up the vertical axis, and 242 (pounds) along the horizontal axis.

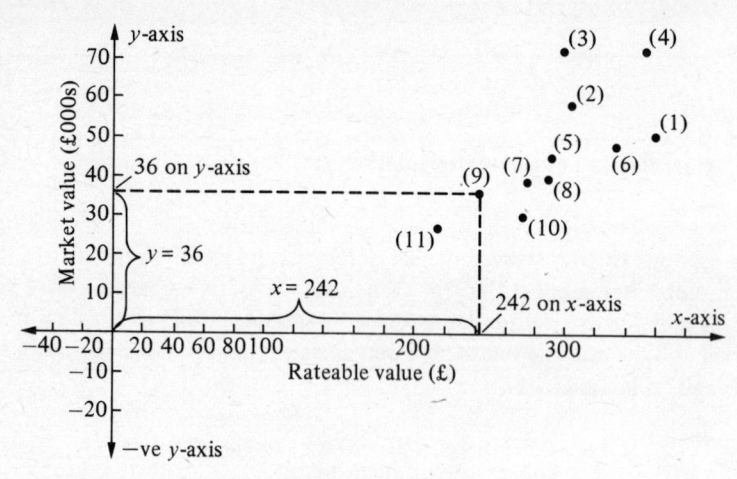

Figure 7.1 *Scattergraph showing market value and rateable value of 11 Fulwood houses*

Many readers may remember drawing graphs in which points were plotted and then joined up with lines. Such graphs are considered later (Fig. 7.3) but in this instance this would not be sensible, for there is no logical order in which to join up the points. Would we join them in order of magnitude of market values, or of rateable values? A graph like this, where it is not appropriate to join the points with lines, is called a *scattergraph* or *scatter diagram*.

There are some important aspects of using coordinate axes in graphs and scattergraphs which may be illustrated from Fig. 7.1:

1. I have used large dots (●), the centres of which mark the points. One may also use smaller dots with circles around them, crosses (X), or crosses around the other way (+) to mark the points.

2. Different scales are used on the two axes. This is common; often they are in different kinds of units altogether (for example grams and centimetres).

3. Both x and y axes also have negative parts where values of x and y will be negative. Obviously, in this case, and in most but not all cases, a negative value for x or y would be meaningless, so we do not need to show the negative axes. In fact, we do not really need to show the section of the x axis between zero and 200 or the section of the y axis between zero and 20, since no points fall in either. We could begin the x axis at 200 and the y axis at 20. Many statisticians, however, like to emphasize the fact that the axes do not start at zero by putting breaks in the axes, as shown in Fig. 7.2, which shows the same data.

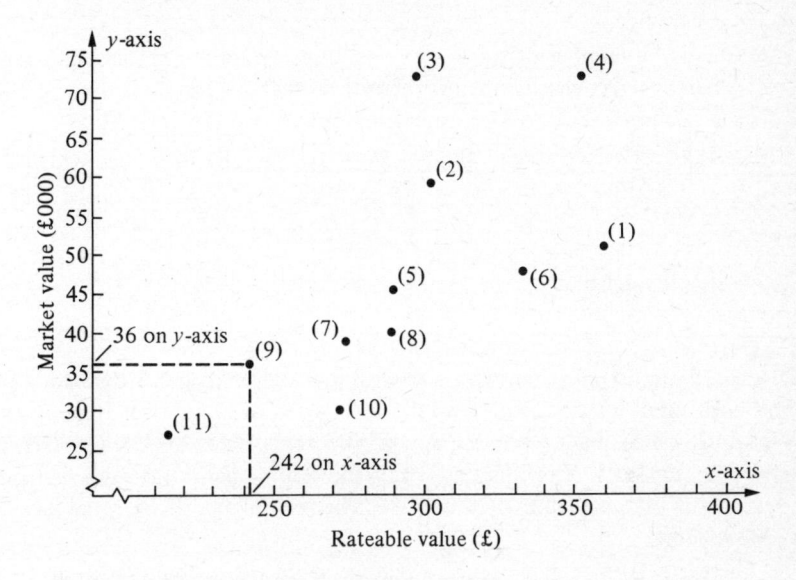

Figure 7.2 *Scattergraph showing market value and rateable value of 11 Fulwood houses*

We should note carefully what is gained and what is lost in doing this. We gain in being able to use a larger scale, and this means that the *differences* between the values on either scale, and the *ratios of differences* between them, are shown more clearly. On the other hand, we lose in that the *ratios of actual values* to each other are no longer obvious. For example, the houses numbered 3 and 4 both have about twice the market value of that numbered 9. This is apparent in Fig. 7.1, but Fig. 7.2 gives no such visual impression.

In general, for a scattergraph or any other graph, an axis starting from zero allows a visual comparison of ratios, while one which has either a break or a non-zero starting point does not. On the other hand, one with a break or a non-zero start may give a better impression of the ratios of the *differences* between values.

Drawing graphs

Though for clarity the grid lines have been omitted from most charts and graphs in this book, graphs will normally be drawn on graph paper initially. The procedure is as follows:
1. Work out the *range* required on each axis. This is the difference in each case between the largest and smallest value required. The smallest will be the start of the axis (zero if it is to start from zero). The largest will be the largest value actually observed or considered possible in a forward projection.
2. Count the number of squares in each direction of the graph paper. The scales on each should then be chosen so that the range of the graph in each direction extends over as much as possible of the graph paper.
3. Label the two scales, and plot the graph.

7.2 Simple time graphs

One of the most common uses of coordinate axes is to put a time scale on the horizontal (x) axis, and some variable which changes over time on the vertical (y) axis. Fig. 7.3 is a graph of the actual sales by UK retailers, as given in Table 5.4. The horizontal axis shows time periods, in this instance quarters of the year from 1979 to 1980. The vertical axis shows retail sales in £000 millions.

There are a number of things which we should note about scales in time graphs:
1. Normally, equal intervals on a scale should represent equal amounts; an exception is the use of log scales, discussed below. It is a misleading practice to use, for example, the same interval for a one-year period in one part of a scale and a quarter-year period in another part of the same scale.
2. The horizontal (x) axis on a time graph very seldom begins at zero, and there is no necessity for the vertical (y) axis to do so. The effects of a non-zero starting point for the y axis on a time graph are analogous to its effects in a scatter diagram. For instance, in a time graph of sales this means that ratios of differences between successive sales figures appear more plainly, but the ratios *of actual* sales figures to each other can no longer be judged at all. This feature of scales may be used deliberately to mislead about a sales 'success'.

Simple graphs—81

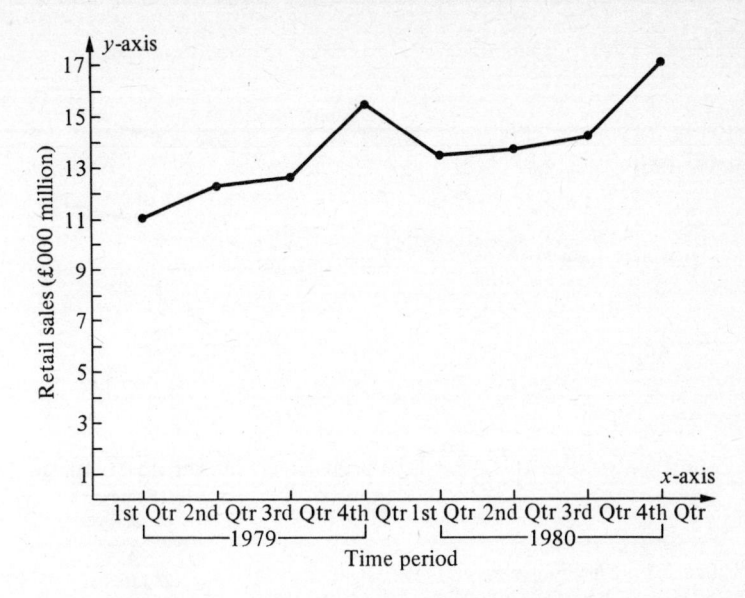

Figure 7.3 *Simple time graph of UK retail sales*

To illustrate this, compare percentage changes over two years in total national food retail sales, giving the national average, with changes in sales by the two foodstore chains 'Smashem Foodstores Ltd' and 'Fidlem Foods Ltd'. First consider Fig. 7.4, which shows changes in national food retail sales from 1979, compared to changes in sales of 'Smashem Foodstores Ltd' and 'Fidlem Foods Ltd'. Each starts in the first quarter of 1979 at 100 per cent. Total national sales rose about 36 per cent over the two years, Smashem sales rose 90 per cent, while Fidlem did poorly with a rise of only about 22 per cent. Fig. 7.4 shows the changes clearly. But now consider Fig. 7.5, which again shows percentage changes in sales since the first quarter of 1979. The percentage changes in sales for Smashem are shown on the same scale as the national ones, but the vertical axis no longer begins at zero. This means that, while a good visual impression may be obtained of how Smashem's increase compares with the national increase, there is no basis at all for a visual impression of the sizes of those increases in absolute terms. In fact, any impression we try to obtain of the sizes of increase will probably tend to exaggerate them.

Turning now to the Fidlem percentage changes in sales, we find that these are actually represented on a different scale altogether from the national figures; this scale is shown on the right of the diagram. The superimposition on one graph of two time series with totally unconnected scales allows no comparison of any kind to be made between the two, and its purpose can only be to mislead. In this it is quite successful, for it makes it look as though Fidlem's

82–*Applied business statistics*

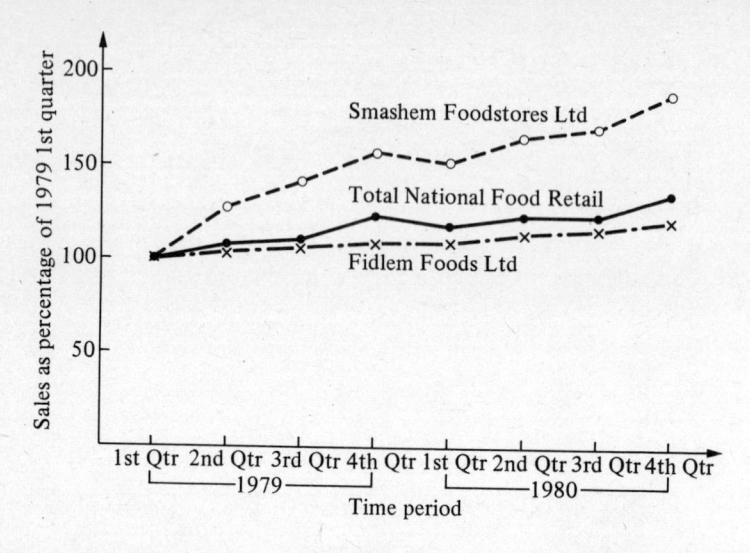

Figure 7.4 *Showing percentage changes in value of retail sales since 1979 1st quarter*

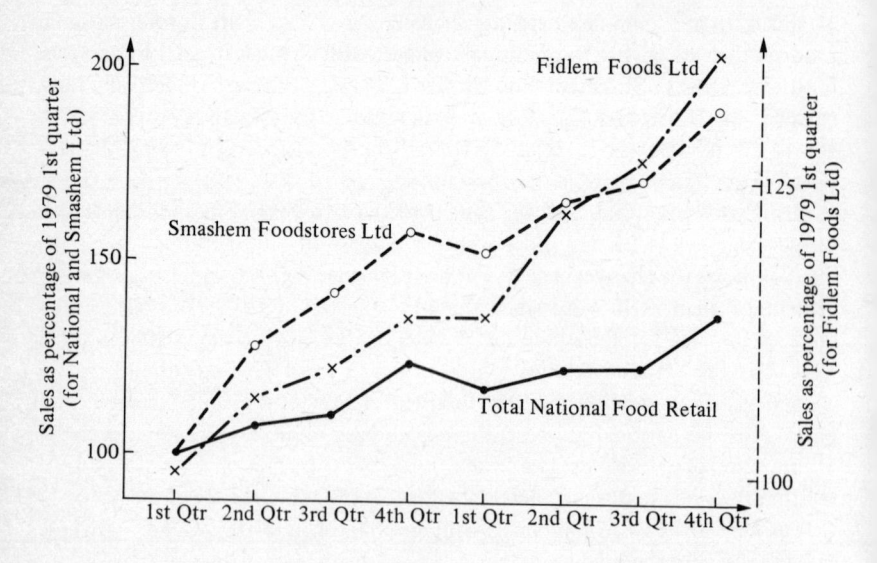

Figure 7.5 *Showing percentage changes in value of retail sales since 1979 1st quarter*

sales have shot up much faster than the national average, whereas Fig. 7.4 showed us that in fact they rose much less. Needless to say, Fidlem's PR department might complete the deception by omitting to mark either scale!

Further examples

Government publications contain many examples of simple time graphs. The current *Social Trends*, for example, includes graphs on divorce rates, consumer expenditure on various items at 1975 prices, the retail price index and average earnings index, among others.

7.3 Band curve charts

Fig. 7.3 showed the quarterly total national retail sales for 1979 and 1980 as given in Table 5.4. But these totals are made up of various different categories of retailers: food, mixed food and non-food, clothing (including footwear), household goods, and other non-food. It can be useful in a time graph to show the breakdown for each period into these sub-categories. Fig. 7.6 does this. The top line of the band curve chart is just the same as the simple time graph of the total sales. But the space between this line and the horizontal axis is divided up into 'bands', the width of each showing the size of sales in that particular category for the period.

For the last quarter of 1980, for example, the figures (in £000 millions) were as follows:

Food	5.02
Mixed food and non-food	5.01
Clothing	1.73
Household goods	2.49
Other non-food	3.01
Total	17.26

An explanatory column (not usual for band curve charts) has been added at the right hand side of Fig. 7.6 to show how this breakdown of figures is given as bands for that last quarter.

This kind of chart also has various other names, such as *layer graph* and *strata graph*.

A variation on the chart shown is to have the bands showing *percentages* of the total which fall into the various sub-categories. The total height of the

84–*Applied business statistics*

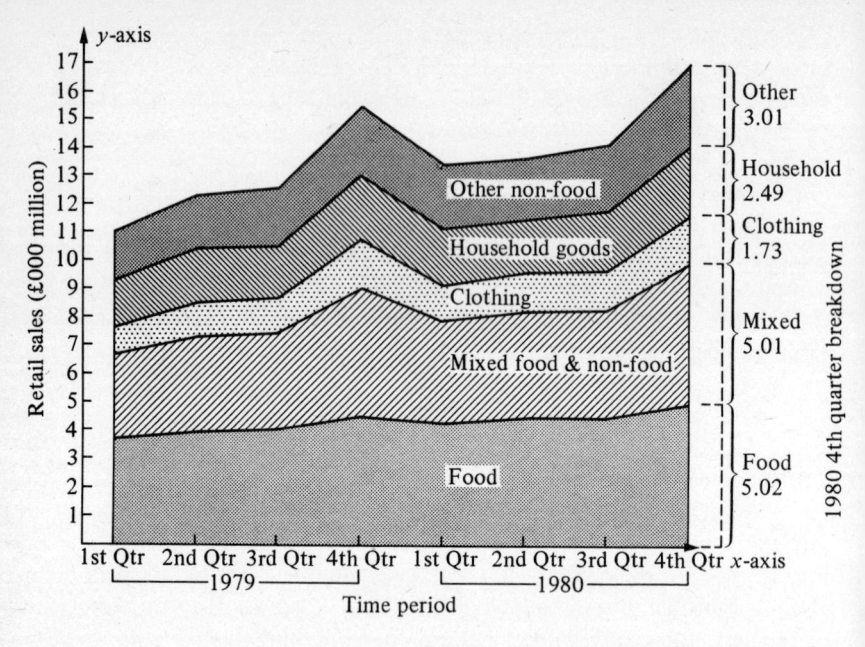

Figure 7.6 *Band curve chart showing the breakdown of total retail sales into various sub-categories*

graph is then the same, representing 100 per cent, for each time period. This variation of the chart stands in the same relation to the kind in Fig. 7.6 as a set of percentage component bar charts stands to a set of ordinary component bar charts (see section 6.3). In fact, a band curve chart can be seen as a kind of development from a series of component bar charts.

Further examples

The band curve chart is often used in Government publications. The current *Social Trends* includes band curve charts on times before first birth in marriages, adoptions, people in employment by sector and industry, government expenditure, causes of death, housing costs, and so on.

7.4 Logarithmic or ratio-scale graphs

There are circumstances in which it is useful to use a logarithmic scale (sometimes called a ratio scale) instead of the ordinary scale on a graph.

Simple graphs—85

Table 7.2 shows the energy consumption (in million tonnes coal equivalent) in the rapidly expanding Slobodian economy. Fig. 7.7 shows these figures on an ordinary graph, with consumption on the vertical axis and the year on the horizontal axis. From Table 7.2 and Fig. 7.7 we can see that coal

Table 7.2 *Slobodian energy consumption (million tonnes coal equivalent)*

Source	1978	1979	1980	1981	1982	1983	1984	1985
Coal	6.7	8.7	10.7	12.7	14.7	16.7	18.7	20.7
Electricity	2.4	3.6	5.4	8.1	12.1	18.2	27.3	41.0
Oil	1.2	1.9	3.1	4.9	7.9	12.6	20.1	32.2

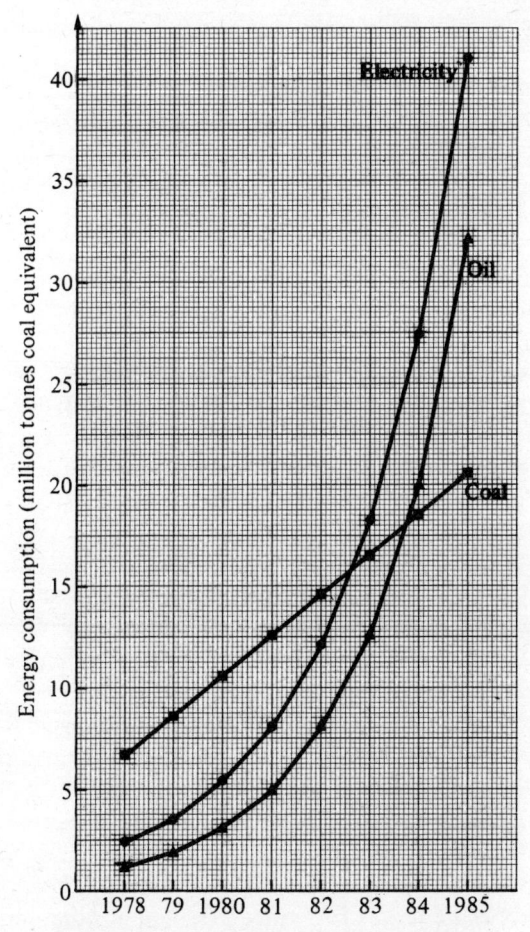

Figure 7.7 *Slobodian energy consumption*

consumption has risen by the same *amount*, two million tonnes, each year. Electricity consumption started off lower, but rose by a constant *factor* of 50 per cent each year; that is, each year's consumption was 50 per cent higher than the previous year's. The oil consumption began even lower, but rose at a rate of 60 per cent per year, so each year's figure was 60 per cent higher than the previous year's.

It is sometimes useful to draw a graph which shows the percentage rate of increase over time, rather than the actual amounts. Such a graph is shown in Fig. 7.8. In Fig. 7.8 the graphs for oil and electricity both appear as straight lines. This is because both are increasing at a constant rate each year. The graph for oil is rising more steeply than that for electricity because the oil consumption figures increase by 60 per cent each year, while the electricity figures increase by only 50 per cent. The *slope* of the line indicates the *rate of increase*.

The coal figures increase by the same *amount* (two million tonnes) each year. But since the consumption is rising, this is a smaller *percentage* of the total for each year. On this *ratio-scale* graph, or *logarithmic scale* graph, the line for coal therefore rises less in the later years, because a smaller percentage rise is being indicated.

Drawing ratio scales

The simplest way to draw this kind of graph would be to obtain some special graph paper. This would need to have an ordinary horizontal scale, with one-centimetre intervals, say. But the vertical scale would need to be such that an increase by a given percentage would cause the same rise wherever on the scale it appeared. For example, a rise of 100 per cent from 2 units to 4 units would give the same interval on the graph as a rise from 16 units to 32 units. Such paper is, in fact, available, and is called *semi-logarithmic graph paper*. It has been used for Fig. 7.8, and its scales are marked on the right of the graph. As originally printed, the first 2.5 cm on the vertical scale represented one unit, the next 2.5 cm two units, the next 2.5 cm four units, and so on.

It may be noted that the vertical scale starts at one and not at zero. This is because the intervals measure rates of increase, and if a figure were to begin at zero, an increase of any size would be an infinite percentage of that starting value. So the scale begins at one unit, though that one unit may be in millions or hundreds if desired. There are no negative values on a scale like this for similar reasons.

It is less easy to manipulate the scales on semi-logarithmic graph paper than on ordinary graph paper. To move a graph up or down on the vertical axis one needs to multiply or divide by a constant – not to add or subtract

Simple graphs—87

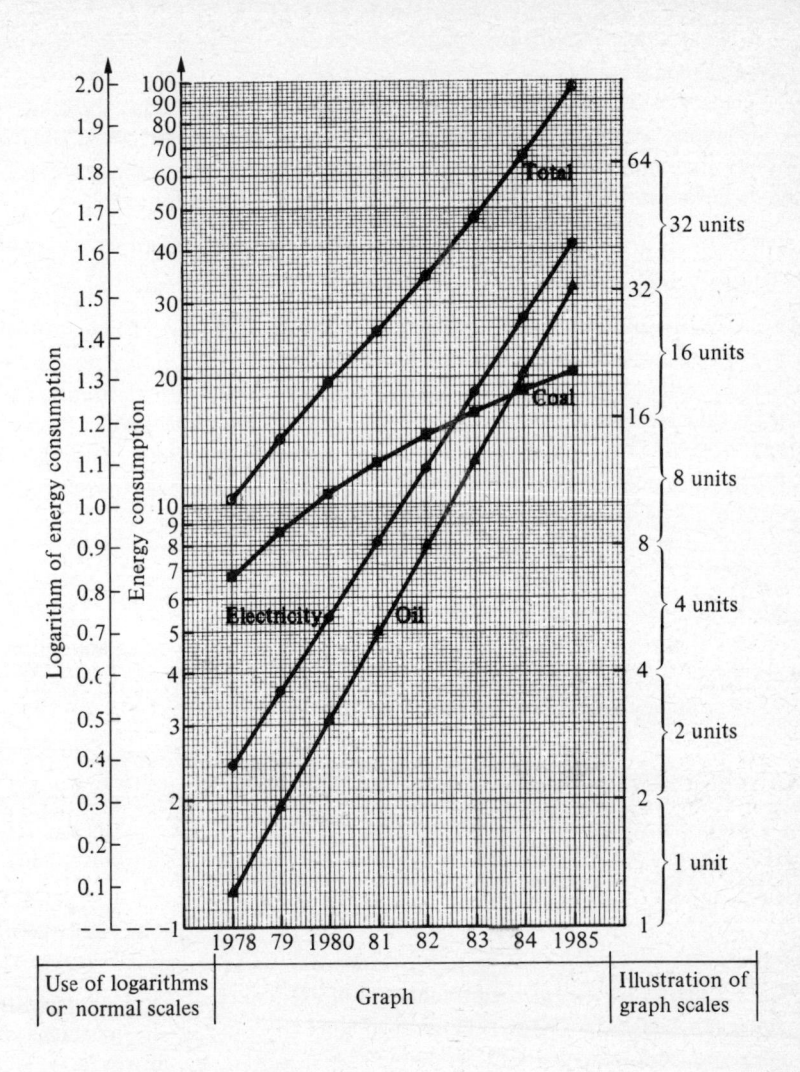

Figure 7.8 *Ratio-scale graph of energy consumption*

as in an ordinary graph. To obtain the right amount of *spread*, that is to use the whole of the available vertical space, means obtaining the *right* semi-logarithmic graph paper, for there are several kinds. A 'one-cycle' paper is suitable for low percentage rates of growth, where in the period concerned figures grow up to ten times the size of the starting value. Using, say, 'three-cycle' paper, one can show data where the figures in the period rise to 1000 times the starting value.

88–*Applied business statistics*

Obviously, semi-logarithmic paper can be used to represent percentage rates of fall, as well as rises.

The special semi-logarithmic paper is easily obtained, and anyone drawing such graphs would be well advised to get some. It is, however, possible to obtain the same effect on ordinary graph paper by plotting the *logarithms* of the figures instead of the actual figures. For the oil figures, for example:

log(1.2) = 0.079   log(1.9) = 0.278   log(3.1) = 0.491   log(4.9) = 0.690
log(7.9) = 0.898   log(12.6) = 1.100   log(20.1) = 1.303   log(32.2) = 1.507

If these were plotted on the ordinary scale shown at the left of Fig. 7.8 they would give the same graph for oil. This is, however, a rather fiddly procedure. First, it will require either a knowledge and availability of logarithm tables, or a calculator which gives logarithms. Second, the final graph must be labelled with the actual figures rather than with logarithms. This means perhaps rubbing out the original scale used, and labelling the vertical scale with actual figures. For serious use the special paper is simpler.

Further example

Table 7.3 shows the January Index of Earnings, Retail Price Index and Wholesale Price Index, for 1974–81. (The Index of Earnings is based on the New Earnings Survey since 1976, and on the old Earnings Survey for 1974 and 1975.)

Table 7.3   *Published January indices*

	1974	1975	1976	1977	1978	1979	1980	1981
Index of Earnings	154.0	205.6	248.1	275.2	301.3	336.8	404.6	480.9
Retail Price Index	100.0	119.9	147.9	172.4	189.5	207.2	245.3	277.3
Wholesale Price Index	209.5	222.1	261.5	337.8	324.9	351.1	451.0	488.5

Fig. 7.9 shows the figures from Table 7.3 plotted on one-cycle semi-logarithmic graph paper. In Chapter 11 we will look in detail at the meanings of index numbers, but obviously a comparison of movements in wholesale and retail prices, or of earnings and prices, can be useful. It may be profitable to generalize at this point and consider several contexts in which ratio-scale graphs are particularly useful.

Uses of ratio scales

1. For single series where the *percentage* change from one period to the next is more important than the actual amount. In particular this is so

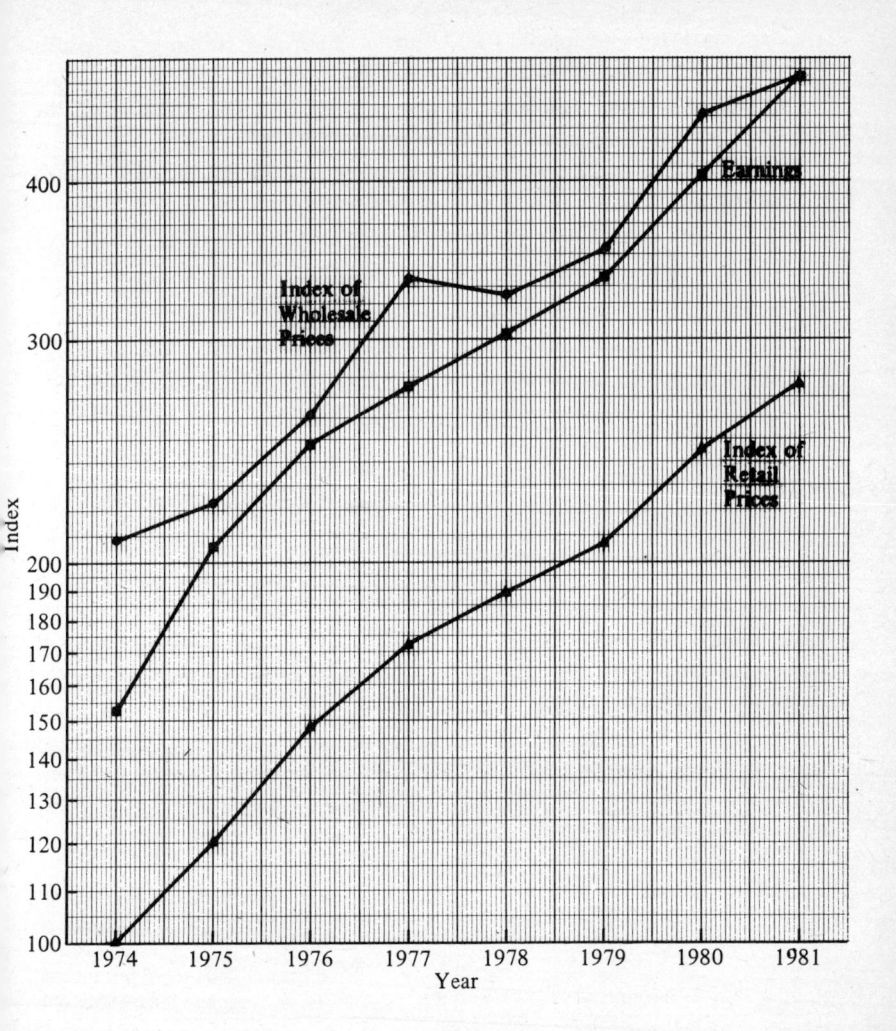

Figure 7.9 *Ratio-scale graph of indices*

for any price indices. A price index rise from 139 to 151, for example, is a higher percentage rise than one from 179 to 191. Using a ratio scale, the same percentage rise of one period's figures compared with the preceding period is represented by the same interval anywhere along the graph. This is shown in the index of retail prices plotted in Fig. 7.9.

2. For comparison of two separate series, even if one begins much higher than the other the same vertical interval represents the same *percentage*

rise. Thus, for example, in Fig. 7.9 the greater rise in earnings than prices between 1980 and 1981 is indicated by a steeper rise on the graph. (Actually a more valid comparison of gross earnings would be to the lesser known Taxes and Prices Index.)

If the units of the two series being compared are not the same, then a ratio-scale graph can still be used to look at percentage changes. However, in this case the distance between the graph and the horizontal axis is of no significance.

3. Where units are the same but one series is very much larger than another, percentage changes can be shown easily on the same axis. Thus in Fig. 7.8 the total energy consumption is shown on the same graph as the consumption of each fuel; this is not really practicable in the ordinary scale in Fig. 7.7.

7.5 Z-charts

There is one other important form of graph which may be found in commercial use, and that is the Z-chart. It consists of three separate time series, drawn on a single set of axes.

Suppose that the Nosmot Travel Company run various holiday package-tours to North America, and keep a record of the monthly receipts. Their figures for the months of 1983 and 1984 are shown in Table 7.4, together with various calculated figures which we shall need.

Table 7.4 *Receipts to Nosmot Travel Company from North America package-tours*

Month	1983 Receipts	1984 Receipts	Total for year so far	Total of this month and previous eleven
January	20	22	22	378
February	8	10	32	380
March	10	13	45	383
April	18	21	66	386
May	18	23	89	391
June	37	39	128	393
July	48	52	180	397
August	70	78	258	405
September	58	59	317	406
October	50	53	370	409
November	24	29	399	414
December	15	16	415	415
Total	376	415		

Simple graphs—91

For 1983, Table 7.4 shows only the monthly receipts. For 1984 monthly receipts are shown, together with a total received in 1984 so far – the 'cumulative total' of the 1984 monthly figures. The last column shows the 'moving 12-month total': for each month in 1984 it shows the sum of that month's figures plus those of the previous eleven months. Until December this includes some months of 1983. The three columns relating to 1984 are plotted in

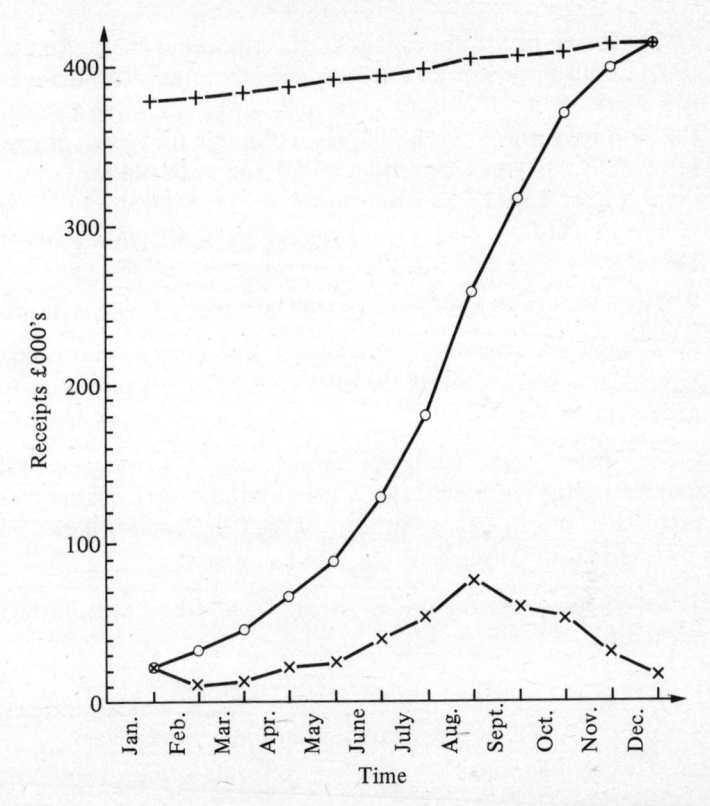

Figure 7.10 *Z-chart of receipts to Nosmot from North American packages*

Fig. 7.10. A chart like this, called a 'Z-chart' for obvious reasons, can be compiled gradually throughout the year. If desired, target figures can be projected on the graph for future periods. In our example we based the 'year' on calendar years, but it could, of course, be based on the financial year of April to March.

Some books advocate plotting the monthly figures in the centres of the months rather than at the right-hand ends like the cumulative figures. I do not see any particular advantage in this, and prefer to plot them underneath the cumulative ones as shown.

92–*Applied business statistics*

Sometimes the monthly figures are plotted on the same horizontal time axis, but on a larger vertical scale. This may perhaps be useful, especially if small differences from month to month are important.

Recall and exercise questions

1. The following figures for ages and selling prices of Range Rovers were taken from a single issue of *Exchange and Mart*: five years old for £5915; two years old for £6500; two years old for £5750; seven years old for £4950; three years old for £4995; seven years old for £4800; one year old for £9450; this year's model for £9750; two years old for £6250; nine years old for £2995; ten years old for £2795; one year old for £8995; four years old for £4995; three years old for £6495; nine years old for £2950; seven years old for £3850.

   Plot these figures on a scattergraph, and comment on what it shows.

2. From Table 5.4 draw a time-series graph, with quarters shown along the *x* (horizontal) axis, showing the total actual quarterly retail sales for the periods given.

3. From Table 5.2, plot the figures for quarterly UK passenger car production on a graph (with quarters on the *x* axis). For each quarter show the proportions of cars falling into each of the four listed categories, making the graph a band-curve chart.

4. Table 7.5 shows expenditure by visitors from various parts of the world to the UK.

Table 7.5 *Overseas visitors to the UK*

Earnings at current prices (£m) – unadjusted

		Total	North America	European Community	Other Western Europe	Other areas
1979	1st quarter	412	72	103	57	180
	2nd quarter	637	128	200	87	222
	3rd quarter	1088	194	347	139	408
	4th quarter	660	117	171	103	269
1980	1st quarter	516	77	143	62	234
	2nd quarter	715	135	230	80	270
	3rd quarter	1124	182	366	132	444
	4th quarter	610	114	152	88	256
1981	1st quarter	472	70	129	60	213

Source: *Business Monitor* MQ6.

Draw a band chart to show proportions of each area's contribution for each quarter.

5.

Table 7.6 *Fuel price indices and the Retail Price Index*

		1973	1977	1978	1979	1980
Domestic Sector	Coal and Coke	128	275	305	357	456
	Gas	115	200	206	213	249
	Electricity	120	301	332	360	458
Industrial sector	Coal	132	317	345	413	520
	Gas	115	354	446	508	690
	Electricity	114	275	303	335	413
Retail Price Index		100	195	211	239	282

Note: Fuel indices from *Energy Trends*, with industrial sector gas rebased for simplicity. Retail Price Index also rebased.

Use one-cycle semi-logarithmic graph paper to plot all these series and comment on what the graphs show.

6.

Table 7.7 *Colour television sets sold by Image Ltd*

				Thousands of sets sold							
	Jan	Feb	Mar	Apr	May	Jun	Jul	Aug	Sep	Oct	Nov	Dec
1983	6.6	8.0	10.8	9.2	10.8	11.9	11.1	10.8	11.9	13.6	18.7	11.6
1984	9.7	9.6	10.5	8.6	10.1	10.5	10.2	9.2	13.3	12.2	16.5	10.7

Draw up a Z-chart for these figures, and briefly explain its meaning.

Problems

1. For one of the groups in the Standard Industrial Classification, obtain information on movements in the last three years in:
   (a) wage rates of manual workers
   (b) normal weekly hours of manual workers
   (c) sales
   (d) output

   Present this in a brief report on the industry.

8
Diagrams for Grouped Frequencies

8.1 The histogram

In an earlier chapter, Fig. 6.6 showed a bar chart of newspaper readers split into various categories by social class and newspaper read. These categories were *qualitatively* different: they were different kinds of thing rather than numbers or measurable variables.

In some situations, however, we have a set of categories which consist of different *sizes* of some measurable variable, rather than things different in kind. In particular, we may have a grouped frequency table of the kind described in section 5.5. The 157 Prunes employees' mean weekly earnings which were listed in Table 5.7 may be grouped as in Table 8.1. Now the bar

Table 8.1   Last year's weekly earnings of 157 employees of Prunes Stretch Covers Ltd on piecework

Earnings (£)	Number of employees
30–33	2
33–36	4
36–39	7
39–42	13
42–45	22
45–48	36
48–51	43
51–54	15
54–57	9
57–60	6
Total	157

chart of Fig. 6.6 had frequencies – numbers of readers in each group – shown on its vertical scale, but its horizontal scale was not a numerical scale at all. Rather, its horizontal scale contained an arbitrarily ordered sequence of categories, for the papers might have been put in a different order.

But suppose that the categories are class intervals along a numerical scale, as in Table 8.1? If we draw a diagram in which the vertical axis represents the frequency in each class interval, it would be sensible to put the class intervals themselves along the horizontal scale. Obviously we will put them in the right order, so that the horizontal scale is now also numeric, being the numerical value of weekly earnings. Fig. 8.1 shows such a diagram, which

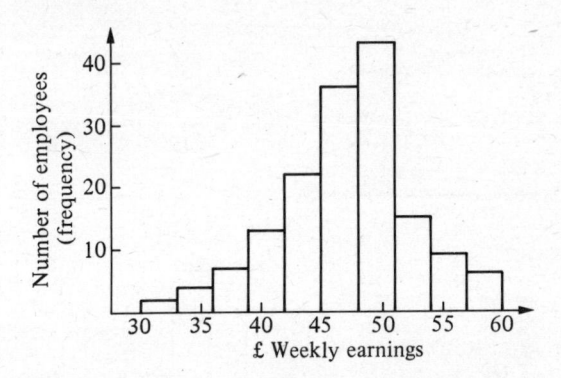

Figure 8.1 *Histogram showing last year's weekly £ earnings of 157 employees of Prunes Stretch Covers Ltd*

is called a *histogram*. Each class interval along the horizontal scale forms the base of a block, and the height of that block shows the number of values which fall within that class interval. Thus, for example, the first class interval £30–33 forms the base of a block whose height is 2 (the frequency in that first interval).

The intention is to convey visually the distribution of frequencies between the different class intervals, and we sometimes refer to 'the distribution' to mean the pattern of this. We can see, for example, that there are nearly twice as many weekly earnings between £48 and £51 as there are between £42 and £45. We can see that most earnings are between £35 and £55 per week, with a peak around £49 per week.

All this is conveyed visually by the histogram. But what exactly forms the basis of our visual impression? In fact it will not be the *heights*, but the *areas* of the blocks. In our table of earnings, Table 8.1, all the class interval lengths were the same; therefore the blocks are all the same width, and the heights of the blocks are proportional to their areas.

But suppose that our table, as often happens, had been as in Table 8.2, with unequal class intervals. In Table 8.2 the first and the last class intervals are longer than the rest (reasons were given in section 5.5 why this is sometimes done). A histogram in which the vertical scale simply showed frequencies in terms of heights would then look like Fig. 8.2.

Table 8.2 *Last year's weekly earnings of 157 employees of Prunes Stretch Covers Ltd on piecework*

Earnings (£)	Number of employees
30–39	13
39–42	13
42–45	22
45–48	36
48–51	43
51–54	15
54–60	15
Total	157

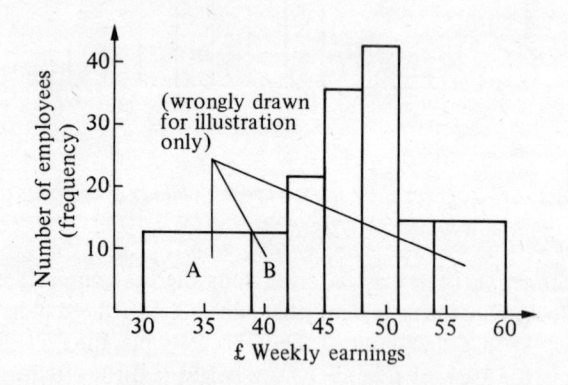

Figure 8.2 *Histogram showing last year's weekly earnings of 157 employees of Prunes Stretch Covers Ltd*

But Fig. 8.2 is misleading! For example, the intervals £30–39 and £39–42 each contain 13 observations. But because of the unequal widths of the blocks the block whose base is £30–39 (labelled A) is three times the area of the block whose base is £39–42 (labelled B). This gives a visual impression which magnifies its importance compared with the number of observations it actually contains. We are similarly misled if we use the blocks whose bases are on the intervals £51–54 and £54–60 to make a visual comparison of frequencies in those two intervals. In short, since our visual comparisons will be a comparison

Diagrams for grouped frequencies–97

of areas, Fig. 8.2 is misleading because the areas of the blocks are not proportional to the frequencies which they represent.

So what is the simplest way to make the heights so that the areas *are* proportional to the frequencies?

Let: area = frequency (for each block)
But: area (for any rectangle) = height × width
And here the width is simply the class interval for the block so:

       area = frequency = height × class interval (for each block)

Dividing both sides of this equation by the class interval we get:

$$\frac{\text{frequency}}{\text{class interval}} = \text{height (for each block)}$$

The height of each block, then, can be made equal to (frequency ÷ class interval). This value (frequency ÷ class interval) is termed the *frequency density*.

Table 8.3 shows frequency densities in its third column, and these have been utilized to draw the histogram in Fig. 8.3, the vertical axis of which is therefore labelled 'frequency density' rather than 'frequency' as before.

Table 8.3   *Last year's weekly earnings of 157 employees of Prunes Stretch Covers Ltd on piecework*

Earnings (£) (CI)	Number of employees (f)	Frequency density (f/CI)
30–39	13	1.44
39–42	13	4.33
42–45	22	7.33
45–48	36	12.00
48–51	43	14.33
51–54	15	5.00
54–60	15	2.50
Total	157	

CI: class interval

The histogram of Fig. 8.3 gives a good visual impression of the numbers of observations in each class interval. It may be compared with the original histogram in Fig. 8.1, which was drawn using the more detailed breakdown of the figures; the two are comparable.

The concept of frequency density not only makes the areas of blocks come out in the correct proportions: in itself it tells us how closely or densely packed the observations are within the interval:

Class £30–39 contains 13 observations in its interval of length £9, so each £1 of that interval contains, on average, 13 ÷ 9 = 1.44 observations.

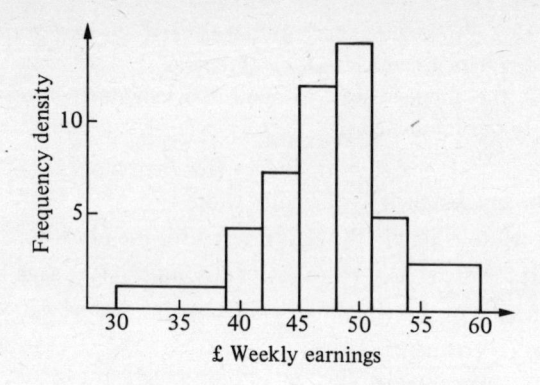

Figure 8.3 *Histogram showing last year's weekly earnings of 157 employees of Prunes Stretch Covers Ltd*

Class £39–42 contains 13 observations in its interval of length £3, so each £1 of that interval contains, on average, 13 ÷ 3 = 4.33 observations.

There are more observations per unit interval (4.33) in the class £39–42 than there are per unit interval (1.44) in the class £30–39.

Frequency density, which is also the height of the histogram block, indicates how closely or densely grouped observations are in each interval: if the frequency density in an interval is 4.33, then observations in that interval are more densely grouped than in an interval for which the frequency density is 1.44. In Fig. 8.3 observations are most densely grouped around the interval £48–51, and then 'tail off' in either direction.

Open-ended classes

We need here to consider the problem posed by *open-ended class intervals*. In Table 8.3 the final class interval could have been '54 and above' rather than the specific interval '54–60'. (Reasons for sometimes doing this were given in section 5.5.) If we have an open-ended interval like this, what length should we use for it when we draw the histogram?

There is really no infallible method to decide on an appropriate length; common sense has to be used. Some authors suggest that if all the other class intervals are equal, then the open-ended interval or intervals should be made the same length. However, if, for example, the observations appear to be tailing off, and the frequency in the last interval is larger, then it would seem reasonable to suppose that the observations ranged over a longer interval. In the end it often boils down to a guess based on common sense. But different people might guess differently: here, for example, one might equally well have guessed that the last interval was twice as long or three times as long as the other intervals. However, while there is some room for variation in the guesses, some guesses, such as ten times as long as the other intervals, would be unreasonable. This kind of situation is not uncommon in statistics, though some people prefer to call such guesses 'estimates' or 'guesstimates'!

Hopefully, the frequency in the open-ended class interval will not be very large, and the exact length we choose for that interval will not be important.

Percentage frequency densities

The third column of Table 8.4 lists the percentage of employees with average earnings in each class interval, rather than the actual frequency. The fourth column shows 'percentage frequency density', that is percentage frequency ÷ class interval, for each group; from this fourth column the histogram in Fig. 8.4 has been drawn.

Table 8.4 *Last year's weekly earnings of 157 employees of Prunes Stretch Covers Ltd on piecework*

Earnings (£) CI	Number of employees f	Percentage of employees %f	Percentage frequency density %f/CI
30–39	13	8.28	0.92
39–42	13	8.28	2.76
42–45	22	14.01	4.67
45–48	36	22.93	7.64
48–51	43	27.38	9.13
51–54	15	9.55	3.18
54–60	15	9.55	1.59
Total	157	100.00	

Are there any advantages in drawing a histogram using *percentage* frequency densities, rather than just frequency densities? Well, we remember that in our earlier histograms, Figs. 8.1 and 8.3, the area of each block represented the frequency of observations within the class interval on which the base of that block stood. Areas of blocks in a histogram drawn on *percentage* frequencies will similarly represent the *percentage* frequencies falling within each class interval. Consider the block (labelled C) which has its base resting on the interval £48–51 on the horizontal scale.

Its area = height × width = height × class interval
= 9.13 × 3
= 27.4

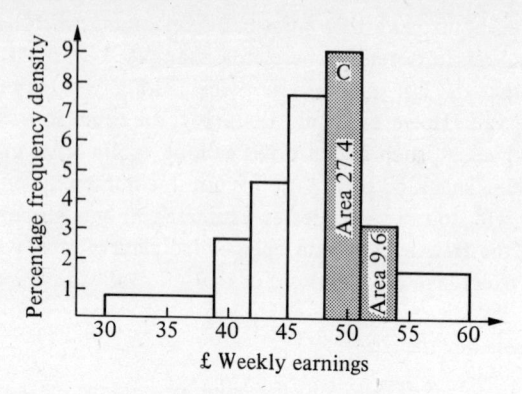

Figure 8.4 *Histogram showing last year's weekly earnings of 157 employees of Prunes Stretch Covers Ltd*

Table 8.4 shows 27.4 per cent of observations to fall in the £48–51 interval which this block represents. In fact, for all the blocks:

$$\text{height of block} = \frac{\text{percentage frequency}}{\text{class interval}}$$

height of block × class interval = percentage frequency = area

This has two important implications:

1. The total area of two adjacent blocks is equal to the total percentage of observations which fall into their combined interval.
2. The total area of all the blocks added together is 100 per cent.

This second point is especially useful if we wish to compare, for example, earnings for workers in two groups of unequal size. Suppose that we wish to compare the earnings of the 157 Prunes workers with those of 63 workers for Windsor Lingerie Ltd. The first two columns of Table 8.5 show the distribution of the earnings of these 63 workers, that is show how many fall into each interval. The last two columns show the relevant calculations for a histogram of Windsor workers' earnings using the percentage frequency densities. Such a histogram is shown in Fig. 8.5, and immediately below it, using the same vertical and horizontal scales, is a similar histogram using percentage frequency densities for the 157 Prunes employees' earnings.

From point 2 above we remember that the total area of the blocks of any histogram drawn on percentage frequency densities is 100 per cent. This is especially useful in a situation like this, where we have two sets of figures, with class intervals in the same units, here weekly earnings in pounds, but unequal totals of observations. In such a situation we *can* make both horizontal

Diagrams for grouped frequencies—101

Table 8.5 Last year's weekly earnings of 63 employees
of Windsor Lingerie Ltd on piecework

Earnings (£) CI	Number of employees f	Percentage of employees %f	Percentage frequency density %f/CI
30-34	3	4.76	1.19
34-38	9	14.29	3.57
38-42	13	20.63	5.16
42-46	25	39.68	9.92
46-50	8	12.70	3.17
50-60	5	7.94	0.79
Total	63	100.00	

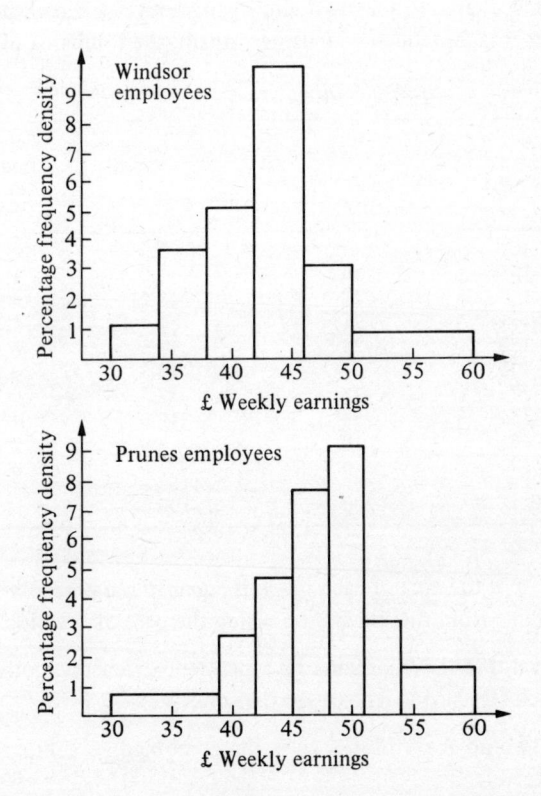

Figure 8.5 *Histograms to compare average weekly earnings of 63 Windsor and 157 Prunes employees*

102–*Applied business statistics*

and vertical axes the same in the two histograms, as we have done here. Since the total area of the blocks in each histogram is 100 per cent, the total areas in the two histograms are physically the same. What this means is that we are not distracted by one histogram being larger in total area than the other. We are free to concentrate visually on the differences in pattern of distribution. For example, from the two histograms in Fig. 8.5 we can see that average earnings as a whole are less at Windsor than at Prunes, and that the peak of the distribution of Prunes workers' earnings is to the right of that for Windsor workers.

Further example

The height in centimetres is measured for a group of 200 adult males in the UK. From this a table (Table 8.6) and histogram (Fig. 8.6) can be drawn up based on percentage frequency densities. Again, the total area of the blocks is

Table 8.6 *Height in centimetres of 200 adult males in the UK*

Height (cm) CI	Number of men f	Percentage of men $\%f$	Percentage frequency density $\%f/CI$
153–161	3	1.5	0.1875
161–165	10	5.0	1.25
165–169	27	13.5	3.375
169–171	21	10.5	5.25
171–173	25	12.5	6.25
173–175	27	13.5	6.75
175–177	25	12.5	6.25
177–181	38	19.0	4.75
181–185	18	9.0	2.25
185–193	6	3.0	0.375
Total	200	100.0	

100 per cent, and the area of each block is equal to the percentage of observations which fall within the interval on which the base of the block stands:

The interval 175–177 contains 12.5 per cent of observations, so the area of the block (D) having its base on that interval is 12.5.

The area of two adjacent blocks gives the combined percentage within their combined interval:

The two intervals 175–177 and 177–181 form a total interval of 175–181, which holds a total percentage of 31.5. The combined area of the two blocks (D and E) is 31.5.

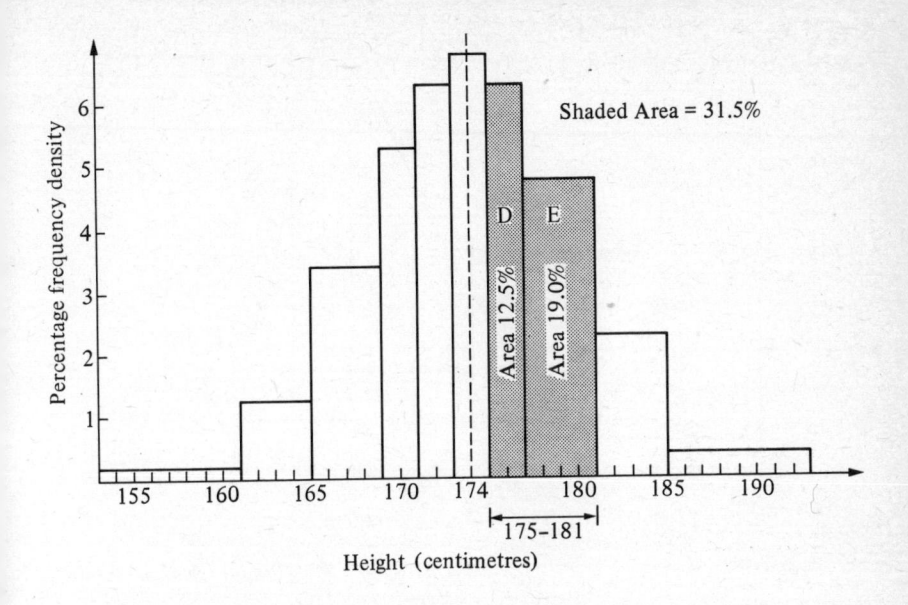

Figure 8.6 *Heights of a group of 200 men from the United Kingdom*

The peak of the histogram in Fig. 8.6 comes at about 174 cm. This implies that observations are most densely grouped around this point: this is the most common kind of height. Numbers then tail off in each direction, and men under 155 cm or above 200 cm are rare.

8.2 Frequency polygons

Sometimes it is thought useful to 'smooth' the outline given by the histogram, to give a better idea of its shape. The most common way to do this is by a frequency polygon. This, as shown in Fig. 8.7, is based directly on the histogram itself. The mid-point of the top of each histogram block is joined to the next, as shown. This, of course, leaves a question about what to do at each end. I have left the polygon as terminating at the top of each end histogram block. Sometimes these points are joined up to the horizontal (*x*) axis, at the mid-points of imaginary class intervals outside each end of the actual ones. If the class intervals are all equal this might make some sense. If they are not, then it is difficult to see any rational basis for choosing an imaginary interval. In practice I am sceptical of the supposed advantages of frequency polygons over histograms, partly for this reason.

104—*Applied business statistics*

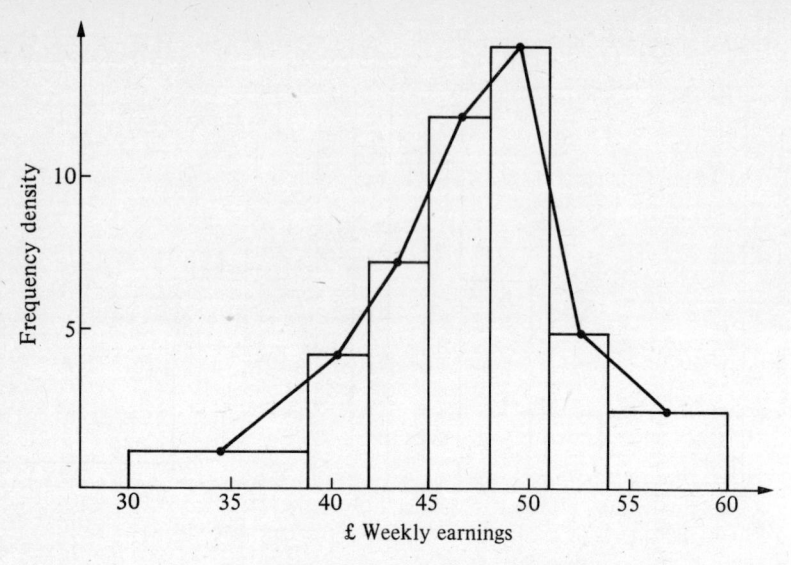

Figure 8.7 *Frequency polygon and histogram showing last year's weekly earnings of 157 Prunes employees*

8.3 Cumulative frequency curves

The cumulative frequency curve (sometimes called the *ogive*) is also based on a frequency table. It can, in fact, be based on an ungrouped or a grouped frequency table, though the latter is more common. It is plotted from a cumulative frequency table, as shown in Table 8.7. The third column shows the

Table 8.7 *Weekly earnings of 157 Prunes employees on piecework*

Earnings (£) CI	Number of employees f	Number in and below this income range (Cumulative less-than frequency)	Number in and above this income range (Cumulative more-than frequency)
30–39	13	13	157
39–42	13	26	144
42–45	22	48	131
45–48	36	84	109
48–51	43	127	73
51–54	15	142	30
54–60	15	157	15
Total	157		

'cumulative less-than frequencies' – the numbers of observations in and below the class interval against which each appears. The last column shows analogous 'cumulative more-than frequencies'. Both are plotted in Fig. 8.8. It should be noted that the 'cumulative less-than frequencies' are plotted at the *ends* of their respective class intervals, while the 'cumulative more-than frequencies' are plotted at the beginnings of theirs.

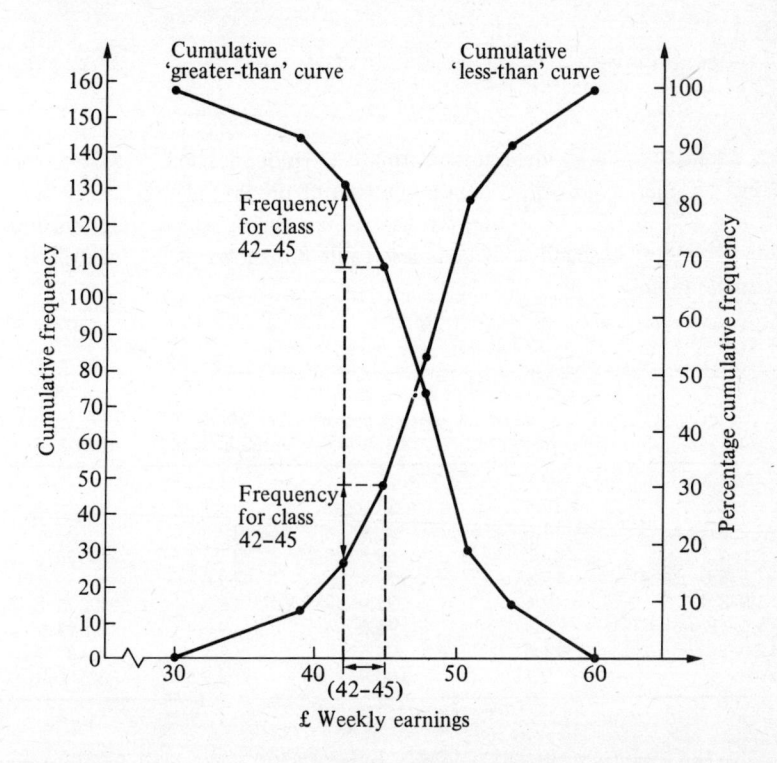

Figure 8.8 *Cumulative 'greater-than' and 'less-than' curves for earnings of 157 Prunes employees*

The visual connection with the actual frequencies is illustrated for the class interval £42–45: its frequency (22) is shown as the difference between two successive figures on each curve.

To draw the two curves only the left-hand vertical axis (showing actual cumulative frequencies) is necessary. It is possible to plot either kind of cumulative curve in terms of *percentage cumulative frequencies*. For this, instead of 'Number in and below this income range', we use 'percentage in and below this income range'. Since, in fact, this would affect only the labelling of the axis and not the shapes of the curves, the percentage cumulative frequency axis is added for demonstration purposes on the right-hand side of Fig. 8.8.

Note that for cumulative frequency calculations or curves we use actual frequencies or percentage frequencies, *never* the frequency densities or the percentage frequency densities.

8.4 Lorenz curves

There are situations in which it is informative to plot one cumulative frequency against another. Figures for UK distribution of pre-tax income are shown in Table 8.8. The UK tax system works on 'tax units', which are either individuals or joint husband-wife incomes. Table 8.8 shows income distribution for such tax units.

Table 8.8   UK pre-tax incomes

Pre-tax income (£000s)	Percentage of tax units	Cumulative percentage of tax units	Percentage of total income	Cumulative percentage of total income
-1.5	17.5	17.5	4.9	4.9
1.5-2.5	19.5	37.0	9.5	14.4
2.5-3.5	14.6	51.6	10.6	25.0
3.5-4.5	12.2	63.8	11.7	36.7
4.5-6.0	15.6	79.4	19.6	53.6
6.0-8.0	12.6	92.0	20.8	77.1
8.0-12.0	6.0	98.0	14.5	91.6
12.0-20.0	1.6	99.6	5.7	97.3
20.0-	0.4	100.0	2.7	100.0

For the year concerned, incomes under £1500 accounted for 17.5 per cent of tax units. If income was equally distributed, these tax units would also account for 17.5 per cent of the total income. In fact, these lowest 17.5 per cent of tax units accounted for only 4.9 per cent of the total income. This is shown in the first point of Fig. 8.9. This shows that a figure of 17.5 on the horizontal (percentage of tax units) axis is associated with one of 4.9 on the vertical (percentage of income) axis. In the range of incomes up to £2500 there are 37.0 per cent of tax units, and only 14.4 per cent of income. This forms the next point on the graph. The line at 45° to the axes shows what the graph would be like if income were equally distributed (if 17.5 per cent of tax units earned 17.5 per cent of income, 37 per cent of tax units earned 37 per cent of income, and so on). This 45° line may be compared with the

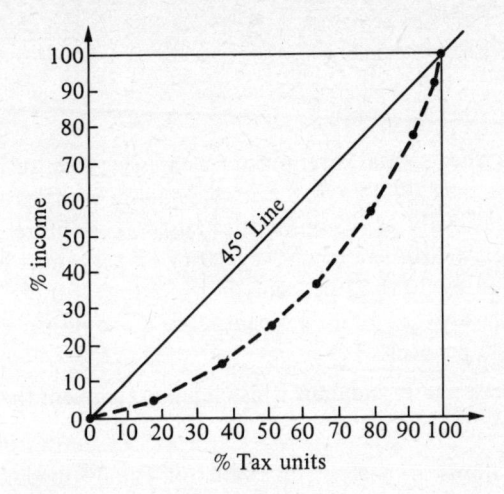

Figure 8.9 *Lorenz curve for UK income distribution*

graph drawn from the cumulative percentages on both axes, which is called a Lorenz curve. The further away the Lorenz curve is from the 45° line, the greater the inequality of income.

Gini coefficients

The Gini coefficient is defined as the ratio of the area between the Lorenz curve and the 45° line to the total area of the triangle formed by that 45° line and the two axes. For perfectly equally distributed incomes the Gini coefficient would be zero, and the higher its value the more unequally distributed are incomes.

Uses

Lorenz curves are mainly used for income distribution analysis (for pre- or post-tax incomes), but may also be used to measure inequalities in other situations, for example number of trade unions *vs* number of members in unions of different sizes, or number of classes *vs* number of pupils in classes of different sizes.

Recall and exercise questions

1. What is a 'frequency density'?
2. In a period when central government was pressing local authorities to cut expenditure, the *CIPFA Journal* carried details of the rises in expenditure over the previous year: 23 authorities raised expenditure by 0 to 5 per cent, 32 by 5 to 10 per cent, 83 by 10 to 15 per cent, 126 by 15 to 20 per cent, 114 by 20 to 25 per cent, 54 by 25 to 30 per cent, 15 by 30 to 35 per cent, 8 by 35 to 40 per cent, 2 by 40 to 45 per cent and 5 by more than 45 per cent.

    Tabulate these values and draw a histogram to represent them. Also draw a cumulative frequency curve.

3. Table 8.9 shows an analysis of awards of Family Income Supplement (current in October 1979) by earnings and age of head of family.

Table 8.9 *Number of awards of Family Income Supplement, analysed by earnings and age of head of family*

Earnings of head of family (£)	Under 21	21–25	26–50	Over 50	All
			(Number of families)		
Nil	—	10	60	20	90
Under 5.00	—	10	500	70	580
5.00– 9.99	—	—	290	30	320
10.00–14.99	10	30	500	50	590
15.00–19.99	20	70	1 070	130	1 290
20.00–24.99	260	250	1 920	120	2 550
25.00–29.99	470	730	4 240	210	5 650
30.00–34.99	910	1470	7 770	420	10 570
35.00–39.99	1020	2100	8 930	420	12 470
40.00–44.99	1000	2800	11 540	660	16 000
45.00–49.99	250	1250	9 650	700	11 850
50.00–54.99	20	240	5 110	210	5 580
55.00–59.99	—	80	1 830	90	2 000
60.00–64.99	—	—	460	40	500
65.00–69.99	—	—	110	—	110
70.00–74.99	—	—	10	—	10
75.00 or more	—	—	—	—	—
All	3960	9040	53 990	3170	70 160

Source: 10 per cent sample: Social Security Statistics.

For each age group draw up a histogram of the earnings of heads of families, and compare the four patterns revealed.

Diagrams for grouped frequencies – **109**

4. Draw a histogram for the data in Question 8 of Chapter 5.

5. Table 8.10 shows a breakdown of readership of women's weekly magazines.

   Table 8.10 *Readership of weekly women's magazines (in thousands)*

	Woman's Own	Woman	Woman's Weekly
15–24	838	796	335
25–44	1216	1145	715
45–64	970	905	905
65+	652	551	752

   Draw a histogram to show the age distribution for each magazine, and comment on any differences these reveal.

6. Table 8.11 shows the distribution of spells of certified incapacity due to industrial injury by duration for Great Britain.

   Table 8.11 *Periods of certified incapacity due to industrial injury (thousands)*

Duration (days)	Men	Women
0–3	19	3
4–6	77	11
7–12	121	18
13–18	82	13
19–24	51	9
25–48	85	13
49–78	33	7
79–150	19	4
151+	8	3

   Draw a histogram for the men, and one for the women, and comment briefly.

7. Table 8.12 shows the New Earnings Survey results for the normal basic weekly hours worked by male and female manual workers.

Table 8.12 Normal basic weekly hours

Hours	Men %	Women %
30–35	1.1	16.0
35–39	6.6	15.9
39–40	30.9	50.2
40–43	9.5	6.0
43–46	14.0	5.9
46–49	11.8	2.8
49–52	8.1	1.4
52–56	7.6	0.8
56–70	8.8	0.8
70+	1.7	0.1
Total	100.0	100.0

Draw histograms for the two distributions and comment on the differences between them.

8. The *Employment Gazette* (January 1981) carried Table 8.13.

Table 8.13 Membership of trade unions at end 1979 (United Kingdom)

Number of members	Number of unions	All membership (thousand)	Percentage of Number of unions	Percentage of Membership of all unions
Under 100	73	4	16.0	0.0
100–499	124	30	27.3	0.2
500–999	47	34	10.4	0.3
1 000–2 499	58	93	12.8	0.7
2 500–4 999	43	154	9.5	1.1
5 000–9 999	24	158	5.3	1.2
10 000–14 999	7	84	1.6	0.6
15 000–24 999	19	364	4.2	2.7
25 000–49 999	17	633	3.7	4.7
50 000–99 999	15	933	3.3	6.9
100 000–249 999	16	2 387	3.5	17.7
250 000 and more	11	8 624	2.4	63.9
All members	454	13 498	100.0	100.0

Draw a Lorenz curve for this data, and briefly explain its meaning.

9
Measures of Average

9.1 The arithmetic mean

We have seen that summarizing is an important function of statistics. One obvious way to do this for a set of figures is to give some kind of 'average'. Some books also refer to this as a 'measure of location' or 'measure of central tendency', but I prefer the simple general term 'average'.

In fact, there are several different measures which could be termed 'averages', but the one used most commonly is perhaps the *arithmetic mean* (often called simply the *mean*). The mean of a group of figures is defined as:

$$\text{Arithmetic mean} = \frac{\text{sum of all the figures in the group}}{\text{number of figures in the group}}$$

Suppose that 'Massive Machinery Ltd' manufacture and service heavy machinery. They run a special bonus scheme to encourage salespeople to find new customers, and in one particular area the nine salespeople have obtained the following numbers of new customers during the last year:

Sidney	0 new customers
Fred	1 new customer
Beryl	1 new customer
Jack	1 new customer
Frank	2 new customers
Jane	2 new customers
John	3 new customers
Mark	4 new customers
Tim	5 new customers

$$\text{Mean} = \frac{\text{sum of all the figures in the group}}{\text{number of figures in the group}}$$

$$= \frac{0+1+1+1+2+2+3+4+5}{9}$$

$$= \frac{19}{9} = 2.11$$

It would be useful to note some commonly used symbols at this point.

The mean. The symbol usually used for this is either $\bar{x}$ (pronounced '*x*-bar') or the Greek letter μ (mu). Sometimes μ is used for the mean of a whole population while $\bar{x}$ is used for the mean of a sample drawn from it. At this point I shall use $\bar{x}$ to signify the mean.

Number of observations. The symbol n is usually used for this.

The sum of. The capital Greek letter sigma (Σ) means 'add together all the values of'. Thus Σx means 'add up all the *x*s'. Note that here *x* is not just one number, but a whole set of numbers like the nine numbers of new customers obtained. If we call the numbers of new customers obtained by each salesperson *x* then:

$$\Sigma x = 0+1+1+1+2+2+3+4+5 = 19$$

Here *x* stands for an array or set of numbers, not just for one number. The sum of the *x*s (Σx) is 19.

$$\text{Mean: } \bar{x} = \frac{\Sigma x}{n} = \frac{19}{9} = 2.11$$

(*Note*: The use of the summation symbol and the formulae in this section is not absolutely essential if the calculation procedures are learned. Practice in the use of formulae is, however, beneficial, as formulae sheets are provided in many examinations.)

Measures of average—113

Now we may note something about the additions we have done.

```
 x           fx
 0   1 × 0 =  0
 1 ⎫
 1 ⎬ 3 × 1 =  3
 1 ⎭
 2 ⎫ 2 × 2 =  4
 2 ⎭
 3   1 × 3 =  3
 4   1 × 4 =  4
 5   1 × 5 =  5
          ───
Σx 19      19
```

The number 1 occurs three times; its 'frequency' of occurrence is three. We had, therefore, to add on a 1 three times. It would have been easier to do this by multiplying the 1 by three and adding on this product. Similarly, the number 2 occurs twice; its 'frequency' is two. To add on two 2s, we could have multiplied the number ($x = 2$) by its frequency ($f = 2$) and added on the product 4 (= 2 × 2).

So rather than add up all the individual x values, we may add up the column giving the cross-products of the xs times the fs: the 'fx' column. Usually this may be set out as in Table 9.1.

Table 9.1

Variable x	Frequency f	fx
0	1	0
1	3	3
2	2	4
3	1	3
4	1	4
5	1	5
Total	9	19

$$\text{The mean } \bar{x} = \frac{\Sigma fx}{n}$$

But the sum of the frequencies is equal to the total number of observations; that is $\Sigma f = n$.

$$\text{Thus: } \bar{x} = \frac{\Sigma fx}{\Sigma f} = \frac{19}{9} = 2.11$$

The first two columns of Table 9.1 simply constitute an ungrouped frequency table (see section 5.4), and we have effectively stated a formula for finding the mean for such a table.

Grouped data

Section 5.5 contrasted ungrouped and grouped frequency tables, and Table 9.2 repeats an earlier example we had of grouped frequencies. Let us consider the 13 employees earning between £30 and £39. What can we estimate as the

Table 9.2 Last year's weekly earnings of 157 employees of Prunes Stretch Covers Ltd on piecework

Earnings (£) x	Number of employees f
30–39	13
39–42	13
42–45	22
45–48	36
48–51	43
51–54	15
54–60	15
Total	157

Table 9.3

Earnings (£) x	Number of employees f	Mid-point of class interval m	Cross-product fm
30–39	13	34.5	448.5
39–42	13	40.5	526.5
42–45	22	43.5	957
45–48	36	46.5	1674
48–51	43	49.5	2128.5
51–54	15	52.5	787.5
54–60	15	57	855
Total	157		7377

total of their weekly earnings (i.e. all their weekly earnings added together)? Some of them are likely to be earning more than £34.5 (the mid-point of the interval) and some less than £34.5; so if we take 13 × 34.5 as our total, this is likely to be a good estimate. Similarly, for the thirteen in the interval £39–42, we could take thirteen times 40.5 as a good estimate of the total of their earnings. Table 9.3 shows this approach generalized into a method of estimating the grand total of all weekly incomes, and hence of finding the arithmetic mean. For each class interval group, we may take the frequency (f) times the mid-point (m) of the interval as a good estimate of the total of earnings for that group. The sum of all these cross-products (Σfm) then estimates the total earnings for all the employees. To find the average we simply divide by the number of employees (n or Σf):

$$\text{Mean: } \bar{x} = \frac{\Sigma fm}{n} = \frac{7377}{157} = £46.987 \text{ (approximately £47)}$$

This is the approach and formula to use when we have data grouped in class intervals. A slight problem arises if the first or the last class interval or both are *open-ended*; for example, suppose that the last interval had been '54 and above'. In such instances we have to estimate (or I might really say 'guesstimate') the upper limit in order to obtain a mid-point for the interval. Some books suggest some set rules for estimating this mid-point, but I mistrust these. If no other information is available, and a common sense guess is impossible, then one would simply have to say that an accurate estimate of the mean was impossible. Hopefully, however, the open-ended interval has been used because there are few observations this high; a low frequency may mean that if the guesstimate is a little out it will not affect the results much.

Handling large numbers*

In section 2.3 we looked at the question of rounding and errors. If we were, for example, compiling a frequency table from a set of numbers which were all in tens of millions, we would tend to round figures to the nearest million and show them in the table as in millions, showing 34 000 000 simply as 34, for example. Suppose, then, that we round the data suitably, and set up a frequency table in appropriate units – tens, hundreds, thousands or whatever. Could the method illustrated in Table 9.3 ever involve numbers too large to handle? Modern electronic calculators normally have a capacity to work to at least eight digits, and desk top computers and many calculators work to more. It is therefore highly unlikely that the calculator capacity will be exceeded. On the assumption that any sensible person having to do such calculations today will be equipped with a calculator, there is no real reason to learn any other method. Nevertheless, I shall illustrate here a traditional method for reducing the magnitude of the numbers.

Its basis may be illustrated thus: suppose that we have four people whose heights are 172 cm, 176 cm, 180 cm, and 175 cm respectively. Then

$$\text{Mean} = \frac{172 + 176 + 180 + 175}{4} = 175.75 \text{ cm}$$

We could equally well, however, find the average *amount by which they exceed 170 cm*:

$$\text{Mean amount by which they exceed 170 cm} = \frac{2 + 6 + 10 + 5}{4} = 5.75 \text{ cm}$$

From this we deduce that the actual mean height is $5.75 + 170 = 175.75$ cm.

---

* For the BEC National level 'Applied Statistics' option module and similar courses, starred sections may be omitted without loss of continuity.

Table 9.4 illustrates the same principle for grouped frequency tables. The fourth column gives the amount by which each mid-point exceeds some constant A. In this case $A = 34.5$, but it may be any number we choose, and

Table 9.4

Earnings (£) x	Number of employees f	Mid-points of class intervals m	$m-A$	$f(m-A)$
30–39	13	34.5	0	0
39–42	13	40.5	6	78
42–45	22	43.5	9	198
45–48	36	46.5	12	432
48–51	43	49.5	15	645
51–54	15	52.5	18	270
54–60	15	57	22.5	337.5
Total	157			1960.5

we choose its value for convenience of arithmetic. In this example I made A 34.5 because it makes the numbers simple. Usually either the smallest mid-point or a round number just below it (such as 30 in this example) will make the arithmetic simplest. The mean will then be given by:

$$\text{Mean} = A + \frac{1}{n} \Sigma f(m-A)$$

$$\doteq 34.5 + \frac{1}{157} \times 1960.5$$

$$= 46.987$$

This gives exactly the same value as before; it is not an approximation, but merely a method to make the arithmetic simpler.

In this particular example this method is unlikely to be any quicker than that shown in Table 9.3, and leaves more room for slips in working. For some examples it may save time, although it is an open question whether this warrants the time spent mastering it. The time might perhaps be better spent in learning how to use the stores on a calculator effectively. Some further comment on alternative methods intended to simplify arithmetic will be made in section 10.5.

9.2 The median

A second common measure of 'average' is the median. If we arrange all the values in ascending order, the median is the middle one. In the example above, of numbers of new orders, there are nine values, corresponding to the nine salespeople. These are shown in ascending order.

Order	1st	2nd	3rd	4th	5th	6th	7th	8th	9th
Value	0	1	1	1	2	2	3	4	5

The middle value here is clearly the fifth, which has the value 2.

In general, it should be obvious that if there are n observations, the middle one will be the $\left(\frac{n+1}{2}\right)$. Here there are nine observations, so $n = 9$ and the median is the $\left(\frac{9+1}{2}\right)$ - the fifth. Note that the median is *not* equal to 5. It is the *fifth observation*, when ranked in ascending order, and is equal to 2.

In our example, n is an *odd* number (9), so there is a middle observation. Obviously, if n is an *even* number, say 10, then there will not, strictly speaking, be a middle observation. In practice in such cases statisticians adopt a rule of thumb to take the middle two observations (the 5th and 6th if there are 10), add them up and divide by two. This, it should be noted, *may* sometimes result in a value being given for a median which is not equal to any of the actual values for the variable. If n is an odd number, or if it is even but the middle two observations are equal, then the median *will* be equal to one of the values actually taken by the variable.

How do we obtain the median if the data is given in the form of a frequency table like that in the first two columns of Table 9.5? We remember that the

Table 9.5

Variable x	Frequency f	Cumulative frequency F
0	1	1
1	3	4
2	2	6
3	1	7
4	1	8
5	1	9

median is the $\left(\frac{n+1}{2}\right)$ when ranked in ascending order, so for nine observations it is the $\left(\frac{9+1}{2}\right)$ - the fifth observation. Fig. 9.1 shows us how the *cumulative frequencies* can help us to locate this fifth observation. There are four observations up to and including the 1s, and six observations up to and including the 2s. The fifth observation must therefore be one of the 2s:

so median = fifth observation = 2. In practice, of course, we need not draw a diagram, but can simply use the last column (cumulative frequencies) in Table 9.5. If there are four observations up to and including the 1s and six up to and including the 2s, then the fifth must be one of the 2s. Needless to say, this approach can be applied to much larger tables of data where frequencies are given for specific values of the variable.

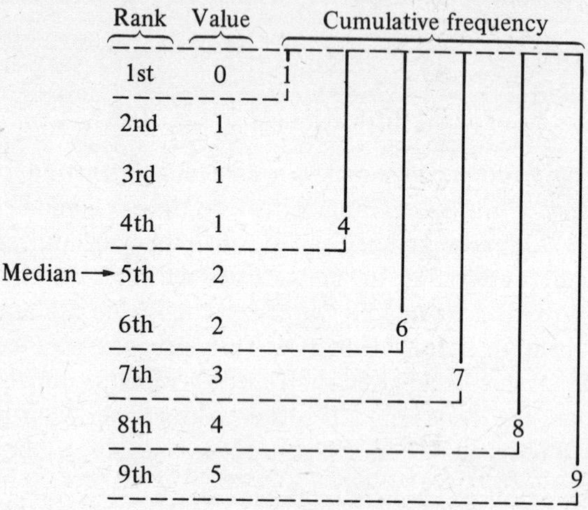

Rank	Value	Cumulative frequency
1st	0	1
2nd	1	
3rd	1	
4th	1	4
Median → 5th	2	
6th	2	6
7th	3	7
8th	4	8
9th	5	9

Figure 9.1 *How cumulative frequencies help find median*

How do we find the median if our data are given grouped in class intervals? Let us look at the example of Prunes employees' wages, given again in the first two columns of Table 9.6. There are 157 observations altogether, so $n = 157$. The median is the $\left(\frac{n+1}{2}\right)$ observation when ranked in ascending order. So here the median is the $\left(\frac{157+1}{2}\right)$ = 79th observation. The cumula-

Table 9.6

Earnings (£) x	Number of employees f	Cumulative frequencies F
30–39	13	13
39–42	13	26
42–45	22	48
45–48	36	84
48–51	43	127
51–54	15	142
54–60	15	157

lative frequency column in Table 9.6 tells us that there are 48 observations up to and including the value 45, and 84 up to and including the value 48. The 79th is therefore somewhere between 45 and 48.

The interval 45-48 is called the *median class interval*. Note that it is *not* called this because it often happens, as in this case, to be the interval mid-way down those listed. It is the distribution of observations which is important, not how the class intervals happen to have been set up. *The median interval is the one which the cumulative frequency column shows to contain the middle observation.* The 79th observation, then, is between 45 and 48. To estimate its exact location in this interval, we must make some assumption about the 36 observations which the interval contains. The simplest assumption is that they are evenly spread within it. If we make this assumption, then the median may be located either using the appropriate formula, given below, or by means of a graph, also given below.

$$\text{Median} = \text{lower boundary of median class} + \left[\frac{\frac{n}{2} - \text{cumulative frequency of class before median class}}{\text{frequency } f \text{ in median class}} \times \text{class interval of median class} \right]$$

In our example the median class has been found from cumulative frequencies to be 45-48. So:

$n = 157$
Lower boundary of median class = 45
Cumulative frequency of class before median class = 48
Frequency f in median class = 36
Class interval of median class = 3

Substituting: $\text{Median} = 45 + \left[\frac{\frac{157}{2} - 48}{36} \times 3 \right]$
$= 47.54$

The graphical method involves drawing the ogive or cumulative frequency curve, as shown in Fig. 9.2. To find the median we draw a horizontal line on the cumulative frequency curve at the point $(n/2)$ up the vertical axis. In our example this is at 78.5. At the point where this horizontal line cuts the curve, we draw a vertical line down onto the horizontal axis. Where it crosses the axis is the median value, £47.54.

The median is the value which has half the observations below it and half above it. It is also useful to define the *first quartile*, which is the value which has 25 per cent of the observations below it and 75 per cent above it. It may be found from the cumulative frequency curve in a similar way to the median, by drawing a horizontal line at $(n/4)$ up the vertical axis. Where this cuts the curve, we draw a vertical line down to find the first quartile. Here it is £43.80. Finally, it is also useful to define a *third quartile*, which is the value which has 75 per cent of the observations below it and 25 per cent above it. This may be found from the cumulative frequency curve by drawing a horizontal line at $(3n/4)$ up the vertical axis, and drawing down from where this line cuts the curve. In our example the third quartile is £50.35.

Both the quartiles may be found using formulae, but for our purposes the graphical method will suffice.

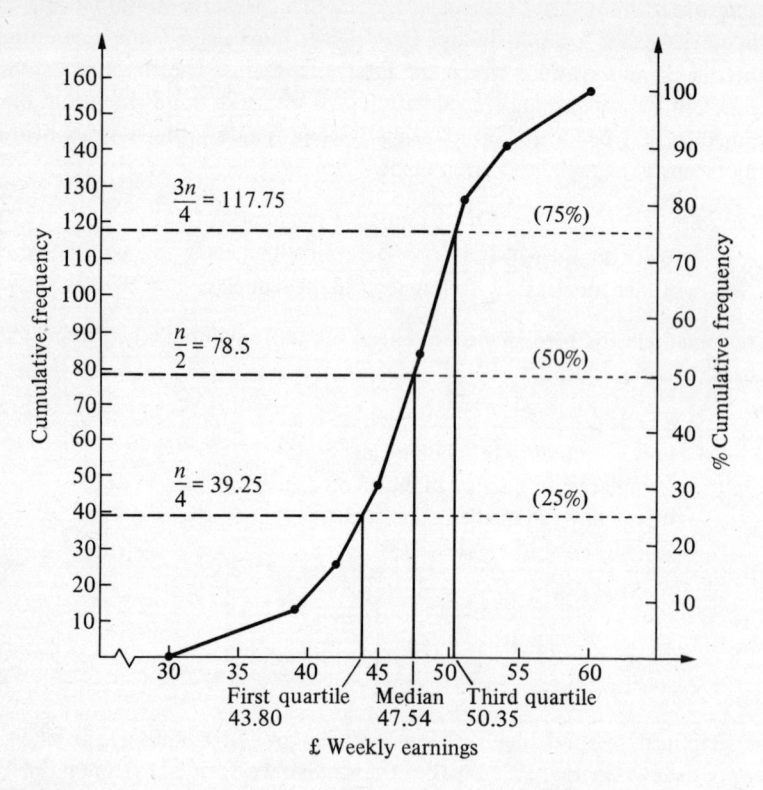

Figure 9.2 *Using cumulative frequency curve to find median and quartiles*

9.3 The mode

The mode is another measure of 'average'. In the simplest cases it may be defined as the 'most frequently occurring value'. Thus, in the example we have used of numbers of new customers, the mode is 1 since this occurs more times (that is three times) than any other value. In general, if we are given a frequency table in which frequencies are listed for various specific values of the variable, then the one which occurs most often (that is, has the highest frequency) is the mode.

What if we have a set of data in which all the values are different? In such a case there is no 'most frequently occurring' value. We may, however, take as the mode the value around which the observations are most thickly clustered or bunched. If we seek a mode in such an instance, we will probably be dealing with data given in class groupings. We turn now, therefore, to consider finding the mode for such an example, taking the weekly earnings of Prunes employees.

Table 9.6 gives the frequency table for the Prunes figures, and the histogram is shown in Fig. 9.3. The histogram, we remember, shows the 'frequency density' within each class interval. But this is just another way of saying that

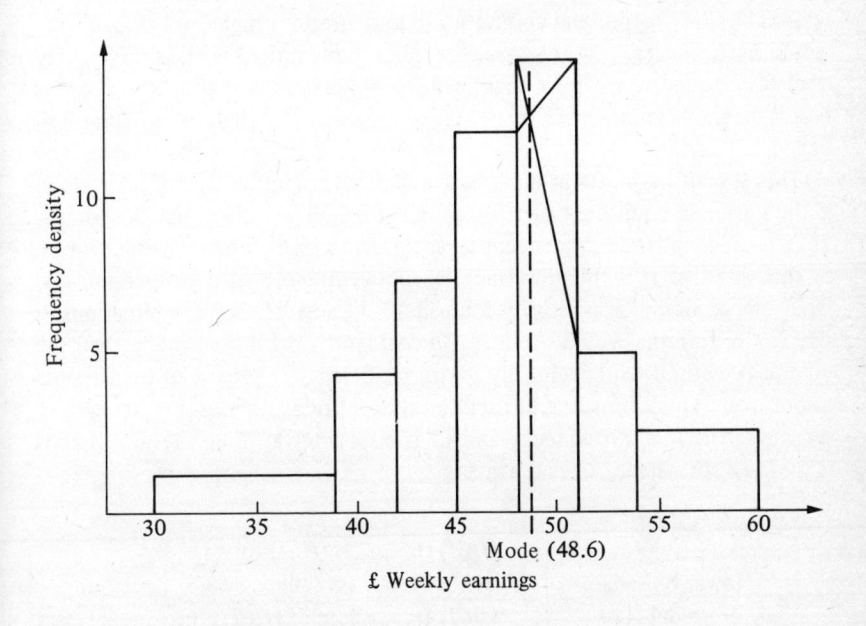

Figure 9.3 *Histogram for Prunes employees' earnings*

it shows how closely bunched the observations are within each class interval. In other words, the tallest histogram block contains the most closely bunched observations. It may be called the *modal class*. Note that the 'modal class' is the class with the highest value of (f/CI), which may *not* be the one with the highest frequency if the class intervals are unequal.

If a single figure estimate for the mode is required, then some assumption has to be made about the distribution of observations within the modal class. Some books just take a simple approach and use the mid-point of the modal class. More commonly, note is made of the frequency densities of the classes either side of the modal class, and observations within the modal class are assumed to be more frequent towards the side with the higher frequency density. This may be done geometrically, as shown in Fig. 9.3. The tops of the histogram blocks have been joined up in a special way, so that the mode can be estimated as at the point where the two lines cross. Note that this is not just a matter of drawing an X across the top of the block for the modal class! The tops of the blocks on either side have been joined. This illustrates the general graphical method of obtaining a mode.

This method, of course, contains some implicit assumptions about the distribution of observations within the modal class; unlike the assumptions made for observations within the *median* class, these are by no means self-evidently reasonable. So, to pretend to give a mode figure to a high level of accuracy would be to mislead. In any case, the mode, unlike the mean, is not usually used for any purpose which might call for a high degree of accuracy. For this reason, a graphical estimate of the mode is accurate enough for any practical purpose.

This should be borne in mind as I now briefly mention the formula which is the algebraic equivalent of the graphical method just described. A formula, of course, can provide more decimal places than a graph, but we should beware of thinking that it is therefore more accurate. As pointed out in section 2.4, 'spurious accuracy' is positively misleading. I know of very few situations in which the formula would produce a *meaningfully* greater accuracy than the graphical method, and personally regard proficiency in the use of the formula as optional. It may, however, be useful to state it briefly here, as some examiners do expect it. If f is frequency and CI is class interval, then the modal class is the class for which f/CI is the greatest:

$$\text{Mode} = \frac{\text{lower boundary}}{\text{of modal class}} + \left[\frac{(f/CI) \text{ for modal class} - (f/CI) \text{ for preceding class}}{2 \times (f/CI) \text{ for modal class} - \text{sum of } (f/CI)\text{s for classes either side}} \times \frac{CI \text{ of}}{\text{modal class}} \right]$$

Applying this to Table 9.6, the modal class is £48–51, for which f/CI is $43/3 = 14.333$, which is the greatest frequency density. From the formula the mode is:

$$48 + \left(\frac{14.333 - 12}{2 \times 14.333 - (12 + 5)} \times 3 \right) = 48.60$$

Since I regard the graphical method as simpler and this formula as optional, I shall omit the formula from the section summary, but it has been included for those who want it.

9.4 Other averages*

Two other 'averages' are given in many books. Both relate to very special situations.

Geometric mean

The geometric mean for a group of n numbers is obtained by multiplying all the n numbers together, and taking the nth root. Suppose, for example, that we have the two numbers 9 and 4:

$$\text{Geometric mean} = \sqrt[2]{9 \times 4} = \sqrt{36} = 6$$

The only practical use for this in business statistics concerns the very special situation where we are dealing with rates of growth or expansion. For example, if the Ruritanian index of retail prices rose as follows:

1983 as a percentage of 1982: 140 per cent
1984 as a percentage of 1983: 130 per cent
1985 as a percentage of 1984: 127 per cent

what is the average rise in prices as percentage of the previous year's?

$$\text{Geometric mean} = \sqrt[3]{140 \times 130 \times 127} = 132.2$$

A rise each year of 132.2 per cent on the previous year's prices would have produced the same rise over the three-year period.

Harmonic mean

The harmonic mean is defined as the reciprocal of the average of the reciprocals for a set of data:

$$\frac{1}{\frac{1}{n}\Sigma\frac{1}{x}}$$

It is even more specialized and much more rarely appropriate than even the geometric mean, and the only examples given in most business textbooks concern averaging car speeds. For our purposes it is probably best to forget it and catch a train!

9.5 Distribution shapes and averages

I have left until this point the introduction of some important terms describing various different shapes of distribution, so that their relationships with the three most important kinds of average might be mentioned.

Unimodal distributions

A unimodal distribution has a histogram with a single 'bump' or high point. Fig. 8.6 showed a unimodal distribution for men's height. Such a 'bump' or high point is, by definition, a mode, and so if there is only one bump it is called 'uni-modal'. In Fig. 8.6 the histogram is also practically symmetrical, so its single mode is in the centre. For such a unimodal and symmetrical distribution the median and the mean will also take the same value as the mode, in the centre of the distribution.

Skewed distributions

A unimodal distribution can also be *positively skewed*, as in the histogram shown in Fig. 9.4. This shows an income distribution for spinsters and widows aged 65 years or more. It is called positively skewed because the tail to the right of its single mode (or bump) is larger than that to the left.

What is the effect of this shape? The mean (£1880) is affected by the few very very large values (one lady's income was over £20 000) and so it is larger than the median (£1356) and the mode (£976). Since the median divides the area in half, and the right-hand tail is larger than the left-hand one, the median is to the right of the mode. This gives the order mode – median – mean shown.

A *negatively skewed* distribution would have a left-hand tail larger than the right-hand tail, either side of a single mode. In this instance the order would be reversed: mean – median – mode.

Measuring skewness*

There are various ways to measure skewness. The commonest uses the concept of 'standard deviation', the definition and method of calculation of which

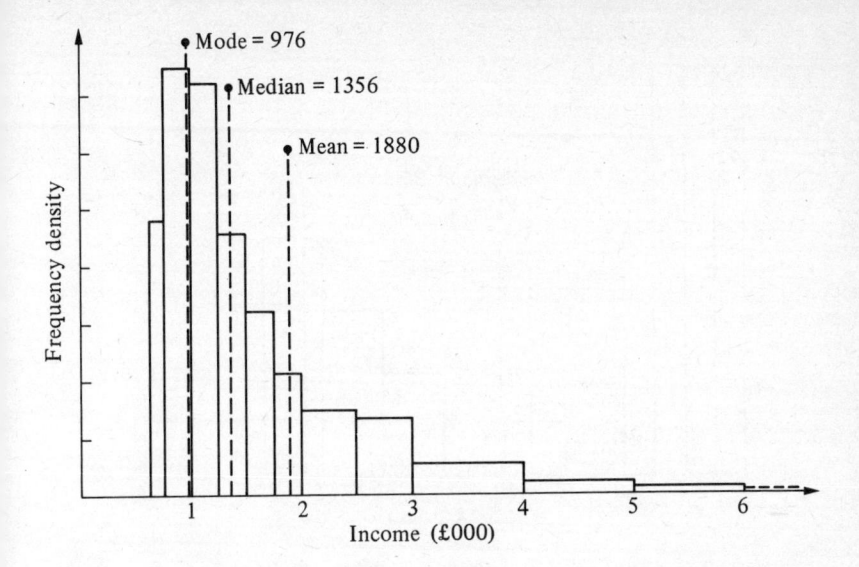

Figure 9.4 *Histogram showing (positively skewed) incomes for widows and spinsters over 65*

are given in section 10.4. The measure is Pearson's coefficient of skewness:

$$\text{Pearson's coefficient of skewness} = \frac{3\,(\text{mean} - \text{median})}{\text{standard deviation}}$$

The sign of this coefficient indicates whether the skew is positive or negative. Theoretically, values of skew as extreme as -3 or $+3$ would be possible, but in practice a value of -1 or $+1$ indicates a high degree of skew. There are often practical difficulties in calculating it, for in many highly skewed distributions the last class interval is given as open-ended, and this makes the calculation of both mean and standard deviation unreliable.

Bimodal distribution

A distribution where the histogram has *two* 'bumps' or modes is called a bi-modal distribution. Fig. 9.5 shows the distribution of ages of female pedestrians killed by road accidents in Great Britain in 1978. It is 'bi-modal' since it has two distinct modes, at just over 10 years and at 73 years respectively. The mean is 40.5 years, and the median is 35.8 years, and in this instance it is noticeable that they both come at an age where there were in fact proportionately *fewest* fatal accidents!

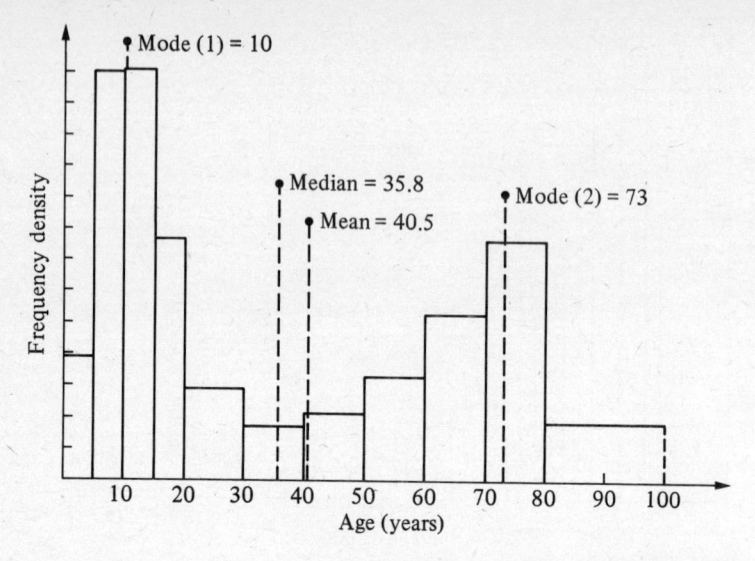

Figure 9.5 *Histogram showing Great Britain female pedestrian road deaths, 1978*

Uniform distribution

Some histograms have a totally flat top, with no kind of 'bump' at all. These are termed *uniform* or *rectangular* distributions. Fig. 9.6 shows a practical distribution which is approximately uniform, the distribution of weights of

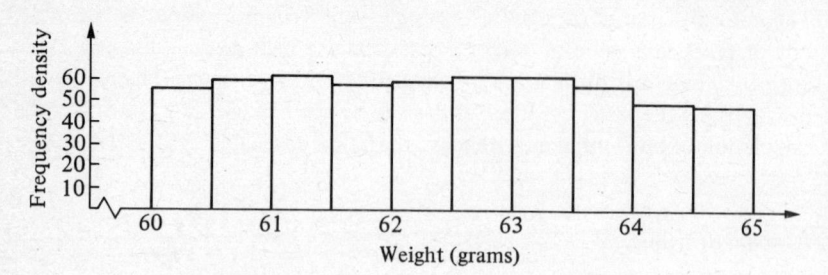

Figure 9.6 *Weights of 288 grade 3 eggs*

288 randomly selected grade three eggs. The nearly 'flat-top', or uniform, nature of this distribution implies that it has no meaningful mode, and its near symmetry makes the mean and median nearly equal (mean = 62.41; median = 62.43). A perfectly uniform distribution would have no mode and an equal mean and median.

9.6 Choosing the right 'average'

How should one decide which of the various measures of average are appropriate? One approach would be simply to list advantages and disadvantages of each form of average, but I do not believe this to be very satisfactory. Averages like harmonic and geometric mean are obscure, and will only be used for very special and clearly defined situations, as described above. The other three are all concepts relatively simple to understand, and in the modern 'calculator age' can all be easily calculated. The issue is not abstract advantages and disadvantages, but the *choice* from the three for a particular purpose. The primary question is purpose, although sometimes the particular features of the data may affect the suitability of a particular average to fulfil that purpose.

The purpose of the average

The two most important questions here are those of arithmetic manipulation and of typicality. If any *arithmetic manipulation* is likely to be required using the average (whether explicit or implicit) then the arithmetic mean is the suitable measure. Only a mean can be used to estimate a total; for example:

1. If the mean number of new customers obtained per salesperson is 2.11, then the estimated total for 15 salespeople would be $15 \times 2.11 = 31.7$ new customers.
2. If the mean weekly earnings for 157 Prunes employees are £47, then the total earnings are $157 \times 47 = £7379$.

Only a mean can be used to obtain a pooled average for two sets of data; for example if the nine salespeople in Area A obtained a mean of 2.11 new customers, and the twelve in Area B obtained a mean of 2.75 new customers, then the overall mean for the two areas together is:

$$\frac{9 \times 2.11 + 12 \times 2.75}{9 + 12} = \frac{52}{21} = 2.48 \text{ new customers}$$

Only the mean is used in probability calculations involving distributions (this will be seen in a later chapter).

Sometimes the purpose of an average is given as obtaining a *typical value*. This, or any similar term is necessarily vague compared with a criterion of suitability for arithmetic manipulation. Various things might be meant by it, but most people will probably have in mind:

1. A value which it is possible for the variable itself to take.
2. A value which often comes up in practice.

In fact the choice of an average will only be important in certain kinds of distribution. In any *symmetrical* distribution the mean and median will be the same and choice between them is not crucial. If the symmetrical distribution is also unimodal then the mode will take the same value, while if it is uniform there will be no mode at all. The choice of a 'typical' average is therefore only crucial if the distribution has more than one mode or is skewed.

In either of these cases, the mean is likely to be less suitable than the other two. If the data is discrete the median and mode will fulfil (1) and the mean will not, while as we saw in section 9.5 the mean is less likely than the other two to be near the 'popular' range of values.

Deciding between the mode and median may be less simple. The mode will fulfil both (1) and (2); while the median will often fulfil (1) but not (2) if the distribution is bimodal or skewed. It may be, however, that by careful thought it can be seen whether we really want the 'most common value' (mode) or the 'middle value' (median).

Examples

To illustrate the various issues of purpose and choice of mean, we may consider the example above of numbers of new customers gained by nine salespeople.

$$\text{mean} = 2.11 \qquad \text{median} = 2 \qquad \text{mode} = 1$$

For arithmetical manipulation we require the mean; for example, to estimate the number of new customers which might be obtained by 20 salespeople take $20 \times 2.11 = 42.2$ (i.e. about 42).

The median and mode both take whole-number values which the variable itself could take: the mean does not. Either median or mode could be seen as 'typical' depending on one's exact interpretation of this term.

The distribution as a whole is positively skewed, so the order of size is mode – median – mean.

Different groups of people may legitimately be interested in different kinds of average for the same data; in the example of weekly earnings of Prunes employees the management may be interested in the *mean* because they may be interested in the overall production and wage bill, which is related to the mean. The unions may be interested in the *median*, because it will tell them what the 'middle-of-the-road' member takes home. A prospective employee may be interested in the *mode* because it is the amount he or she is most likely to finish up earning.

Calculation problems

The choice of which kind of average to use should really be determined by the purpose of the average, not by the form in which we happen to have been given the data. Nevertheless, it should be noted that to find a mean accurately one needs to have exact values, or closed ranges of values for *all* the observations. For the median and mode this is not necessary.

In many published tables the final class interval may be open-ended, and some of the variables put into these open-ended intervals may be very much higher than the rest of the figures. For example:

1. In many official publications an income distribution may state '£20 000 and above'. As we know - and remember that the top 10 per cent of incomes account for 25 per cent of the total income - *some* incomes are very high indeed.
2. A table of trade union membership may have a final interval of '250 000 or more'. Yet we know that some unions are very large: the TGWU has over two million or around 16 per cent of *all* union members.

The mean is affected by the numerical values of such very high figures, and if we do not know them accurately then our estimate for the mean may be inaccurate. In such instances we may be forced to cite the median, even if it is not ideal, as it can be found without knowing the precise figures.

Lastly, we may note a second implication of the mean being affected by very high or 'freak' values. Where samples are being used as representative of larger groups, the mean will be affected by any freak values in the sample. This may mean that it is less 'stable' than the median. Usually, however, in such circumstances the necessity for the average to be arithmetically manipulated will require the mean in any case.

9.7 Summary

Arithmetic mean

The sum of all the values divided by the number of values. Formulae for calculating it are:

Mean $\bar{x} = \dfrac{\Sigma x}{n}$ if all individual values are given

Mean $\bar{x} = \dfrac{\Sigma fx}{n}$ if frequencies are given for stated exact values of the variable

Mean $\bar{x} = \dfrac{\Sigma fm}{n}$ if frequencies are given for the values falling in stated ranges or class intervals

Especially appropriate for arithmetic manipulation like pooling samples or reconverting to estimate totals. It uses all data values, and is therefore affected by extreme values.

Median

The middle observation [the $\frac{1}{2}(n+1)$ observation] when they are ranked in ascending order. Its location may be found from the cumulative frequency, or if the data is in class intervals the median class may be found.

For data given as individual values the median may be found directly.

For data where frequencies are given for various stated exact values the median may be found from the cumulative frequency.

For grouped data with frequencies given for each class interval, the median class should be found from the cumulative frequency; it is that class which contains the $\frac{1}{2}(n+1)$ observation. The correct formula is then:

$$\text{Median} = \text{lower boundary of median class} + \left[\dfrac{\frac{n}{2} - \text{cumulative frequency of class before median class}}{\text{frequency}(f) \text{ in median class}} \times \text{class interval of median class} \right]$$

Alternatively it may be found graphically from the cumulative frequency curve.

The median gives a 'typical' or 'middle-of-the-road' value, and is usually a value which the variable itself may take.

Mode

The mode is the most frequently occurring value for discrete data, or is the point around which the observations are the most closely clustered.

For discrete data given either individually or with frequencies for each stated value of the variable the mode is obvious by inspection; it is the one with the highest frequency of occurrence.

If the data is given with frequencies for each group or class interval, then the mode is best estimated graphically from the histogram. The modal class is the one with the tallest histogram block, and a value for the mode is obtained from this.

The mode is the most likely value to occur if one is taken at random.

Skewness

A distribution which has a long tail to the *right* of a single mode is *positively* skewed. A distribution which has a long tail to the *left* of a single mode is *negatively* skewed.

Recall and exercise questions

1. Define in general terms the arithmetic mean, the median and the mode.
2. What is the meaning of Σx?
3. What is the shape of a negatively skewed distribution?
4. During the month of February the sales records for washing machines each day in a particular store were as follows:

    2 3 1 2 2 5 3 0 1 1 2 4
    0 4 3 2 1 1 2 3 2 4 2 1

    Find the arithmetic mean, the median and the mode, and explain briefly in everyday terms the meaning of each.
5. From a frequency table constructed from data in Exercise 6 of Chapter 5, find the arithmetic mean, the median and the modal household size. If the researchers are looking for a 'typical' household, what would you recommend?
6. From the data in Exercise 2 of Chapter 8, estimate the percentage rise which was most common amongst local authorities, and what a 'typical' or 'middle of the road' authority might raise its expenditure by in the period.

7. Using the data of Table 8.9, find the mean, the median and the modal earnings for each of the four groups of family heads of different ages.

8. Find measures of average duration for the incapacities of the men and of the women in Table 8.11. Briefly comment on the problems posed by the open-ended class intervals. What can be said about the shapes of the distributions?

9. From Table 8.12:
    (a) Find the mean, median, and modal number of hours worked by male and by female manual workers.
    (b) What is typical of each group?
    (c) Is there any difference between the hours of a typical male and a typical female worker?
    (d) Is there any difference between the most common hours for men and for women?
    (e) If we took 1000 male manual workers at random from the whole population, what would you estimate to be the *total* number of weekly hours worked by this group?
    (f) What kind of shape are the two distributions of hours?

10. From the data in Exercise 8 of Chapter 5:
    (a) Estimate the mean number of passengers carried per flight.
    (b) Estimate the most probable number of passengers to be carried on a randomly chosen flight.
    (c) The airline estimate that a charge of £90 per passenger gives an adequate profit on a *charter* flight over the same route; but on a charter flight all seats are filled. What would the airline have to charge per passenger to make the same overall profit on the regular flight?

11. Use the data in Table 8.13 to obtain a figure for an 'average' size of union, and comment briefly on why you chose the particular measure of average that you did.

Problems

1. A bitter industrial dispute has been in progress between a group of male manual workers and the management over their basic number of hours. The present number is 44 hours per week, and the management claim that this is below the national average. The men, on the other hand, claim to have figures to prove that the average male manual worker in fact works less than this figure.

Assuming the hours shown in Table 8.12 to relate to the period concerned, prepare a brief statistical report to present to the arbitration committee explaining the real situation in terms of working hours.

2. Obtain copies of a professional journal (e.g. *The Certified Accountant*) over several months. Extract details of advertisements for qualified persons below director level over this period, avoiding any duplication of readvertised posts. Use your figures to present a very brief report for the student paper of the body, analysing the average salary which a practitioner might expect when qualified.

3. Obtain figures for the ages at marriage of males and females over recent years (e.g. from *Annual Abstract of Statistics*), and analyse whether or not there has been any change in the average age at which people marry.

4. From Table 5.3 present an analysis of the age patterns for different kinds of accidental death in the home for males and females in Great Britain.

10
Measures of Dispersion

10.1 The concept of dispersion

In Chapters 8 and 9 we looked at last year's weekly earnings of 157 Prunes employees. We found that for the 157:

$$\text{Mean earnings} = £46.99$$
$$\text{Median earnings} = £47.54$$
$$\text{Modal earnings} = £48.60$$

These figures also indicated a slightly negatively skewed distribution (see section 9.5).

But there is a further feature of such a wage distribution which might be of interest to us. To illustrate this, consider the weekly earnings last year of 54 typists working for the same company. Apart from the very occasional bit of overtime, they all received the same basic wage, so there was little difference between the weekly earnings of the different typists. The mean for their weekly earnings may also turn out to be £46.99 per week; but the *variation* between the earnings of different individuals would be much less than for the employees on piecework.

The question of the degree of 'variation' (or 'dispersion' or 'spread') between a set of figures is an important one. Various ways have been devised to measure it. We will now look at these different measures, using the examples from Chapter 9 of the nine salespeople and the 157 Prunes pieceworkers.

10.2 The range

In section 9.1 we referred to the number of new customers found by each of nine salespeople: Sidney got none, Fred, Beryl and Jack each got one, Frank and Jane got two each, John got three, Mark got four, and Tim got Five. The *range* is one measure of dispersion, where 'range' is defined as:

$$\text{Range} = \text{biggest value} - \text{smallest value}$$

For our salespeople the biggest number of new customers is five, and the smallest is zero, so the range is $(5-0) = 5$.

The range is a measure which is easily understood and calculated. It depends, however, on the numerical values of just two of the observations; in a sample of any size a 'freak' value at one end or the other of the scale is quite likely to occur and so affect the range. The range is really only a suitable measure, therefore, for small samples. In this context the range has been widely used in quality control schemes.

Example

The weights of packets of cheese are printed automatically. As a check, every half hour a sample of five packs is taken from the flowline production, and weighed accurately. The *mean* weight of the five is plotted on one chart, and the *range* of weight on another. If either goes outside certain set limits the process is checked.

The range has been used in preference to standard deviation (section 10.4) in such situations because it has been easier to calculate. However, with the proliferation of cheap calculators preprogrammed to calculate standard deviation, this advantage of the range disappears, and the range may fall into disuse.

10.3 The mean deviation*

One obvious way to measure dispersions is mean deviation, defined thus:

Mean deviation
 = Average over all values of (distance between the value and the mean)

In this context the appropriate average is the arithmetic mean of distances. These 'distances' must, of course, be taken as positive, whether the value is above or below the mean. Table 10.1 applies this to the salespeople example:

Table 10.1 *Finding mean deviation for numbers of new customers obtained by nine salespeople*

Name	Variable value	Positive distance from mean of 2.11
Sidney	0	2.11
Fred	1	1.11
Beryl	1	1.11
Jack	1	1.11
Frank	2	0.11
Jane	2	0.11
John	3	0.89
Mark	4	1.89
Tim	5	2.89
Total		11.33

The mean deviation is the average of the deviations taken as positive values. This will be their total divided by the number of observations:

$$\text{Mean deviation} = \frac{11.33}{9} = 1.26$$

Formulae for the mean deviation when the data are in grouped or ungrouped frequency tables are given in Appendix B. They have not been given in the text because the mean deviation is not much used. The standard deviation (section 10.4) turns out to be a much more useful measure of dispersion.

10.4 Variance and standard deviation

The mean deviation was the average of the (positive) deviations between each value and the mean. This obviously measures how spread out the data are. But suppose that instead we find the average of the deviations *squared*. If the deviations are themselves big, when squared they will certainly be very big. Table 10.2 illustrates this. The second column shows the deviations of each variable value from the mean, which is 2.11. The third column contains the deviation squared in each case; this will, of course, always be a positive number whether the actual deviation is positive or negative in sign. The *variance* is defined as the average of these squared deviations. Again, the most appropriate average is an arithmetic mean.

Variance

= Average over all values of (square of deviation of value from the mean)

$$= \frac{\text{Total of squared deviations from mean}}{\text{Number of observations}}$$

$$= \frac{\Sigma(x-\bar{x})^2}{n} = \frac{1}{n}\Sigma(x-\bar{x})^2$$

$$= \frac{1}{9} \times 20.8898 = 2.3211$$

Just as the average of deviations (taken as positive) of values from the mean is a measure of how spread out the values are, so the mean of these deviations *squared* also measures how spread out they are; it too is a measure of dispersion.

Table 10.2 Standard deviation calculation for numbers of new customers

Variable value x	Deviation from mean $(x-\bar{x})$	Deviation squared $(x-\bar{x})^2$	Frequency f	$f(x-\bar{x})^2$
0	−2.11	4.4521	1	4.4521
1	−1.11	1.2321 ⎫		
1	−1.11	1.2321 ⎬	3	
1	−1.11	1.2321 ⎭		3.6963
2	−0.11	0.0121 ⎫	2	
2	−0.11	0.0121 ⎭		0.0242
3	0.89	0.7921	1	0.7921
4	1.89	3.5721	1	3.5721
5	2.89	8.3521	1	8.3521
Total		20.8898	9	20.8898

Suppose that our data is in an ungrouped frequency table. The fourth and fifth columns of Table 10.2 illustrate calculation of the variance for such data:

Variance $= \frac{1}{n}\Sigma f(x-\bar{x})^2$ where f is the frequency in each instance

Which gives us, of course, exactly the same value as before.

Standard deviation

The variance does measure dispersion, and for some purposes has useful properties, but it does have a great disadvantage in that its units are *squared* units. For example, if the original units are in years then the variance is measured in square years! The most useful measure of all for dispersion turns out to be the standard deviation, which is derived directly from the variance:

Standard deviation $= \sqrt{\text{Variance}}$

Thus in our example:

Standard deviation $= \sqrt{2.3211} = 1.5235$

Now if the variance is big, then its square root, the standard deviation, will also be big, so standard deviation also measures dispersion. Its units, however, are not squared units, but are the same as those of the original data: if the variables are in years then so is the standard deviation.

The standard deviation proves to have various properties which make it a very useful measure of dispersion; we shall see some of these in later chapters. To find a standard deviation one always has to find the variance, and then take its square root. However, since our only interest in the variance in this book is as a stage on the way to finding the standard deviation, no further special mention will be made of it.

Notation

Various letters are used to denote standard deviation. The most common are 's' for the standard deviation within a sample, and the lower-case Greek letter sigma σ (not to be confused with the upper-case sigma Σ) for the standard deviation within a whole population. For clarity I have decided to use 's.d.(x)' to indicate the standard deviation of a set of x values. The notation s.d.( ) enables us to signify standard deviation for any set of values as specified within the brackets. However, since some readers will be using this book for externally set exam papers which provide formula sheets, I shall sometimes give key formulae using the σ notation as well. It should be noted that the standard deviation always applies to a *set* of values and cannot be applied to a single value or observation.

We then have the formulae for the standard deviation:

Standard deviation $=$ s.d.(x)

$\qquad = \sqrt{\dfrac{1}{n}\Sigma(x-\bar{x})^2}\quad$ for individual data

$\qquad = \sqrt{\dfrac{1}{n}\Sigma f(x-\bar{x})^2}\quad$ for ungrouped frequency tables

Alternative formulae*

In the pre-calculator era it was useful to invent procedures to make arithmetic as simple as possible. There are several alternative formulae for the standard deviation:

For individual data:

$$\text{s.d.}(x) = \sqrt{\frac{1}{n}\Sigma x^2 - \bar{x}^2} \quad \text{or} \quad \sqrt{\frac{1}{n}\Sigma x^2 - \left(\frac{1}{n}\Sigma x\right)^2}$$

For ungrouped frequency tables:

$$\text{s.d.}(x) = \sqrt{\frac{1}{n}\Sigma fx^2 - \bar{x}^2} \quad \text{or} \quad \sqrt{\frac{1}{n}\Sigma fx^2 - \left(\frac{1}{n}\Sigma fx\right)^2}$$

Since, for most purposes, it is doubtful whether these have any significant advantage today, when the assumption is that arithmetic will be done on a calculator, they will not be illustrated here. Alternatives for *grouped* frequency tables *will* be illustrated below, for the benefit of anyone still wishing to use them.

Grouped data

Table 10.3 illustrates the method of finding the standard deviation for data which is given in a grouped frequency table. The first four columns are the

Table 10.3 *Calculation of standard deviation for grouped frequency tables – Prunes pieceworkers*

Earnings (£) x	Frequency f	Mid-points of class intervals m	fm	$m - \bar{x}$	$f(m - \bar{x})^2$
30–39	13	34.5	448.5	−12.5	2031.25
39–42	13	40.5	526.5	−6.5	549.25
42–45	22	43.5	957	−3.5	269.5
45–48	36	46.5	1674	−0.5	9
48–51	43	49.5	2128.5	2.5	268.75
51–54	15	52.5	787.5	5.5	453.75
54–60	15	57.0	855	10.0	1500.00
Total	157		7377.0		5081.50

same as in Table 9.3, and are used to calculate the mean $\bar{x}$

$$(\bar{x} = \Sigma fm \div \Sigma f = 7377.0 \div 157 = 46.9873 = 47)$$

In section 9.1 we saw how, if the exact values are not known, the mid-points of the class intervals can be used to estimate the mean. Similarly here we can take the formula:

$$\text{s.d.}(x) = \sqrt{\frac{1}{n}\Sigma f(x - \bar{x})^2} \quad \text{for ungrouped frequencies}$$

And instead of the unknown xs use the mid-points (m):

$$\text{s.d.}(x) = \sqrt{\frac{1}{n}\Sigma f(m-\bar{x})^2} \quad \text{for grouped frequencies}$$

This formula has been applied in the last two columns of Table 10.3. From each mid-point m the mean $\bar{x}$ is subtracted (column 5), and the remainder is squared and multiplied by the frequency f (column 6).

$$\text{s.d.}(x) = \sqrt{\frac{1}{n}\Sigma f(m-\bar{x})^2}$$

$$= \sqrt{\frac{1}{157}(5081.50)} = 5.689$$

So the standard deviation is about £5.69.

A check. When doing a calculation of mean, median, mode or standard deviation it is useful to make a rough mental check that the figures arrived at are in the right order of magnitude. The mean, median and mode should all be around the middle of the range of figures, and even in a skew distribution will obviously not be *outside* that range. The standard deviation will usually be such that five or six standard deviations will just cover the range of the variable. In our example, the variable ranges from 30 to 60, so a standard deviation of £5.69 is of the right order of magnitude.

Coefficient of variation

Suppose we wish to compare the variability in the earnings of stretch covers workers with those of workers in some entirely different industry, or with workers in the USA. This other industry might have earnings with a standard deviation of, say, £7.86, compared with £5.69 at Prunes. But although this standard deviation is larger, if the *mean* earnings in the other industry were, say, £63.54 per week, compared with Prunes' £46.99, this might be misleading.

We define a measure called the coefficient of variation:

$$\text{Coefficient of variation} = \frac{\text{Standard deviation}}{\text{Arithmetic mean}} \times 100$$

For Prunes

$$\text{Coefficient of variation} = \frac{5.69}{46.99} \times 100 = 12.1$$

For the other industry

$$\text{Coefficient of variation} = \frac{7.86}{63.54} \times 100 = 12.4$$

Thus the variations are not really very different.

Measures of dispersion–141

The concept of the coefficient of variation might also be of use where, for example, one wished to compare UK workers with workers in a similar industry in the USA. One set of figures would be in pounds and the other in dollars, and to try to convert on an exchange rate might be misleading – should one use exchange rate or comparable purchasing power? The coefficient of variation enables variation to be looked at in percentage terms, irrespective of currency.

10.5 Alternative formulae and large numbers*

The formulae for standard deviation given in section 10.4 will enable a fairly rapid calculation of standard deviation to be made for practically any set of data, given a calculator. There is really no reason to learn to use more complicated formulae, and it may be hoped that increasing numbers of courses will recognize this.

Traditionally, however, textbooks have reflected a dislike for the formulae stated in section 10.4 for actual calculations. The reason for this is that the mean $\bar{x}$ is seldom a whole number, so $(x-\bar{x})$ is likely to be a long decimal. This makes very little difference today, but in days when all arithmetic had to be done using long multiplication or tables it was time-consuming. Various alternatives were therefore developed, which were easier to use (at least for the contrived textbook problems) in pre-calculator days, but are of dubious advantage today. They will be dealt with here because some examiners or professional bodies may still wish to use them.

For grouped frequency tables, the following formula will give *exactly* the same results as before, assuming no rounding errors:

$$\text{s.d.}(x) = \sqrt{\frac{1}{n}\Sigma fm^2 - \bar{x}^2}$$

Sometimes this may be written substituting the formula for mean $\bar{x}$ as well:

$$\text{s.d.}(x) = \sqrt{\frac{1}{n}\Sigma fm^2 - \left(\frac{1}{n}\Sigma fm\right)^2}$$

The use of this is illustrated in Table 10.4.

It may be noted that using the formula in section 10.4 we could afford to round the mean from £46.9873 to £47 and still obtain three decimal place accuracy on the standard deviation. With this revised formula this is not possible: check for yourself.

Table 10.4 Alternative formula for standard deviation of grouped data

Earnings (£) x	Frequency f	Mid-points of class intervals (m)	(fm)	(fm²)
30–39	13	34.5	448.5	15 473.25
39–42	13	40.5	526.5	21 323.25
42–45	22	43.5	957	41 629.50
45–48	36	46.5	1674	77 841
48–51	43	49.5	2128.5	105 360.75
51–54	15	52.5	787.5	41 343.75
54–60	15	57	855	48 735
Total	157		7377.0	351 706.50

$$\text{s.d.}(x) = \sqrt{\frac{1}{n}\Sigma fm^2 - \left(\frac{1}{n}\Sigma fm\right)^2}$$

$$= \sqrt{\frac{1}{157}(351\ 706.50) - (46.9873)^2}$$

$$= 5.689 \quad \text{as before}$$

This alternative formula may sometimes be quicker than the one in section 10.4, although the difference will often be fairly marginal. It has, however, the disadvantage that the numbers involved are larger; unlike the other formula we end up by subtracting one very large number from another very large number to get the comparatively small number in which we are interested. At the time of writing, most pocket calculators have a display of no more than ten digits, many only eight. If a calculated number exceeds the number of possible digits, then one of two things may happen:

Either 1. The calculator will signify that the calculation cannot be done, by displaying E or ERROR.

or 2. It will round the number, and tell you where the decimal point goes. For example:

75648 × 5698713 = 4.3109   11

The 11 tells us to move the decimal point eleven places to the right.

The first of these makes the calculation impossible; the second may leave us with a large error in the final answer. This implies that if the numbers are large this method is unsuitable.

Table 10.5 *Alternative method of calculating mean and standard deviation - earnings of Prunes pieceworkers*

Earnings (£) x	Frequency f	Mid-points of class intervals m	d $\left(\dfrac{m-A}{c}\right)$	fd $f\left(\dfrac{m-A}{c}\right)$	fd^2 $f\left(\dfrac{m-A}{c}\right)^2$
30–39	13	34.5	0	0	0
39–42	13	40.5	2	26	52
42–45	22	43.5	3	66	198
45–48	36	46.5	4	144	576
48–51	43	49.5	5	215	1075
51–54	15	52.5	6	90	540
54–60	15	57	7.5	112.5	843.75
Total	157			653.5	3284.75

To overcome this problem of large numbers, a third method has traditionally been used. The calculations for this, for mean and for standard deviation, are illustrated in Table 10.5. The method involves calculating a fourth column (d) as follows:

$$d = \frac{\text{mid-point} - \text{a constant }(A)}{\text{another constant }(c)} = \frac{m-A}{c}$$

These constants A and c can be any values we like, although obviously we choose ones which make the arithmetic simpler. Some books call A the 'assumed mean'. (I usually explain to my students that they do this because we *don't* assume it is the mean – and this confuses people so that they pay statisticians to come and explain it to them!) Actually it can be any constant we like. It will be best, however, to choose one of the mid-points themselves. If we use one which is half-way down the table, then some of the values in the fourth column will be negative. To avoid this here I use the first one, 34.5, for A.

The value of c will normally be the class interval if all classes are equal. In this case nearly all classes are 3 or multiples of 3, so I have made c equal to 3. If the class intervals have no common multiple, then we would make c equal to 1.

Having got our fourth column of ds, columns five and six will be fd and fd^2, respectively. The formulae for mean and for standard deviation are then:

$$\text{Arithmetic mean } \bar{x} = A + \frac{c\Sigma fd}{n} = A + \frac{c\Sigma f\left(\dfrac{m-A}{c}\right)}{n}$$

$$= 34.5 + \frac{3 \times 653.5}{157}$$

$$= £46.987 \quad \text{as before}$$

Standard deviation
$$\text{s.d.}(x) = c \times \sqrt{\frac{1}{n}\Sigma fd^2 - \left(\frac{1}{n}\Sigma fd\right)^2}$$
$$= c \times \sqrt{\frac{1}{n}\Sigma f\left(\frac{m-A}{c}\right)^2 - \left[\frac{1}{n}\Sigma f\left(\frac{m-A}{c}\right)\right]^2}$$
$$= 3 \times \sqrt{\frac{1}{157} \times 3284.75 - \left(\frac{1}{157} \times 653.5\right)^2}$$
$$= 3 \times \sqrt{20.9220 - 17.3257}$$
$$= 5.689 \quad \text{as before}$$

The sole purpose of this procedure is to make the arithmetic simpler, which in this example it succeeds in doing. Its disadvantage is that students find it complicated, and even when mastered slips in working out the values of d are easy to make. Now that any such calculations would be sensibly attempted only with a calculator, the time it saves is negligible - and it may even take longer. In my view the procedure is now obsolete, but it has been illustrated here for reasons stated above.

10.6 Dispersion and quartiles

In section 9.2 (and especially Fig. 9.2) we looked at the calculation of the median and the quartiles; the lower quartile Q_1 has 25 per cent of the observations below it, while the upper quartile Q_3 has exactly 25 per cent above it. It is clear that the more spread out the data are, the further apart these two quartiles will be.

We define the *interquartile range* as $(Q_3 - Q_1)$; it is the interval which contains 50 per cent of the observations. The *semi-interquartile range*, sometimes called the *quartile deviation*, is defined as half this figure:

semi-interquartile range $= \frac{1}{2}(Q_3 - Q_1)$

For Fig. 9.2 $\qquad\qquad\qquad = (50.35 - 43.80)/2 = 3.275$

The important point about this as a measure of dispersion is that it will not be affected by any 'freak' or extreme values. So for some situations where one is comparing distributions which are very heavily skewed, or else the information about the upper limits of values (for example, in income distributions) is incomplete, the semi-interquartile range can be useful. For any more usual situation the standard deviation has better properties.

The semi-interquartile range is in the same units (pounds, tonnes or whatever) as the original data. If we wish to obtain a measure of the kind of *percentage* variation in absolute terms, then we need to divide the interquartile range $Q_3 - Q_1$ by some measure of the original values. To do this we define:

$$\text{Quartile coefficient of dispersion} = \frac{Q_3 - Q_1}{Q_3 + Q_1}$$

For Fig. 9.2

$$\text{Quartile coefficient of dispersion} = \frac{50.35 - 43.80}{50.35 + 43.80} = 0.0696$$

Assuming Q_3 and Q_1 positive, this must always give an answer less than one, and it is a number without any units like pounds or tonnes.

10.7 Summary

Range

The difference between the largest and the smallest values in a set of data is called the range.

Mean deviation*

Each value of a data set will be a certain distance from the arithmetic mean. If we take the average (i.e. mean) of these distances (all taken as positive) this is the mean deviation.

Variance

The variance is the mean of the *squares* of distances between the values and the arithmetic mean:

$$\text{Variance} = \frac{1}{n}\Sigma(x - \bar{x})^2$$

Standard deviation

The standard deviation is defined as the square root of the variance:

$$\text{s.d.}(x) = \sqrt{\frac{1}{n}\Sigma(x-\bar{x})^2} \quad \text{if all individual values are given}$$

$$= \sqrt{\frac{1}{n}\Sigma f(x-\bar{x})^2} \quad \text{if frequencies are given for various stated exact values for the variable}$$

$$= \sqrt{\frac{1}{n}\Sigma f(m-\bar{x})^2} \quad \text{if frequencies are given for the values falling in various stated ranges or class intervals}$$

Coefficient of variation

This is defined as:

$$\text{Coefficient of variation} = \frac{\text{Standard deviation}}{\text{Arithmetic mean}} \times 100$$

Quartile deviation or semi-interquartile range

This is defined as: $\quad \text{Semi-interquartile range} = \dfrac{Q_3 - Q_1}{2}$

Quartile coefficient of dispersion

$$\text{Quartile coefficient of dispersion} = \frac{Q_3 - Q_1}{Q_3 + Q_1}$$

Recall and exercise questions

1. What is the 'range'?
2. What is the relationship between the variance and the standard deviation?
3. Define the coefficient of variation.
4. Define: (a) the interquartile range
   (b) the semi-interquartile range
   (c) the quartile coefficient of dispersion
5. For the figures in Exercise 6 of Chapter 5, find:
   (a) the range
   (b) the variance
   (c) the standard deviation
   (d) the coefficient of variation

6. From the data in Exercise 2 of Chapter 8, estimate the standard deviation of local authority expenditure increases.
7. Using the data of Table 8.11, compare the variation in lengths of incapacity among men with those among women, using:
   (a) the quartile deviations
   (b) the standard deviations (try this with an assumed upper limit for the open-ended intervals of 300, and again with 500, and see if it makes much difference).

Problems

1. You have been asked by the Equal Opportunities Commission to produce a report comparing the hours worked by men and by women in full-time manual work (use Table 8.12 or more recent figures if obtainable). This should include comment on the average hours, general pattern, variability of hours and so on.
2. Table 10.6 shows various figures relating to the UK footwear industry for a particular year.

Table 10.6 *Footwear manufacturers*

Number of employees	Number of establishments	Number of employees	Percentage of gross output
1– 10	125	622 ⎫	
11– 24	142	2 451 ⎪	11.3
25– 49	42	1 577 ⎬	
50– 99	78	5 729 ⎭	
100– 199	93	13 095	13.9
200– 299	38	8 936	9.8
300– 399	20	6 547	6.7
400– 499	11	5 103	4.9
500– 749	19	10 081	12.1
750– 999	5	4 375	4.3
1000–2499	9	14 216	15.6
2500 and +	4	17 535	21.4
Total	586	90 267	100.0

Produce a report analysing the distribution of sizes of establishments, their importance to the industry in terms of output and employment, their efficiency, and so on.

Note. This example relates to a number of previous sections of this book.

11
Index Numbers

11.1 Price indexes

If we were asked to find the average retail price of a litre of ordinary pasteurized milk in 1985 as compared with that in 1975, this would be a clearly defined task. We would compile an ungrouped frequency table of prices (in pence) for 1975, another table for 1985, and find the arithmetic mean of each table. But suppose instead that we were asked to make a comparison of average prices in the two years for dairy products *as a whole*? We would now want to compare 'averages' of items which were of different kinds. Index numbers are designed to help us in this situation. Since indexes (often written as 'indices') are usually used, in our field of interest, to compare 'averages' in two time periods, I will restrict the discussion to this aspect.

Index numbers are used very widely in administration, business and commerce, and many official published statistics take this form. They measure not only 'average' prices, but also other monetary levels, such as wage rates and earnings, and various kinds of volume, such as industrial production, export volumes and retail sales. In this section we will begin with the basic principles of price indexes.

Price index construction

What has happened to the prices of dairy products in Ruritania in recent years? Table 11.1 reveals all: it shows the actual price in Ruritanian pence for each item counted as a 'dairy product', for 1975, 1980 and 1985. It also shows later prices as a percentage of those for the same items in 1975:

> The price of a tub of butter in 1975 was 8p and in 1985 was 42p, so the price in 1985 as a percentage of the 1975 price is: $\frac{42}{8} \times 100\% = 525\%$.

These percentages are called *price-relatives*, and they enable us to compare easily the proportionate changes in price of different items since 1975.

Table 11.1 Prices of Ruritanian dairy products 1975–1985

Item	Price in 1975 In pence	As a percentage of 1975 price	Price in 1980 In pence	As a percentage of 1975 price	Price in 1985 In pence	As a percentage of 1975 price
1 litre of milk	10	100	20	200	37	370
250 g block of cheese	20	100	50	250	80	400
250 g tub of butter	8	100	20	250	42	525
150 g carton of yoghurt	6	100	16	267	38	633

Note: Prices are in Ruritanian pence
Source: Our Special Correspondent

But suppose that we wish to consider the changes in prices between 1975–85 in the 'dairy industry in general', or, put another way, the changes in 'average price for dairy products as a whole'? One suggestion might be simply to compare the average prices (arithmetic means) for the listed items in 1975 with those of later years:

$$\frac{\text{Average for 1985}}{\text{Average for 1975}} = \frac{\frac{1}{4}(37+80+42+38)}{\frac{1}{4}(10+20+8+6)} = 4.477 \text{ (or } 447.7\,\%)$$

This figure, however, is virtually meaningless. Because it arbitrarily takes one each of the units in which the items happen to be given, it assumes that a price rise of 1p on a 250-gram block of cheese is of equal importance to a rise of 1p on a litre of milk. This is unrealistic, since Ruritanian families buy many more litres of milk than blocks of cheese.

What about averaging the price relatives? Price-relatives in 1985 for the four items are 370, 400, 525 and 633. These average to:

$$\tfrac{1}{4}(370+400+525+633) = 482$$

This figure, however, assumes that a 10 per cent rise in cheese prices, for example, is of the same importance as a 10 per cent rise in milk prices; this is again unrealistic because much more is spent on milk than on cheese, so rises in milk prices are more important.

We have been looking for a way to measure 'average' price rises on dairy products. But *both* the approaches examined so far have involved considering price rises of dairy products mixed in arbitrary quantities. It would make more sense to look at the price rises in the total cost of a 'typical' or 'average' shopping basket of dairy products.

We need to know, therefore, what goes into the 'average' Ruritanian family family shopping basket each week. Fortunately, the Ruritanian government, with great foresight, did a massive survey in 1975 to find out family consumption of the different items. Obviously, it found great differences between families: some eat no butter, only margarine, some dislike cheese, some can afford to buy more of everything, and so on. So the 'shopping basket' used

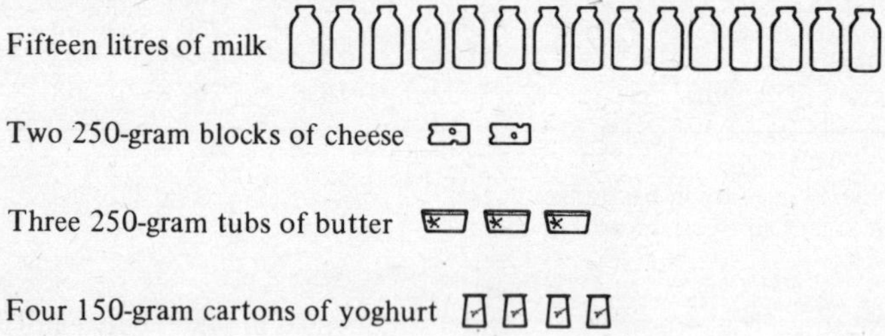

Fifteen litres of milk

Two 250-gram blocks of cheese

Three 250-gram tubs of butter

Four 150-gram cartons of yoghurt

Figure 11.1 *Average weekly dairy product shopping basket for Ruritanian families, 1975*

has to be an average, say an arithmetic mean, of all families' consumption. Fig. 11.1 shows what this worked out to be from the 1975 survey. Thus in 1975 the average weekly bill paid for dairy products by Ruritanian families was as follows:

$$1975 \text{ quantity} \times \frac{1975}{\text{price}} = \text{Amount spent}$$

15 litres of milk @ 10p	1.50
2 blocks of cheese @ 20p	0.40
3 tubs of butter @ 8p	0.24
4 cartons yoghurt @ 6p	0.24
Total	£2.38

The average family expenditure on dairy products in 1975 was £2.38. By 1985 average expenditure had risen, but its rise was due both to price increases *and* to increases in consumption of everything except butter. To compare *only* prices, we need to consider the 1985 bill for a similar basket of goods to that purchased in 1975. For 1985 this would be:

$$1975 \text{ quantity} \times \frac{1985}{\text{Price}} = \text{Amount spent}$$

15 litres of milk @ 37p	5.55
2 blocks of cheese @ 80p	1.60
3 tubs of butter @ 42p	1.26
4 cartons yoghurt @ 38p	1.52
Total	£9.93

The percentage rise in the cost of the same basket of goods from £2.38 to £9.93 represents the average rise in dairy product prices in the period:

$$\frac{9.93}{2.38} \times 100 = 417.2 \text{ per cent}$$

This figure of 417.2% is the *price index* for 1985 based on 1975, and it may be compared with the individual price-relatives for items in Table 11.1.

Some terms basic to index numbers may be introduced here. 1975 is the *base year*, for the price relatives give individual item prices as percentages of 1975 prices, and the index gives general prices as a percentage of those in 1975. Price-relatives and indexes for the base year will therefore always be 100 (i.e. 100 per cent).

The amounts of each item to include in our 'basket' were taken from *base year quantities*, and this system is sometimes called *base weighting* (these weights are, of course, not physical weights; the term is used here mathematically). When used for a price index it may also be called a *Laspeyres price index*. A formula for a Laspeyres Price Index may be expressed in words or in symbols:

$$= \frac{\text{Total for all items of}\left[\text{price}\binom{\text{in year for which}}{\text{index is required}} \times \text{quantity (in base year)}\right]}{\text{Total for all items of}\left[\text{price (in base year)} \times \text{quantity (in base year)}\right]} \times 100$$

or $= \dfrac{\Sigma p_i q_b}{\Sigma p_b q_b} \times 100$ Where p_i is price in year for which index is required
p_b is price in base year and
q_b is quantity in base year

The year for which an index is being calculated is called the 'current year', the '*n*th-year' and the 'given-year' in different books. Note that the term 'current year' does *not* mean the one we are now in: it means the year for

152–Applied business statistics

which an index is being calculated. The current year above is 1985, but if we were calculating the index for 1980, then 1980 would be the current year.

Further example. If we require the Laspeyres price index for 1980 based on 1975:

$$\text{Index} = \frac{\text{Total for all items of [price (in 1980)} \times \text{quantity (in 1975)]}}{\text{Total for all items of [price (in 1975)} \times \text{quantity (in 1975)]}} \times 100$$

$$= \frac{(20 \times 15) + (50 \times 2) + (20 \times 3) + (16 \times 4)}{(10 \times 15) + (20 \times 2) + (8 \times 3) + (6 \times 4)} \times 100$$

$$= \frac{524}{238} \times 100 = 220.2$$

Current-year weighting

So far our basket of goods has been based solely on average quantities consumed in the base year 1975. But, as already mentioned, consumption

Table 11.2 *Average weekly Ruritanian household consumption of dairy products 1975-1985*

| | | Year | |
Dairy product	1975	1980	1985
Number of litres of milk	15	16.2	18.6
Number of 250 g blocks of cheese	2	2.3	2.7
Number of 250 g tubs of butter	3	2.6	2.3
Number of 150 g cartons of yoghurt	4	4.9	6.2

changed over time. Table 11.2 shows this for both 1980 and 1985. Now if we want a price index for 1985 based on 1975, there is no reason why we should not compare bills for shopping baskets based on the 1985 rather than the 1975 consumptions. Thus:

1975				*1985*		
18.6 litres milk	@ 10p	1.86		18.6 litres milk	@ 37p	6.88
2.7 blocks cheese	@ 20p	0.54		2.7 blocks cheese	@ 80p	2.16
2.3 tubs butter	@ 8p	0.18		2.3 tubs butter	@ 42p	0.97
6.2 cartons yoghurt	@ 6p	0.37		6.2 cartons yoghurt	@ 38p	2.36
Total		£2.95		Total		£12.37

$$\text{1985 price index is } \frac{12.37}{2.95} \times 100 = 419.3$$

This figure is slightly higher than the one of 417.2 which we obtained by comparing prices for the basket containing 1975 quantities.

It should be noted that 1975 is still the base year. The difference is just that we use the quantities consumed in the year for which we want the index to compare prices with those in the base year. This is called *current-year weighting* (again remember that 'current year' means the year for which we want an index, not necessarily the one we are in now). When these are used for a price index it is called a *Paasche price index*.

A formula for a Paasche price index may be expressed in words or in symbols:

$$= \frac{\text{Total for all items of}\left[\text{price}\binom{\text{in year for which}}{\text{index is required}} \times \text{quantity}\binom{\text{in year for which}}{\text{index is required}}\right]}{\text{Total for all items of}\left[\text{price (in base year)} \times \text{quantity}\binom{\text{in year for which}}{\text{index is required}}\right]} \times 100$$

or $\quad = \dfrac{\Sigma p_i q_i}{\Sigma p_b q_i} \times 100 \quad$ Where p_i is price in year for which index is required
p_b is price in base year, and
q_i is quantity in year for which index is required

Further example. If we require the Paasche price index for 1980 based on 1975, using formula, the index is:

$$\text{Index} = \frac{(20 \times 16.2) + (50 \times 2.3) + (20 \times 2.6) + (16 \times 4.9)}{(10 \times 16.2) + (20 \times 2.3) + (8 \times 2.6) + (6 \times 4.9)} \times 100$$

$$= \frac{569.4}{258.2} = 220.5$$

It may be noted that the Laspeyres index uses the same base year quantities for however many years the index is calculated; the Paasche, on the other hand, requires the different quantities used in each year for which an index is calculated. Thus our Paasche example used 1980 quantities to calculate the 1980 index, and 1985 quantities to calculate the 1985 index. This means that a Paasche index requires more data collection. On the other hand it does consider price changes based on up-to-date consumption patterns for each year — and most people are more concerned with changes in the cost of maintaining their *present* consumptions than with the cost of past patterns of consumption.

Laspeyres and Paasche will differ markedly in their results only when *two* conditions *both* apply:

(a) Price rises are markedly larger for some items than for others

154–*Applied business statistics*

and (b) the ratios of quantities of different items to each other have changed in pattern over the period concerned.

If the first does not apply, and all items have risen by about the same, then the index will be the same irrespective of what quantities are used as weights. If the second condition does not apply, and the pattern of quantities is about the same for all years, then using base or current weights will make no difference.

Weighting price-relatives

Any realistic price index is compiled along the lines shown above. Prices for each year are weighted by being multiplied by quantities in a standard 'basket' of goods, and the total is compared with that for the same basket in the base year. Mathematically, however, it can sometimes be easier to work by weighting the *price-relatives* rather than the actual prices for each year. If suitable weights are chosen then this gives *exactly* the same results as the equivalent Laspeyres or Paasche index already described.

In our Laspeyres index, for example, the suitable weights would be the item expenditures in the base year (remember that 'weights' here are not physical; the term is being used mathematically):

$$\text{Weight for milk} \quad (15 \times 10) = 150$$
$$\text{Weight for cheese} \quad (2 \times 20) = 40$$
$$\text{Weight for butter} \quad (3 \times 8) = 24$$
$$\text{Weight for yoghurt} \quad (4 \times 6) = 24$$

The index for any year is then given by the formula (in words or symbols):

$$\text{Index} = \frac{\text{Total for all items of } [\text{price-relative} \binom{\text{in year for which}}{\text{index is required}} \times \text{weight for item}]}{\text{Total for all items of } [\text{weight for item}]}$$

or $= \dfrac{\Sigma R_i W}{\Sigma W}$ where R_i is price-relative in year for which index is required, and W is weight (for each item).

(Summed over all items)

The price-relatives for 1985 for each item are given in Table 11.1, so the index based on the weights just given will be:

$$\frac{(370 \times 150) + (400 \times 40) + (525 \times 24) + (633 \times 24)}{150 + 40 + 24 + 24} = \frac{99292}{238} = 417.2$$

Since our weights in this case were founded on base-year expenditures this has given *exactly* the same result as a Laspeyres price index using base-year quantities. If we had wished to obtain a Paasche index, then we could have used current-year expenditures for weights.

In our example the weights added up to 238. It can be more convenient to have them add up to 100 or 1000. If we multiply each weight by

$$(1000/238 = 4.2)$$

then the weights will add up to 1000, but the effect of using them will be the same. If we do this:

Weight for milk: = 630.25
Weight for cheese: = 168.07
Weight for butter: = 100.84
Weight for yoghurt: = 100.84

You may like to verify that using these weights gives *exactly* the same result as before, the index is 417.1932 or approximately 417.2 (to one decimal place).

Interpreting index numbers

Various points about index number interpretation will be made later, but at this stage it is important to note one very basic thing about the kind of index series we have looked at. Suppose that we have the Laspeyres series:

1975	1980	1985
100	220.1	417.2

The difference between 1985 and 1980 is $(417.2 - 220.1) = 197.1$. But what units is this in? It is 197.1 'percentage points of the 1975 prices'. If we wish to find the actual percentage rise in prices of 1985 compared with 1980, then this is:

$$\frac{417.2 - 220.1}{220.1} \times 100 = 89.6 \text{ per cent}$$

So the 1985 prices are 317.2 per cent up on those of 1975, but only 89.6 per cent up on those of 1980 (i.e. they are at 189.6 per cent of their 1980 levels).

This point is of special importance in comparing two series, as shown in Table 11.3. Prices rose by 40.2 percentage points, compared with a 31.2 percentage point rise in earnings. When, however, these rises are shown as percentages *of 1979 levels*, it is clear that earnings rose more than prices in the period.

Table 11.3 UK rises in prices and earnings 1979-1980

Index	1979	1980	Rise in percentage points of base year	Rise as percentage of 1979 levels
General Index of Retail Prices (Jan 74 = 100)	223.5	263.7	40.2	18.0
Index of Average Earnings (whole economy) (Jan 76 = 100)	150.9	182.1	31.2	20.7

A second point is that to find the *annual rate* of inflation which would raise prices from 100 to 417.2 in ten years is a problem involving sequences and series. As shown in Chapter 22, the annual rate would be given by x in:

$$100(1+x)^{10} = 417.2$$
$$x = 0.154$$

so the annual rate of inflation would be 15.4 per cent. Further problems and points of interpretation will be discussed below in the practical context of the General Index of Retail Prices.

Purchasing power

Another very important aspect of interpreting price indexes concerns 'purchasing power'. The basic meaning of the dairy product index is that what could be bought for £100 in 1975 would cost £417.20 in 1985 (assuming an average mix of products). This means that each £1 spent on dairy products in 1985 would buy only:

$$\frac{100}{417.2} = 24 \text{ per cent of what it bought in 1975}$$

In terms of 1975, the pound in 1985 had a purchasing power of only 24 per cent, for dairy products.

The concept of purchasing power can be used to translate a monetary expenditure into real terms. Suppose, for example, that Ruritanian hospitals spent the following on dairy products:

	1975	1980	1985
Expenditure (£0 000s)	60	130	220

By 1985 the pound was worth only 24 per cent of its value in 1975 for dairy

Index numbers–157

products. Therefore the 1985 expenditure would have bought only the same as 24 per cent of 220 in 1975: 52.8 (£0 000s). In real terms, then, there has been a *reduction* in expenditure on dairy products, assuming, of course, that the mix of products bought by hospitals is 'average'.

Purchasing power is a useful concept either for particular industries or for prices generally. The Government publishes tables of purchasing power of the pound in the UK, based on the General Index of Retail Prices.

11.2 Linking and chaining

One of the major problems in price indexes is that consumption patterns change over time. In Ruritania, for example, so little cream was consumed in 1975 that it was not worth including it in the index (the Ruritanians had always preferred yoghurt to add to sweet food). But in 1982 an enterprising dairy company introduced 'Whippy', a specially packaged and seasoned cream (with nuts!), which caught on with the public. By 1985 average weekly sales had reached such high levels (1.4 packs per household at 62p each) that it was felt that it should really be included in the dairy products index since it was basically cream.

But this produces problems. Even if we use a Paasche index, and current weightings therefore *include* Whippy, a *price* for the product in 1975 is still required. But Whippy did not exist in 1975! So what can be done? A meaningful index including the Whippy can be found for 1986 based on 1985. This is found to be 113.2, and we thus have two series:

	1975	1980	1985	1986
Laspeyres with 1975 as base year:	100	220.1	417.2	–
Laspeyres with 1985 as base year:	–	–	100	113.2

It is really more convenient to link two separate series like this into one continuous series, this is done as follows: The new index says that in 1986 dairy prices were 13.2 per cent higher than in 1985. To continue the old series, one would therefore expect the 1986 figure to be 13.2 per cent up on the 1985 figure of 417.2:

$$1986 \text{ figure} = 417.2 \times \frac{113.2}{100} = 472.3$$

This gives one series:

1975	1980	1985	1986
100	220.1	417.2	472.3

158–*Applied business statistics*

This provides a single mathematical sequence of prices as a percentage of 1975 prices, but it does it *without* assuming that a basket of goods relevant in 1975 is also relevant in years 1986 and later.

Chain-base

A technique sometimes usefully combined with linking is the chain-base method. The first part of Table 11.4 shows a chain-base index for the Ruritanian textile industry, 1983 to 1986. In a chain-base index, each year's

Table 11.4   *Chain-base and chain-linked price index numbers for Ruritanian textiles*

Content	1983	1984	1985	1986
Chain-base index number	100	113.4	111.2	114.1
Explanation of the chain base — 1st pair, 2nd pair, 3rd pair	100	113.4 / 100	111.2 / 100	114.1
Chain-linked index number	100	113.4	126.1	143.9
Explanation of the chain-linking system for indexes	100	113.4 up 11.2% to / 100 ......↑.....	126.1 up 14.1% to / 111.2 ↑ / 100	143.9 / 114.1
Procedure for the calculation of the chain-linked system	100 / 100	113.4 × $\frac{111.2}{100}$ = / 113.4 / 100	126.1 × $\frac{114.1}{100}$ = / 111.2 / 100	143.9 / 114.1

figure is given as a percentage, not of a fixed base year, but rather of the year immediately before it, for example: in 1984–5 prices rose 11.2 per cent so index for 1985 is 111.2. The complete series is useful, since by 'rate of inflation' we usually mean the rise in prices for one year as a percentage of prices in the previous one. This series gives exactly that, whereas differences between consecutive indexes in an ordinary fixed-base series are percentages of base-year prices, not of those in the previous year.

The second part of Table 11.4 shows the conversion of a chain-base index, where each year's index is a percentage of the previous year's, to a fixed-base index, where each is a percentage of the same year (here 1983). For example:

1985 prices were 11.2 per cent up on those of 1984, according to the chain-base index. In 1984 the index was 113.4, so in 1985 it will be 11.2 per cent up on 113.4:

$$113.4 \times \frac{111.2}{100} = 126.1$$

Following this procedure for all years we get a chain-linked index.

It is possible to use the *same* weights for each index in the chain-base series; for example, one could use the 1983 expenditures as weights. If the latter are used, then when the series is converted to a chain-*linked* series, it will give us a simple Laspeyres price index (you might like to try this).

It may, however, be beneficial to change the weights for each pair of years compared in a chain-base series. In many instances this can solve the problem of consumer habits changing over time. In Laspeyres or Paasche indexes a single basket of goods, using base- or current-year quantities, is presumed to be at least a reasonably realistic purchase in both the base year and the year for which the index is calculated. If these are far apart and social or consumer habits have changed, then this brings problems. But the chain-linked index only presumes that any given basket of goods is a reasonably realistic purchase for two consecutive years. In so short a period a massive change in consumer habit is very unlikely, so this is usually feasible. The only situation where chain-linking would not help us to overcome problems of change would be, for example, an index for some specific area of industry where there was a revolutionary technological breakthrough or a complete change in the way of collecting data changed the pattern within one year.

In a Laspeyres or Paasche index the year shown as 100 (the base year) does have particular importance: it influences the value for every year after, and in a Laspeyres index it also provides the weights. In a chain-linked system with annual revision of weights the year shown as 100, though it may still be *called* the base year, has no particular importance or influence. The series can be changed immediately to have any other year as base year, simply by multiplying or dividing all the values by an appropriate factor. For example, the Ruritanian textile index can have 1985 shown as 100 simply by dividing all values by 1.261.

It is sometimes claimed that it is misleading to compare later and earlier figures in a linked or chain-linked series, because later figures are based on a greater level of consumption (and so presumably customer satisfaction) than earlier ones. If, for example, a retail price index once included only black and white TV sets, but now includes colour sets, it is claimed that the figures are not comparable. This claim is based on a misunderstanding. A chain-linked price index measures neither *expenditure* (which is the product of consumption *and* price), nor *satisfaction*; it measures nothing more than *changes in price*. The actual price of colour TV sets is, moreover, irrelevant.

Only changes in their price affect the index, not how expensive they are compared with other items.

11.3 Practical price indexes

Though even recent textbooks contain incorrect information about it, the General Index of Retail Prices – RPI to its friends – is one of the most commonly cited price indexes. For this reason it will be useful to discuss general problems of indexing while outlining the construction of the RPI, although for reasons of space other indexes will then have to be sketched in very rapidly. (References to Government and other publications are listed in section 11.6).

To begin with its general aims: the index (with monthly and annual figures) measures changes in retail prices for an average 'basket' of consumer items. The term 'basket of items' is, of course, figurative, for the index includes consumer items like electricity and transport which are not usually carried in baskets! What should be included on this basis?

Obviously any kind of saving or investment expenditure would have to be excluded, as these are not consumer items. The RPI also excludes most insurances since these are seen as a kind of 'saving' against particular eventualities, and are held to be included in repair or replacement costs. Other items may have to be excluded on pragmatic grounds, because of the difficulties in defining a standard 'unit' for which price changes might be measured. The RPI excludes betting payments and church and charity donations for this reason; it also excludes educational and medical fees (see refs. 2 and 12), even though some might question whether this is reasonable.

The RPI includes items open to consumer choice from actual disposable income. Involuntary expenditures, in particular income tax and National Insurance, are rightly excluded.

One of the most difficult items to deal with in such an index is mortgage repayments by owner-occupiers. The basic problem is that the index is to measure prices of consumer items; a house is lived in but is also an investment. Thus one of two alternative lines of argument might be taken:

1. House purchase is investment in an asset. The actual *consumption* is the occupancy of a house which could be rented out, so the 'price' of that occupancy is its potential rent. A change in house prices or mortgage interest rates alters the costs of the investment, but there is no obvious tendency for rents to fluctuate with either. There is, then, no obvious

connection between house prices or mortgage interest rates and potential rent (the 'price of occupancy').

2. Most owner-occupiers, if they wish to occupy a house of the kind occupied, have no choice but to pay the prevailing mortgage interest rate. To them the mortgage repayment does represent the 'cost' of being able to 'consume' their living accommodation, and a rise in mortgage rates increases that cost. However, they also realize that the cost *is* an investment. Therefore only the interest element of the mortgage payments is counted for the index, not the repayment of capital, which is considered an investment.

Until 1975 the compilers of the RPI accepted the first argument, and included in their shopping basket a theoretical rent for owner-occupiers, calculated from the rateable value of their houses. As they were presumed to pay this rent to themselves it was also added to their income. Since 1975 argument 2 has been accepted, owner-occupation has been priced according to the mortgage interest paid (see ref. 5). The latter is estimated from an index of house prices, from which outstanding mortgage debts are estimated, and from prevailing mortgage interest rates.

Household divergence

The RPI is based on the average household 'shopping basket' of consumer items, but this conceals great differences between households. One source of difference between households is household composition; for example consumption patterns of pensioners differ greatly from those of young families. The RPI excludes pensioners, and separate index series are produced for one- and two-person pensioner households. The differences in weights between the general and the two-pensioner indexes are shown in Table 11.5, which also shows the price indexes for 1981 for each of the eleven categories listed.

There are also differences due to income levels. The RPI weightings exclude the top 3 to 4 per cent of incomes; this involves 10 to 15 per cent of the total income. Expensive consumer items are therefore under-represented. But even of those items included some, of course, will never affect poorer households, for whom food might appropriately be given a higher weighting (see refs. 8, 9 and 10).

There are some geographical differences, since rents and rates are higher in some areas than others.

Lastly, and unavoidably, there are differences due to choice of lifestyle. Some of us, for example, consume no alcohol or tobacco, but Table 11.5 shows that weightings for these are included in the RPI. Obviously it would

Table 11.5 *RPI item weights and price indexes*

Item	General index weights 1968	1975	1981	Two-pensioner index weights 1981	Price Index Jan 1981
Food	263	232	207	416	266.7
Alcohol	63	82	79	44	277.7
Tobacco	66	46	36	52	296.6
Housing	121	108	135	0[a]	285.0
Fuel and light	62	53	62	142	355.7
Durable household goods	59	70	65	44	231.0
Clothing and footwear	89	89	81	71	207.5
Transport and vehicles	120	149	152	63	299.5
Miscellaneous goods	60	71	75	84	293.4
Services	56	52	66	72	289.2
Meals out	41	48	42	12	307.5

Source: *Employment Gazette*

[a] The two-person pensioner household index excludes housing as being too difficult to calculate meaningfully.

be impracticable to produce a separate index series for each combination of area, income and lifestyle. To produce an average is sensible enough – as long as its limitations are recognized.

Index type and weights

A misleading emphasis is placed in some books on choosing a base year which has 'normal' prices. In fact the annual rate of inflation is a function of the ratios of successive annual index figures, not their size; thus whether prices were low or high in the base year is unimportant. If base-year (Laspeyres) weightings were used, then it would be important for *quantities* in the base year to be 'normal', and if current-year (Paasche) weightings were used then a year with 'abnormal' quantities could distort the index just for that year, but whether base-year prices were abnormally high or low is unimportant.

In fact, the RPI uses a chain-linked system with, since 1962, annually revised weights. The weights of most items are based, since 1975, on the latest year's figures alone, although weights of a few subject to fluctuation or high sampling error are based on the average of the last three years. For such a chain-linked system the period given as 100 (Jan 1975 at the time of writing) exerts no influence on ratios of successive index number values.

Obviously, to set weights average consumption levels are needed. A large-scale *Household Expenditure Survey* in 1953-4 has been followed since by a *Family Expenditure Survey* (FES), issued annually since 1962. Some 10 or

11 thousand households are selected randomly, using a stratified four-stage sampling scheme. Interviewers, who make initial contact at various times throughout the year, seek cooperation in provision of information by all members of these households over 16 years old, and achieve complete cooperation from about two-thirds of households. The interviewer completes a *Household Schedule* of expenditure on rent, rates, utility costs, insurances, season tickets, car purchase, and so on, and an *Income Schedule* on income for anyone over 16. In addition each 'spender' keeps a *Diary Record Book* of all expenditure for 14 consecutive days. Information is confidential, and is analysed by computer.

The weights are in the eleven broad categories shown in Table 11.5, with further subdivision into a total of 95 sections. It would be plainly impossible to follow prices of every consumer item purchased in these 95 sections. In each section, therefore, 'typical' items have been selected which will, it is argued, give a good indication of price changes in the section as a whole. Some kind of selection like this is inevitable for most price indexes. The RPI utilizes price changes on some 350 items in all.

Price information

Prices of items obviously vary with brand, quality and place of purchase. To find price changes between two periods, the same *quality* of goods must be priced, and the prices averaged over the kinds of shops from which people actually purchased in the periods concerned.

For the RPI, prices are collected by Department of Employment staff in some 200 local offices. On a predetermined Tuesday near the middle of each month, they find retail prices actually charged (excluding any credit charges) for the same goods in the same, several thousand cooperating retailers. These retailers are representative of the kind and location of shops found nationally. Those prices that are standardized may be found centrally or directly. The 150 000 individual prices obtained in this way are averaged into prices for each of the 350 items, and it is hoped that the various procedures ensure comparability of quality between periods as far as possible. It should, perhaps, be noted that the crucial point is not that the average price arrived at should be a true reflection of the actual averages paid nationwide, only that the percentage *changes* in the prices are accurate.

RPI and the cost of living

There is still much confusion over whether or not the RPI measures changes in the cost of living. Much of the confusion is over the term 'cost of living' itself. To some people it means 'costs of survival' – of the 'bare necessities'. Personally, I doubt if it has much meaning if defined in this sense; certainly the RPI does not attempt to measure this. A second meaning would be the cost of purchasing the average consumption levels in any period. This would rise *both* with price and with quantities consumed: it would be a measure of average expenditure in each period. Paradoxically, if defined in this way, should prices stay constant and average consumptions rise, then the 'cost of living' would go up!

If pressed, most people would probably accept a third definition. They would mean by a rise in the cost of living that the costs of maintaining *the same level of lifestyle* rose between one period and the next. Does the RPI measure this? We must, of course, distinguish between the general deficiencies common to all index numbers, and any which are specific to the RPI as a measure of the 'cost of living' in this third sense. All indexes are averages, and the costs of maintaining any individual lifestyle, such as that of a teetotal non-smoker living in the North West, may vary in ways different from the average. But this is a general point to bear in mind in index *interpretation*, not a specific criticism of the RPI. The real question is whether, if 'cost of living' in this sense is a meaningful concept at all, anything better than the RPI might be found to measure it. Well, the RPI measures the changes in the costs of purchasing a given (average) consumption of items out of disposable income. In practice, and especially now that mortgage repayments are included, this *is* likely to be allied to the 'cost of living' in the sense that we are now using the term. Thus, whereas older government publications (followed by most books) simply said that the RPI was 'not a cost-of-living index' (ref. 2); more recent official comment recognizes the vagueness of this term, but adds that there is 'no great harm' in calling the RPI a 'cost-of-living' index as long as its sense is understood (ref. 6).

Other price indexes

A monthly index of average earnings is compiled by the Department of Employment. Their information is provided direct by samples (and in some instances by complete returns) from public and private sector concerns. An old series (begun in January 1963) included only manufacturing industry, and was discontinued in December 1980. The present series includes virtually all sectors of the economy, and began from January 1976. It covers gross earnings paid monthly or weekly to manual and non-manual employees.

If the index of average earnings (IAE) measures average income, and the RPI consumer prices, then does a comparison of the two say how well off we are? No: the RPI measures items purchased with *disposable income*, and the IAE the *gross* earnings. Changes in income tax or national insurance affect neither, but clearly affect how well off we are. For this reason from January 1978 a new 'tax and price index' (TPI) was begun. This takes RPI and adjusts it to allow for changes in direct taxation (income tax and national insurance). This should, theoretically, be suitable to compare with the average earnings index.

References to other 'price' indexes may be found in the recommended books about sources; these indexes include the index of wholesale prices, index of wage rates, and the *Financial Times* index of share prices.

11.4 Quantity indexes

Often we may wish to measure changes in quantities over time. For single items this is no problem. For example, we might look at changes in average household consumptions of milk in terms of litres. But what if we wanted to consider changes over time in average household consumption of dairy products *as a whole*?

Table 11.6, like Table 11.2, shows changes in household consumption in Ruritania for each individual item; it shows the actual quantities consumed, and also shows consumption for each item as a percentage of the 1975 consumption. These are 'quantity-relatives', analogous to the 'price-relatives' mentioned in section 11.1. As with the prices, a simple total or average volume for each year (say in grams) would be meaningless: a gram of milk is not

Table 11.6 *Average Ruritanian household consumption of dairy products, 1975-85*

Item	Volume in 1975 Number	As a percentage of 1975 number	Volume in 1980 Number	As a percentage of 1975 number	Volume in 1985 Number	As a percentage of 1975 number
Litres of milk	15	100	16.2	108.0	18.6	124.0
250g blocks of cheese	2	100	2.3	115.0	2.7	135.0
250g tubs of butter	3	100	2.6	86.7	2.3	76.7
150g cartons of yoghurt	4	100	4.9	122.5	6.2	155.0

equivalent to a gram of cheese. Similarly, to average out the quantity-relatives would be to assume that a 10 per cent increase in milk consumption was of equal importance to a 10 per cent increase in cheese consumption; there is no reason to believe that this is so.

What we obviously need to do is again to *weight* the quantities or their quantity-relatives in some way. Now, in this particular case it might be open to us to use, say, calorie content. This might be appropriate if our interest was nutritional. But, much more commonly, the only or most appropriate weighting is, alas, in terms of prices. We could measure relative consumption in 1985 as:

$$\frac{\text{Monetary value of 1985 consumption if bought at 1975 prices}}{\text{Monetary value of 1975 consumption as bought at 1975 prices}}$$

This is a base-weighted (Laspeyres) quantity index, and the formula is:

$$\text{Index} = \frac{\text{Total for all items of } \left[\text{price (in base year)} \times \text{quantity}\binom{\text{in year for which}}{\text{index is required}}\right]}{\text{Total for all items of } [\text{price (in base year)} \times \text{quantity (in base year)}]} \times 100$$

or

$$\text{Index} = \frac{\Sigma p_b q_i}{\Sigma p_b q_b} \times 100$$

Where p_b is price of each item in the base year
q_b is quantity of each item in the base year
q_i is quantity of each item in the year for which the index is required.

If 1975 is the base year and 1985 the current year (the year for which we want an index):

$$\text{Index} = \frac{(10 \times 18.6) + (20 \times 2.7) + (8 \times 2.3) + (6 \times 6.2)}{(10 \times 15) + (20 \times 2) + (8 \times 3) + (6 \times 4)} \times 100$$

$$= \frac{295.6}{238} \times 100 = 124.2$$

(You might like to check that the Laspeyres quantity index for household consumption in 1980 comes to 108.5).

Again, as with the price indexes, it is possible to use current-year (Paasche) weightings instead of base-year ones. The formula for a Paasche quantity index is:

$$\text{Index} = \frac{\text{Total for all items of } \left[\text{price}\binom{\text{in year for which}}{\text{index is required}} \times \text{quantity}\binom{\text{in year for which}}{\text{index is required}}\right]}{\text{Total for all items of } \left[\text{price}\binom{\text{in year for which}}{\text{index is required}} \times \text{quantity (in base year)}\right]} \times 100$$

or

$$\text{Index} = \frac{\Sigma p_i q_i}{\Sigma p_i q_b} \times 100$$

Where p_i is price of each item in the year for which the index is required
q_b is quantity of each item in the base year
q_i is quantity of each item in the year for which the index is required.

Applying this formula to the prices and volumes given above, the Paasche quantity indexes for household consumption of dairy products in Ruritania work out to: 1975, 100; 1980, 108.7; 1985, 124.5; these are similar in this instance to the Laspeyres indexes.

As with price indexes, it is possible to present quantity indexes in a *chain-base* format. It is also possible to link, or to chain-link quantity indexes, but since the procedures for this are exactly the same as those for price indexes I shall not show any quantity examples here.

As with price indexes, it is possible to weight the quantity-relatives rather than actual quantities. The formula used is identical except that R_i now stands for the quantity-relative for the year indexed:

$$\text{Index} = \frac{\Sigma R_i W}{\Sigma W}$$

Since the weights will be in ratios of expenditures they will in fact be the same weights as were used for the price indexes.

The interpretation of quantity indexes in terms of percentage points of the base year, and the meaning of the ratios of successive index values, are again very similar to price indexes. There are, however, some further important points to which I now progress.

Price, quantity and expenditure

By 'expenditure' we usually mean expenditure in monetary terms, and this is the product of price and quantity bought. Now we remember that between 1975 and 1985 the Laspeyres quantity index for Ruritanian household dairy product consumption rose from 100 to 124.2, and the Laspeyres price index rose from 100 to 417.2. What does this imply?

The quantity index rise means that for every 1 unit purchased in 1975 the average family purchased 1.242 units in 1985. Had they been able to purchase at 1975 prices, their expenditure would have been multiplied by 1.242. But the price index rise means that for every pound spent at 1975 prices, they would spend £4.172 in 1985; thus their expenditure would be multiplied by a *further* factor of 4.172. These two things taken together should mean a rise in average *expenditure* by a factor of: $1.242 \times 4.172 = 5.1816$.

But was average household expenditure on these products in 1985 actually 518.16 per cent of its 1975 level? No: in 1985 the average expenditure was £12.36 per week, as against £2.38 per week in 1975, a rise to 519.33 per cent of the 1975 level. The fact that these figures are not the same, although they are quite close, is a fault of the Laspeyres index system. It is, incidentally, a fault shared by the Paasche system, which in this case would give 522.0.

This may be put another way. We looked earlier at ideas of purchasing power, and found using a Laspeyres index that the 1985 Ruritanian pound bought 24 per cent of what it bought in 1975. The average expenditure of £12.36 in 1985 should then have bought what 12.36 × 24 per cent = £2.966 would buy in 1975. The ratios of 1985 and 1975 quantities should therefore have been 2.966/2.38 = 1.246 or 124.6 per cent. This *should* give us the quantity index, but in fact it is slightly different (although again not much out).

Consumption, sales and production

The quantity index which I chose to use for illustration was one of household consumption. But average household consumption is directly related to the total retail sales: the total retail sales must be the average household consumption in a year multiplied by the number of households. If the number of households remains constant over time, then percentage changes in average household consumption will exactly match percentage changes in total retail sales. The same quantity index, then, will measure either. But suppose that the number of Ruritanian households rose by five per cent during 1975 to 1985? The Laspeyres quantity index for total retail dairy sales would then be:

$$124.2 \times \frac{105}{100} = 130.41$$

Alternatively, of course, we could have used the national sales volumes for each item to compute the quantity index for total retail sales directly. Retail sales indexes, which are really just quantity indexes whether we call them indexes of value or of volume, are published in the UK in various Government publications, such as *Business Monitor*.

The other major use of quantity indexes is for production levels. For the Ruritanian dairy industry, one presumes that for each item:

production = consumption + exports + use for processing − imports

Thus it may be different from the index for retail sales. It will, however, be found in a very similar way by weighting production quantities or quantity-relatives by price (base-year weighting or current-year weighting). It is, however, not obvious whether to use retail or wholesale prices. In the UK indexes of industrial production (see section 11.5) weights are based on 'value added'.

11.5 Practical quantity indexes

Perhaps the most important UK quantity index is the Index of Industrial Production (IOIP), published monthly in various amounts of detail in *British Business*, the *Monthly Digest*, *Economic Trends*, *Business Monitor* and other publications. This is a Laspeyres (base-year weighted) index series, with the base year being changed about every five years. The Central Statistical Office compiles it by weighting together about 300 index series which each represent output in an industry or part of an industry. Most of these 300 are compiled by other government departments by weighting together activity indicators, which number some 700 in all.

The weight given to each industry is proportional to its estimated contribution to gross domestic product at factor cost, that is the value it adds to purchased materials, fuel and services. These values-added, used to calculate weights, are estimated by adjusting in various ways figures from the *Annual Census of Production*. Most changes in price-relatives, however, are estimated from changes in gross outputs, which are easier to monitor.

Table 11.7 shows the changes in weights for the main industries covered between 1970 and the rebasing on 1975. These are based on the 1968 Standard Industrial Classification (SIC). The new revised SIC forms the basis for the new IOIP with 1980 as base year (published from 1983). When the base year is changed, the sub-series are linked in the kind of way we looked at earlier, but the combination of sub-series into one index may be recalculated using the revised weights. The dangers of exaggerated belief in the accuracy of such indexes are illustrated by the fact that under the old weights production in 1973-77 fell by 6.7 per cent, but recalculated with the revised weights (and a few other improvements) it fell by only 3.7 per cent (see ref. 24).

Other indexes

The Department of Industry publishes monthly in *British Business* and in the *Business Monitor* two index series: one for *value* and one for *volume* of retail sales. The value index measures the actual monetary expenditures on all retail sales (including VAT). These will change both with quantities bought and

Table 11.7 Weights for Index of Industrial Production: 1970 and 1975

Order of the 1968 Standard Industrial Classification		Weight (parts per 1000) 1970	1975
II	Mining and quarrying	37	41
III	Food, drink and tobacco	84	77
IV	Coal and petroleum products	7	9
V	Chemicals and allied industries	58	58
VI	Metal manufacture	57	47
VII	Mechanical engineering	100	92
VIII	Instrument engineering	15	12
IX	Electrical engineering	67	66
X	Shipbuilding and marine engineering	16	15
XI	Vehicles	72	68
XII	Other metal goods	48	46
XIII	Textiles	49	40
XIV	Leather	3	3
XV	Clothing and footwear	24	24
XVI	Bricks, cement, pottery and glass	27	28
XVII	Timber and furniture	22	26
XVIII	Paper, printing and publishing	64	58
XIX	Other manufacturing	31	31
XX	Construction	146	182
XXI	Gas, electricity and water	72	80

with the prices. The volume indexes are produced by taking the estimated value of sales each month for each kind of business, and deflating it using a specially constructed price index so as to obtain an estimate of sales at 1976 prices. The price indexes are compiled mainly from materials used for the RPI. The ideas of purchasing power, and relationship between price, quantity and expenditure were looked at above, and the method of deflating expenditure using a price index was outlined.

References to other quantity indexes, such as volume of exports, may be found in the recommended books about sources.

11.6 Index number references

From 1983 all official indexes are to be rebased on 1980. Details of these, and any other changes, are usually carried in publications like the *Employment Gazette*, *Economic Trends*, *British Business*, *Statistical News*, and the latest editions should be consulted. At the time of writing, the following references may be useful; key publications are 5 and 6.

Retail price index

1. *Report on Proposals for a New Index of Retail Prices by the Cost of Living Advisory Committee* March 1956 (Cmnd 9710).
2. *Method of Construction and Calculation of the Index of Retail Prices* HMSO, 1966 (due for republication).
3. *A Report of the Cost of Living Advisory Committee* July 1968 (Cmnd 3677).
4. *Studies in Official Statistics* Research Series
    No. 3: *Some Problems of Index Number Construction* HMSO, 1970;
    No. 5: *Further Problems of Index Number Construction* HMSO, 1973.
5. *Housing Costs, Weighting and Other Matters Affecting the Retail Prices Index: Report of the Retail Prices Index Advisory Committee* February 1975 (Cmnd 5009).
6. The unstatistical reader's guide to the retail prices index. *Employment Gazette* October 1975 p. 971.
7. Technical improvements in the retail prices index. *Employment Gazette* February 1978.
8. Impact of rising prices on different types of household. *Employment Gazette* February 1979 p. 122.
9. Effect of rising prices on low income households. *Employment Gazette* March 1979 p. 250.
10. Recent developments in the Retail Prices Index. A. G. Carruthers et al, *The Statistician*: Vol 29, No. 1 (March 1980).
11. *Family Expenditure Survey Handbook* W. F. F. Kemsley et al, HMSO, 1980.
12. Retail prices in 1980, *and* Retail price indices – annual revision of weights. *Employment Gazette* March 1981 pp. 127 and 137.
13. *Family Expenditure Survey* (see Introduction in latest edition).

Other price indexes

14. The Tax and Price Index – sources and methods. *Economic Trends*, August 1979 p. 81.
15. The New Tax and Price Index. *Department of Employment Gazette* August 1979.
16. On Index of Wholesale Prices. *Trade and Industry*, 22 September 1978, p. 668.
17. The monthly index of average earnings. *Employment Gazette* May 1975 p. 410.

18. Monthly Index of Average Earnings: extension to all industrial sectors. *Employment Gazette*, April 1976 p. 350.
19. Seasonal adjustment of earnings. *Employment Gazette*, July 1980 p. 755.

A source which also describes uses of price indexes is:

20. *Price Index Numbers for Current Cost Accounting* 1979, volume 13; later issued in *Business Monitor*.

Quantity indexes

21. *The Measurement of Changes in Production*: Studies in Official Statistics, Number 25 (HMSO 1976).
22. Recent improvements to output statistics. *Statistical News*, Feb 77 (36) p. 36.30 and Nov 77 (39) p. 39.31.
23. Measuring value added from the census of production. *Statistical News*, May 78 (41) p. 41.4.
24. The rebased estimates of the index of industrial production. *Economic Trends*, May 1979, p. 93.
25. Revisions to index numbers of production. *Economic Trends*, January 1980, p. 99.
26. Retail sales: index to be rebased on 1976 = 100. *British Business*, 29 February 1980, p. 336.

11.7 Summary

Laspeyres price index

$$\text{Index} = \frac{\text{Total for all items of} \left[\text{price} \binom{\text{in year for which}}{\text{index is required}} \times \text{quantity (in base year)} \right]}{\text{Total for all items of [price (in base year)} \times \text{quantity (in base year)]}} \times 100$$

Paasche price index

$$\text{Index} = \frac{\text{Total for all items of} \left[\text{price} \binom{\text{in year for which}}{\text{index is required}} \times \text{quantity} \binom{\text{in year for which}}{\text{index is required}} \right]}{\text{Total for all items of} \left[\text{price (in base year)} \times \text{quantity} \binom{\text{in year for which}}{\text{index is required}} \right]} \times 100$$

Price index using weights and price relatives

$$\text{Index} = \frac{\text{Total for all items of}\left[\text{price relative}\binom{\text{in year for which}}{\text{index is required}} \times \text{weight for item}\right]}{\text{Total for all items of [weight for item]}}$$

$$= \frac{\Sigma R_i W}{\Sigma W}$$

Where R_i is price relative (for each item) in year for which index is required, and
W is the weight (for each item)

Laspeyres quantity index

$$\text{Index} = \frac{\text{Total for all items of}\left[\text{price (in base year)} \times \text{quantity}\binom{\text{in year for which}}{\text{index is required}}\right]}{\text{Total for all items of}\left[\text{price (in base year)} \times \text{quantity (in base year)}\right]} \times 100$$

Paasche quantity index

$$\text{Index} = \frac{\text{Total for all items of}\left[\text{price}\binom{\text{in year for which}}{\text{index is required}} \times \text{quantity}\binom{\text{in year for which}}{\text{index is required}}\right]}{\text{Total for all items of}\left[\text{price}\binom{\text{in year for which}}{\text{index is required}} \times \text{quantity (in base year)}\right]} \times 100$$

Chaining

Chaining is a method of joining two index series, so that a new series started in a particular year forms a continuous series with one previously in operation. Doing this for each year is called 'chain-linking'.

General Index of Retail Prices RPI

The RPI is a chain-linked, Laspeyres (base-weighted) index, with weights changed annually according to the findings of the *Family Expenditure Survey.*

Recall and exercise questions

1. Why do we need index numbers?
2. What does 'current-year weighting' mean?
3. Briefly explain the concept of a chain-linked index, and why it might be used.
4. Name and briefly describe the survey upon which the weights for the General Index of Retail Prices (RPI) is based.

5. What problems are there in using the General Index of Retail Prices as a 'cost-of-living index'?
6. Briefly explain the meaning of a 'quantity index' and name *two* published in the UK.
7. What does 'purchasing power of the pound' mean?
8. Table 11.8 shows movements over a certain period of the quantities and costs of UK animal compound feedingstuffs.

Table 11.8 *Compound animal feedingstuffs, UK*

Livestock	1968/9 Qty	1968/9 Cost	1970/71 Qty	1970/71 Cost	1972/73 Qty	1972/73 Cost	1974/5 Qty	1974/5 Cost
Cattle	3361	36.2	3530	42.1	4017	46.8	3676	77.1
Calf	392	48.7	385	52.5	458	61.6	363	86.2
Pig	2211	36.4	2554	44.0	2554	49.6	2269	83.1
Poultry	4011	38.9	3832	47.3	3759	53.3	3276	88.0
Other	214	39.3	195	46.2	211	50.7	220	75.9

Quantities in thousand tonnes; costs (prices ex-mill and merchants store) in £ per tonne.

Source: Ministry of Agriculture, Fisheries and Food.

(a) Calculate Laspeyres and Paasche price indexes for the period, and briefly explain their meaning.
(b) Calculate Laspeyres and Paasche quantity indexes for the period and briefly explain their meaning.
(c) Comment briefly on any changes in the pattern of price and consumption in animal feedingstuffs over the period.

9. Table 11.9 shows the prices (pence per gallon) and inland deliveries (thousand tonnes) of petroleum products.

Table 11.9 *January prices and deliveries of petroleum products*

	1974	1975	1976	1977	1978	1979	1980	1981	Deliveries (thousand tonnes) 1974	Deliveries (thousand tonnes) 1979
2 star	40.5	71	75	78	74	78	118	130	2830	2182
4 star	42	73	77	80	76	79	120	132	13654	16503
Derv	41.5	56	62	78	84	84	126	140	5518	6057

Delivery figures for 4 star include 3 and 5 star.
Source: Department of Energy

(a) Draw graphs to show the price-relatives of individual products based on 1974.

(b) Superimpose a graph of the Laspeyres price index for all products.
(c) What difference, if any, would it make to use 1979 as base year instead of 1974?
(d) Begin a new series from 1979 and link this to the earlier one based on 1974.

10. Table 11.10 shows British landings of shellfish.

Table 11.10 *Shellfish landings*

Shellfish	Total landings (thousand tonnes) 19-5	19-6	Total value (thousand pounds) 19-5	19-6
Cockles	16.4	18.5	302	356
Mussels	6.9	7.5	193	212
Crab	6.6	7.7	1331	1917
Norway lobster	9.4	12.6	4452	7345
Lobster	0.8	0.9	2514	3414

(a) Find the price-relatives for each type of shellfish of 19-6 based on 19-5.
(b) Use the 19-5 total values (i.e. total expenditures) as weights to calculate a Laspeyres price index for shellfish, using the price-relatives.
(c) Calculate a Paasche price index for shellfish for 19-6 based on 19-5.

11. See Table 11.5; use the 1981 weights provided to weight the 1981 indexes for each of the eleven items and form:
(a) an overall general index of retail prices,
(b) an index of retail prices for two-person pensioner households.

12. Table 11.11 shows the changes in sales and prices of a small company which imports a number of items of office equipment.

Table 11.11 *Sales of Floggit and Dodge Ltd: office equipment*

Item	1976 Price	Sales	1980 Price	Sales	1984 Price	Sales
Calculator FB31	12	150	8	200	5	310
Calculator SB55	65	18	55	72	20	112
Manual typewriter	75	15.2	120	12.1	190	9.3
Electric typewriter	210	7.1	330	6.2	490	7.5
Telephone answering machine	80	0.4	110	0.5	145	0.4

(a) Calculate Laspeyres and Paasche price and quantity indexes.
(b) Briefly explain any differences you find between Laspeyres and Paasche results.
(c) A new calculator is to be added to the range from 1984. What problems will this pose in calculation of any future indexes?

13. Table 11.12 shows the indexes of house prices and the RPI.

Table 11.12 *Indexes of house prices and the RPI, 1973-1981*

Quarter	Secondhand properties Modern	Secondhand properties Older	New properties	All properties	General Index of Retail Prices (all items)
1973 4th qtr	100	100	100	100	100
1974 4th qtr	104	106	105	105	118
1975 4th qtr	114	116	119	116	148
1976 4th qtr	123	125	131	125	170
1977 4th qtr	132	134	145	135	192
1978 4th qtr	171	171	183	172	208
1979 4th qtr	221	229	236	225	244
1980 1st qtr	227	234	249	232	255
2nd qtr	234	242	259	239	270
3rd qtr	236	245	264	242	276
4th qtr	235	244	266	241	281
1981 1st qtr	239	245	275	243	288
2nd qtr	243	249	280	248	302
3rd qtr	245	249	278	248	306

Source: Nationwide Building Society mortgage approvals, *Department of Employment Gazette*.

(a) Use the General Index of Retail Prices to calculate the purchasing power of the pound in later periods in terms of that for the fourth quarter of 1973.
(b) Use this to find the cost *in real terms* of house purchase in later periods compared with 1973.

Problems

1. Use any sources available to you to find consumption figures for different types of alcoholic drink over recent years. Obtain estimates of the retail prices of each category (averaged out), and also the alcohol content (% vol.). It is not necessary to consume the products!
    (a) Produce a quantity index, weighting according to price.
    (b) Produce a quantity index, weighting according to alcohol content.

From these figures, produce a report on the movements in pattern of alcohol consumption over recent years, for each category and overall.

2. Obtain figures for recent years for the circulation (i.e. sales) of the national daily newspapers (try *Social Trends*, or better still *Benn's Press Directory* for several years). Using the current prices of papers, produce a weighted index of total daily paper circulation. Draw up your figures in a brief report:

   (a) indicating any changes in market share of circulation,
   (b) indicating the effects of any new newspapers (the *Daily Star*, for example) and suspensions of publication of others which occur during the period, and
   (c) indicating the overall movements in the industry.

12
Elementary Probability

12.1 The importance and meaning of probability

The concept of probability is vital in many areas of decision-making in industry, commerce and administration. Its spheres of application include quality control, reliability, market research surveys, opinion surveys, forecasting, stock control, insurance and investment appraisal.

The meaning of probability

Consider Table 12.1 which shows the numbers (to the nearest hundred) of various kinds of dwellings found within the suburban area of East Lumstead.

Table 12.1 *Types of dwelling in East Lumstead*

	Private dwellings	Council dwellings	Total
One-bedroom	600	1000	1600
Two-bedroom	3000	1700	4700
Three-bedroom	4400	1800	6200
Total	8000	4500	12500

Suppose that we write the addresses of all the 12 500 dwellings on identical cards and thoroughly mix all these address cards together in a large container. If we then close our eyes and take out one card we have *selected at random* one dwelling. We know, of course, that it must be one of the six categories of dwelling listed above, but we do not know in advance which of the six it will be.

Suppose we make the reasonable assumption that any card is as likely to be selected as any other. This will enable us to assign a *probability* to each of the six categories of dwelling. For example

Probability that dwelling is private two-bedroom

$$= \frac{\text{Number of (equally likely) cards for private two-bedroom dwellings}}{\text{Total number of (equally likely) cards}}$$

$$= \frac{3000}{12\,500} = 0.24$$

Put another way, if we repeatedly *selected at random* one card from the 12 500, in the long run we should expect the proportion of cases in which we get a private two-bedroom dwelling to be 0.24 (or 24 per cent).

This kind of approach may be applied in many situations where we have the following four conditions:

(A) A test situation or experiment
(e.g. selecting at random one card from 12 500 well-mixed cards).

and (B) A definable set of possible *outcomes* to this test situation
(e.g. the 12 500 different cards which could be selected).

and (C) A result or *event* consisting of some of these outcomes
(e.g. the event of getting a private two-bedroom dwelling, which consists of 3000 of the cards).

and (D) A lack of knowledge in advance about which outcome will occur
(e.g. no prior knowledge of which card would be selected).

The simplest definition of the probability of a particular event would be the proportion of times in a very long run of similar test situations that that particular event was the result:

$$Pr(\text{Event } A) = \begin{bmatrix} \text{The proportion of times in a very long run} \\ \text{of similar test situations in which Event } A \\ \text{is the result.} \end{bmatrix}$$

In practice, as in the above example, we may have no information about a long run of past test situations. We may, however, reason from the basic symmetry of the situation that in a long run each of the outcomes will come up as often as the others – they are all equally likely. We may then define:

$$Pr(\text{Event } A) = \frac{\text{Number of equally likely outcomes in which Event } A \text{ occurs}}{\text{Total number of equally likely possible outcomes}}$$

This is the most useful definition of probability for our purposes.

180—*Applied business statistics*

Further examples

1. Probability that a randomly selected card is a three-bedroom council dwelling:

    Pr(3-bedroom council)
    $$= \frac{\text{Number (equally likely) three-bedroom council cards}}{\text{Total number (equally likely) cards}}$$
    $= (1800/12\,500) = 0.144\ (14.4\ \text{per cent}).$

2. Probability that a randomly selected card is a one-bedroom dwelling:

    Pr(1-bedroom dwelling)
    $$= \frac{\text{Number (equally likely) 1-bedroom dwellings}}{\text{Total number (equally likely) dwellings}}$$
    $= (1600/12\,500) = 0.128\ (12.8\ \text{per cent})$

3. If a batch of 1000 components contains 62 faulty, what is the probability that a randomly selected component is faulty?

    Pr(faulty component)
    $$= \frac{\text{Number (equally likely) faulty components}}{\text{Total number (equally likely) components}}$$
    $= (62/1000) = 0.062\ (6.2\ \text{per cent})$

12.2 Basic features of probability

We may note, first that
(a) a number of outcomes could never be negative, and
(b) the number of equally likely outcomes in which any particular event occurs must always be less than or equal to the total number of outcomes.

This implies that a probability is a kind of proportion and is *always a number between zero and one.*

Second, the probability figure may be cited as a percentage. Thus:

0.05 is written as 5 per cent: a probability of 0.05 means a 5 per cent chance.

0.73 is written as 73 per cent: a probability of 0.73 means a 73 per cent chance.

In a sense, per cent means that one should divide by 100 to get a proportion, so in this book no distinction will be made between 5 per cent and 0.05 in probability. However, where a figure is quoted as a percentage, 'per cent' or the % sign will always be used.

Third, if in a probability calculation you obtain an answer which is *not* between zero and one (or, equivalently, between 0 and 100 per cent), your answer is wrong. If this happens in an examination and you are unable to locate your mistake then you should at least put a note on the answer to indicate that you recognize its error.

Complementary events

Two events are 'complementary' if:

(a) one or other must happen

and (b) they cannot both happen

If we take one address card randomly from our 12 500 above:

Event (that we get a one-bedroom dwelling) is complementary to *Event* (that we get a dwelling with more than one bedroom).

It is worth noting that:

$$\text{number of dwellings with more than one bedroom} + \text{number of dwellings with one bedroom} = \text{total number of dwellings}$$

If we now divide both sides by the total number of dwellings we get:

$$\frac{\text{number of dwellings with more than one bedroom}}{\text{total number dwellings}} + \frac{\text{number of dwellings with one bedroom}}{\text{total number dwellings}} = 1$$

But by our definition of probability this means:

Pr(dwelling has more than one bedroom) + Pr(dwelling has one bedroom) = 1

This is generally true.

If events A and B are complementary:

$$Pr(A) + Pr(B) = 1$$

or we could write: $Pr(A \text{ occurs}) + Pr(A \text{ does not occur}) = 1$

which implies: $Pr(A \text{ occurs}) = 1 - Pr(A \text{ does not occur})$.

This last equation can sometimes help us to find the probability that event A occurs more simply than we otherwise could.

Conditional probability

Further information about the known features of a test situation can alter the probabilities that particular results will be obtained from it. Suppose, for example, that we randomly select an address card. What is the probability that it is a one-bedroom dwelling?

$$Pr(\text{dwelling has one bedroom}) = \frac{\text{Number of one-bedroom dwellings}}{\text{Total number of dwellings}}$$

$$= \frac{1600}{12\,500} = 0.128 \ (12.8 \text{ per cent}).$$

But suppose now that we select a card at random, but are told that it is a private dwelling. What *now* is the probability that it has one bedroom? We now know that it is one of the 8000 private dwellings, of which only 600 have one bedroom. Thus:

$Pr(\text{dwelling has one bedroom given that it is private})$
$$= \frac{\text{Number of one-bedroom private dwellings}}{\text{Total number of private dwellings}}$$

$$= \frac{600}{8000} = 0.075 \ (7.5 \text{ per cent})$$

The probability of an event *given* such additional information is called *conditional probability*. In general it is represented in words as: 'probability of event B given A' and it is written variously as:

$Pr(B|A)$

$P_A(B)$

$Pr(B \text{ given } A)$

I shall adopt the last of these three (the simplest) and write:

$$Pr(\text{one bedroom } given \text{ private}) = 0.075.$$

It is important to note that in general $Pr(B \text{ given } A)$ is not the same as $Pr(A \text{ given } B)$. For example:

$Pr(\text{one bedroom } given \text{ private}) = 600/8000 = 0.075 \ (7.5 \text{ per cent})$

$Pr(\text{private } given \text{ one bedroom}) = 600/1600 = 0.375 \ (37.5 \text{ per cent})$

Further examples

1. If one address card is taken at random from the 12 500:
    Pr(three bedrooms) = 6200/12 500 = 0.496 (49.6 per cent)
    Pr(three bedrooms *given* council) = 1800/4500 = 0.4 (40 per cent)
    Pr(council *given* three bedrooms) = 1800/6200 = 0.29 (29 per cent)

2. If two components are taken at random from a box which contains 488 good and 12 faulty components:
    Pr(first component is faulty) = 12/500 = 0.024 (2.4 per cent)
    Pr(second faulty *given* first was good) = 12/499 = 0.024048 (2.4048 per cent)
    Pr(second faulty *given* first faulty) = 11/499 = 0.022044 (2.2044 per cent).

12.3 Addition law*

Suppose that for some purpose we wished to know the probability of selecting a dwelling which was *either* private *or* had three bedrooms *or* both. How many dwellings come under this description?

The number of private dwellings is 8000 and the number of three-bedroom dwellings is 6200. But if we add these two numbers together we will be *double counting* the dwellings which are *both* private and have three bedrooms since they are included in both counts. So we can see that:

Number of dwellings which are private or have three bedrooms
 = number which are private + number which have three bedrooms
   − number which are both

If we now divide both sides by the total number of dwellings we get:

$$\frac{\text{Number private or three-bedroom or both}}{\text{Total number of dwellings}}$$

$$= \frac{\text{number private} + \text{number three-bedroom} - \text{number both}}{\text{Total number of dwellings}}$$

$$= \frac{\text{Number private}}{\text{Total number}} + \frac{\text{Number three-bedroom}}{\text{Total number}} - \frac{\text{Number both}}{\text{Total number}}$$

Which, by definition means:

Pr(private *or* three-bedroom *or* both)

$$= Pr(\text{private}) + Pr(\text{three-bedroom}) - Pr(\text{both})$$
$$= 8000/12\,500 + 6200/12\,500 - 4400/12\,500$$

This illustrates an important general law, the *addition law*:

$$Pr(A \text{ or } B \text{ or both}) = Pr(A) + Pr(B) - Pr(\text{both})$$

Further examples

1. If one card is drawn what is: Pr(two-bedroom *or* council *or* both)?

$$= Pr(\text{two-bedroom}) + Pr(\text{council}) - Pr(\text{both})$$
$$= 4700/12\,500 + 4500/12\,500 - 1700/12\,500$$
$$= 0.376 + 0.36 - 0.136 = 0.6 \text{ (60 per cent).}$$

2. If one card is drawn what is: Pr(two-bedroom *or* three-bedroom *or* both)?

$$= Pr(\text{two-bedroom}) + Pr(\text{three-bedroom}) - Pr(\text{both})$$
$$= 4700/12\,500 + 6200/12\,500 - 0/12\,500$$
$$= 0.376 + 0.496 = 0.872 \text{ (87.2 per cent).}$$

Mutually exclusive events

The last example will strike you as a special case, for the probability of both events (two-bedroom *and* three-bedroom) occurring is zero. Two events which cannot both happen are called *mutually exclusive*. For mutually exclusive events the addition law of probability reduces to:

$$Pr(A \text{ or } B \text{ or both}) = Pr(A) + Pr(B)$$

In fact, for mutually exclusive events this can easily be extended to more than two events:

$$Pr(A \text{ or } B \text{ or } C) = Pr(A) + Pr(B) + Pr(C)$$

Examples

1. What is Pr(one-bedroom private *or* two-bedroom council)?

   $= Pr$(one-bedroom private) $+ Pr$(two-bedroom council)
   $=$ 600/12 500 $+$ 1700/12 500
   $= 0.048 + 0.136 = 0.184$ (18.4 per cent).

2. What is Pr(private *or* two-bedroom council *or* three-bedroom council)

   $= Pr$(private) $+$ Pr(two-bedroom council) $+$ Pr(three-bedroom council)
   $= 8000/12\ 500$ $+$ $1700/12\ 500$ $+$ $1800/12\ 500$
   $= 0.64 + 0.136 + 0.144 = 0.92$ (92 per cent).

One of the most frequent errors in probability is that of using the simplified addition law when the events concerned are *not* mutually exclusive. So if you use the simplified addition law for two or more events make sure that no two of them can happen together.

12.4 Multiplication law*

Suppose that we wish to find the probability of two events *both* happening: for example that a randomly selected dwelling is *both* private *and* has three bedrooms. In this instance it is obvious that the answer is (4400/12 500), since there are 4400 dwellings which are both private and have three bedrooms, and 12 500 altogether. In many instances, however, it is not so simple, and it will be useful to develop a general law to be used.

Let us look first at an even simpler example, illustrated in Fig. 12.1. A departmental section contains 30 executive officers. Half of them are men and half women, (15 men and 15 women). Of the women two-thirds (10) are married, and of the men one-third (5) are married. Suppose that (as an improvement on existing promotion policy!) one is selected at random for promotion:

$\frac{1}{2}$ are men (15)
$\frac{1}{3}$ of the men are married
so $\frac{1}{2} \times \frac{1}{3}$ are married men (5)
The proportion who are *both* men *and* married is $\frac{1}{2} \times \frac{1}{3} = \frac{1}{6}$
Now we may note that in probabilistic terms: $\frac{1}{2} = Pr$(man)
$$\frac{1}{3} = Pr(\text{married } given \text{ man})$$
Thus: Pr(man *and* married) $= Pr$(man) $\times Pr$(married *given* man)
$$= \quad \tfrac{1}{2} \quad \times \quad \tfrac{1}{3}$$
$$= \tfrac{1}{6} \text{ or } 0.16666 \text{ (16.667 per cent)}.$$

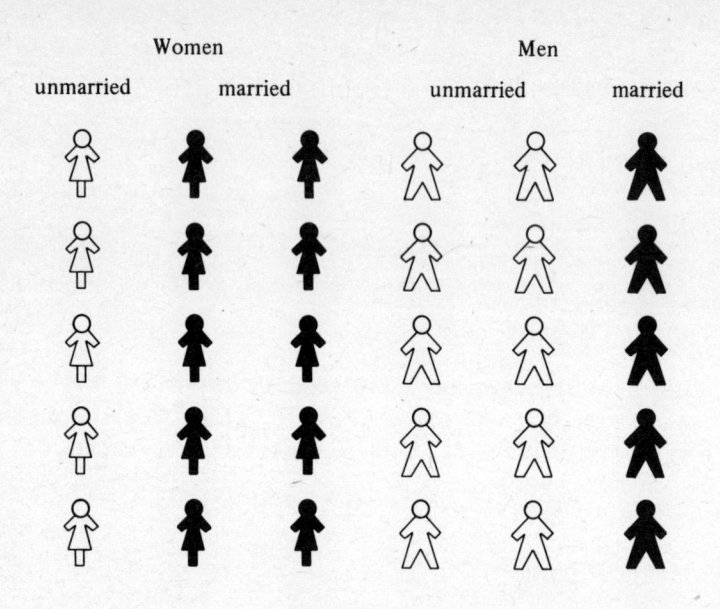

                    Women   Men
      unmarried   married   unmarried   married

Figure 12.1 *A departmental section of 30 people*

This can be similarly applied to our random selection of an address card:

Pr(private *and* three-bedrooms) = Pr(private) × Pr(three bedrooms *given* private)
                                 = 8000/12 500 ×   4400/8000
                                 =   0.64   ×   0.55
                                 = 0.352 (35 per cent).

In general: If there are two events *A* and *B*:

$$Pr(A \text{ and } B) = Pr(A) \times Pr(B \text{ given } A)$$

Further examples

1. If one address card is selected randomly what is the probability that the dwelling is both council and two-bedroomed?

    Pr(council *and* two bedrooms)
        = *Pr*(council)   × *Pr*(two bedrooms *given* council)
        = 4500/12 500 ×   1700/4500
        = 0.136 (13.6 per cent).

2. In a group of 30 people as detailed above, two are randomly selected for promotion. What is the probability that they are both married men?

Pr(first is a married man *and* second is a married man)
$= Pr$(first is married man) $\times Pr$(second is married man *given* first was)
Pr(first is married man) $= 5/30$.

When the second is selected there will be only 29 left, and if the first was a married man only four of these 29 will be married men.

Thus: Pr(second is married man *given* first was) $= 4/29$

Pr(first is married man) $\times$ Pr(second is married man *given* first was)
$=$ 5/30 $\times$ 4/29
$=$ 0.16666 $\times$ 0.137931
$= 0.023$ (2.3 per cent).

In some circumstances we may find that the occurrence or non-occurrence of the event A has no effect on the probability of the event B. In this case we would say that A and B are *independent* events. If this is the case then clearly: $Pr(B$ given $A)$ is just $Pr(B)$ since A makes no difference to it. We then get a simpler form of multiplication law:

$$Pr(A \text{ and } B) = Pr(A) \times Pr(B) \text{ (for } A \text{ and } B \text{ independent)}$$

For example, if one address card is taken at random, noted and replaced, then a second address card is taken at random, noted and replaced, what is the probability that the first was private *and* the second council?

Pr(first private *and* second council)
        $= Pr$(first private) $\times Pr$(second council)
        $=$ 8000/12 500 $\times$ 4500/12 500
        $= 0.2304$ (23.04 per cent).

This simpler form may easily be extended to more than two independent events.
If A and B and C are all independent of each other:

$$Pr(A \text{ and } B \text{ and } C) = Pr(A) \times Pr(B) \times Pr(C)$$

For example, if one takes an address card at random three times, replacing the card between successive samplings, what is the probability of getting all three as private dwellings?

Pr(first private *and* second private *and* third private)
        $= Pr$(first private) $\times$ Pr(second private) $\times$ Pr(third private)
        $=$ 8000/12 500 $\times$ 8000/12 500 $\times$ 8000/12 500
        $= (8000/12\ 500)^3 = 0.261$ (26.21 per cent).

12.5 Using the laws*

In some instances we may use both laws, sometimes more than once, in the course of a problem. For example, if one address card is taken at random, noted and replaced, and then a second address card is taken at random, noted and replaced, what is the probability that of the two cards one was private and the other a council dwelling?

Pr(one of each type)
 = Pr [(first private *and* second council) *or* (first council *and* second private)]
 = Pr(first private *and* second council) + Pr(first council *and* second private)

Here we have used the addition law, and have been able to use the simpler form of it since the probability of both (first private *and* second council) *and at the same time* (first council *and* second private) is zero – it is a contradiction in terms. Each part of the resulting sum can be broken down further:

Pr(one of each type)
 = [Pr(first private) × Pr(second council)] + [Pr(first council) × Pr(second private)]
 = [(8000/12 500) × (4500/12 500)] + [(4500/12 500 × (8000/12 500)]

12.6 Probability distributions and expectancy*

Consider the following example: A docker has an agreement with his employers that he report every day for work. If work is not available that day then he is paid £10 and sent home. If work is available then he receives the full day rate of £20 irrespective of whether it takes the whole day or not. There is no seasonal variation. Out of 480 past working days, there has been work on 288 and on the rest there has been none. What is the probability that there will be work tomorrow? In this instance we cannot refer to our definition involving equally likely events, but we have the other definition:

$$Pr(\text{Event } A) = \begin{bmatrix} \text{The proportion of times in a very long run} \\ \text{of similar test situations in which Event } A \\ \text{is the result.} \end{bmatrix}$$

If we assume that 480 is a long enough run, we may estimate the probability:

Pr(docker gets one day's work) = (288/480) = 0.6 (60 per cent)

We could then ask what his average earnings per day are over a long run. This may be simplest seen by setting out his earnings as in Table 12.2. We remember

Table 12.2 *Earnings of a docker over 480 days*

Earnings x	Frequency f	fx	Probability p	px
20	288	5760	0.6	12
10	192	1920	0.4	4
		7680		16

that the formula for the arithmetic mean for a set of figures may be given as:

$$\text{mean } \bar{x} = \frac{\Sigma fx}{\Sigma f} = \frac{7680}{480} = 16$$

$$\bar{x} = \frac{288 \times 20 + 192 \times 10}{480}$$

$$= \frac{288 \times 20}{480} + \frac{192 \times 10}{480}$$

$$= \frac{288}{480} \times 20 + \frac{192}{480} \times 10$$

But, as you can see, this is the same as multiplying each value of x by its probability and then adding these cross products. This is done more directly in the last column of the above table. This average figure would be termed the *expected earnings*.

The expected earnings is the mean earnings per day *in the long run* of days. Although termed 'expected earnings' it does not mean that £16 is what we would 'expect' for a particular single day. In fact, in this example it is impossible for a docker to receive a figure of £16 in a single day, since he must get either £10 or £20.

Probability distributions

In sections 5.4 and 5.5 we first looked at frequency tables – lists of values of a variable against the frequency with which each value was observed. The first two columns of Table 12.2 form a similar frequency table. In an analogous way we can set out a list of values of a variable against the *probability* of getting each value. The first and fourth columns of Table 12.2 form such a list, called a *probability distribution*. Usually, a probability distribution, like a frequency table, would be set out with the possible values for the variable in ascending order of size.

Further example

A company has records of the number of breakdowns on a particular kind of machine on each day over a 520-day period. The frequency table for this is shown in the first two columns of Table 12.3. The frequencies given have

Table 12.3   Machine breakdowns over 520 days

Number of breakdowns x	Frequency f	Estimated probability p	px
0	173	0.33269	0
1	190	0.36538	0.36538
2	106	0.20385	0.40770
3	38	0.07308	0.21924
4	11	0.02115	0.08460
5	2	0.00385	0.01925
Total	520	1.00000	1.09617

been used to estimate probabilities for a probability distribution. This has then been used to calculate the expected number of breakdowns per day, which is found from:

$$\text{Expected number} = \Sigma px = 1.09617$$

Note, again, that the 'expected number of breakdowns' is the average number of breakdowns per day *in the long run*, not the number which we would 'expect' on any individual day.

12.7   Summary

Definition

Probability of a particular result = Pr(result occurs)
$$= \frac{\text{Number of equally likely outcomes where result occurs}}{\text{Total number of equally likely outcomes}}$$

About probability

Probability is a number between zero and one (or 0 and 100 per cent).

Complementary events

$Pr(A \text{ occurs}) + Pr(A \text{ does not occur}) = 1$

Conditional probability

$Pr(B$ given $A)$ is the probability that B will occur given that A has occurred.

Addition law*

$Pr(A$ or B or both$) = Pr(A) + Pr(B) - Pr($both$)$
Which for *mutually exclusive* events simplifies and extends to:
$Pr(A$ or B or $C) = Pr(A) + Pr(B) + Pr(C)$

Multiplication law*

$Pr(A$ and $B) = Pr(A) \times Pr(B$ given $A)$
Which for *independent events* simplifies and extends to:
$Pr(A$ and B and $C) = Pr(A) \times Pr(B) \times Pr(C)$

Probability distributions

A list of the possible values which a variable could take, and the probability that it will take each value.

Expected value

The mean value that a variable will take in a long run of repeats of the same situation.

Recall and exercise questions

1. Define 'probability'.
2*. Give the meaning of the term:
    (a) independent events;
    (b) mutually exclusive events;
    (c) conditional probability.

3. From Table 5.2 if a passenger car were taken at random from the 1980 fourth quarter production, what would be the probability that it would be:

   (a) an export model over 1600 cc?
   (b) any car over 1600 cc?
   (c) any export model?
   (d) an export model *given that* it was over 1600 cc?
   (e) over 1600 cc *given that* it was an export model?

4. A survey has been done of 3500 households in East Lumsden. Of these, 600 contained car owners who had changed their cars during the last 12 months. They were asked the engine capacity of both their previous and their present cars; their answers are summarized in Table 12.4. This

   Table 12.4 *Results of a survey of 600 car owners in East Lumsden*

	Present engine capacity			
Previous engine capacity	Small (1300cc or less)	Medium (over 1300cc but not over 2000cc)	Large (over 2000cc)	Total
Small (1300cc or less)	180	60	10	250
Medium (over 1300cc but not over 2000cc)	25	155	20	200
Large (over 2000cc)	15	25	110	150
Total	220	240	140	600

   shows, for example, that 60 out of the group of 600 changed their car in the previous 12 months from one with a small engine (1300 cc or less) to one with a medium engine (over 1300 cc but not over 2000 cc).

   An East Lumsden resident is found to be a car owner who changed his car in the last 12 months. From the survey results, what is the probability that:

   (a) he has a medium-sized car at present?
   (b) he has changed from a medium-sized to a large car?
   (c) his present car is from a larger category than his previous?
   (d) his present car is from a smaller category than his previous?
   (e) he has owned a small car either now or previously?
   (f) he has a medium car now given that he had a medium car previously?
   (g) he has changed to a smaller car in the last 12 months given that he previously had a large one?

5. A company marketing greenhouses buys large numbers of a standard size bolt. Because of the ever-present dangers of industrial disruption they prefer not to order all bolts from one single supplier, so they buy 30 per cent of their needs from Ace Bolt Co., 25 per cent from Screwit Ltd, and the rest from U. R. McNutt. From past experience they have found that McNutt's are the best quality, while Screwit supply 4 per cent faulty and Ace 6 per cent faulty bolts.

Complete the table below, showing the sources and numbers defective of 1000 typical bolts received:

	Good	Defective	Total
Ace Bolt Co			300
Screwit Ltd			250
U. R. McNutt			
Total			1000

A bolt is taken at random; what is the probability that:

(a) it comes from Screwit Ltd?
(b) it is defective?
(c) it is a defective Ace bolt?

A bolt is taken at random and found to be defective; what is the probability that:

(d) it was supplied by Ace Bolt Co?
(e) it was not supplied by Screwit Ltd?

6*. A company has three lathes of identical model but different ages. From past records it has been found that the chances of each of them being out of action at any given time are as follows: Machine A is out of action 6 per cent of the time, Machine B for 5 per cent of the time, and Machine C for only about 2 per cent of the time.

Find the probability that at any given time:

(a) Machine A is out of action but the other two are not;
(b) Machine B is out of action but the other two are not;
(c) exactly one machine is out of action;
(d) no machines are out of action;
(e) at least one machine is out of action;
(f) no more than one machine is out of action.

7*. A company manufactures an electronic unit which contains two diodes. If a unit malfunctions it is usually because one or both of these diodes has a manufacturing fault. From past experience it is known that about 2 per cent of diodes manufactured have such faults.

(a) What percentage of units malfunction (i.e. have one or both diodes faulty)?
(b) What percentage have two faulty diodes?
(c) If a unit malfunctions what is the probability that it has only one faulty diode? (Hint: consider 1000 typical units.)

8*. An insurance company offers married couples four alternatives of life insurance:

Policy M1: payable if husband dies within ten years
Policy W1: payable if wife dies within ten years
Policy A1: payable to surviving partner if either dies within ten years
Policy B1: payable to Estate if both die within ten years

Premiums for the policy are to be calculated on a basis of the probability of payment having to be made. The company knows that the probability of a husband aged 28 dying within ten years is 0.03, and the probability of his wife aged 24 dying within ten years is 0.02.

(a) Find the probabilities of payment being made to such a couple under each of the four kinds of policy, on the assumption that their probabilities of dying are independent.
(b) Briefly comment on the implications of the assumption of independence, and say how valid you think it would be.

9*. In a large batch of components it is claimed that about 5 per cent are faulty. Two components are taken at random; what is the probability that:

(a) neither is faulty?
(b) the first is faulty and the second good?
(c) exactly one is faulty?
(d) at least one is faulty?

Three components are taken at random; what is the probability that:

(e) all are good?
(f) exactly one is faulty?

10*. A new leisure centre is to be opened, containing squash courts, swimming baths and a gymnasium. The official opening date has been arranged and Sir Peasmold Runtfuttock has been booked for the opening ceremony. Unfortunately, the builders are unsure if the facilities will be finished in time as this will depend upon the weather. After consulting the Meteorological Office they estimate the following probabilities:

Pr(squash court finished) = 0.8
Pr(swimming baths finished) = 0.6
Pr(gymnasium finished) = 0.9

(a) What is the probability that all facilities will be completed in time?
(b) The council are prepared to go ahead if at least two facilities are going to be ready; otherwise they wish to postpone the opening. What are the chances that they will be able to go ahead as planned?

11*. A building company has submitted tenders for three major projects:

A Government contract worth £250 000 to them if successful
A new supermarket worth £150 000 to them if successful
An office block worth £160 000 to them if successful

In their estimation their chances of successfuly obtaining these contracts are: 60 per cent for the Government contract, 50 per cent for the new supermarket, and 80 per cent for the office block.

(a) Cash-flow problems would result if no contracts are obtained. What is the chance of this happening?
(b) Resource problems would result if all contracts were obtained. What is the chance of this happening?
(c) What is the chance of obtaining exactly two of the contracts?
(d) What is the chance of obtaining contracts worth at least £200 000?
(e) What is their expected return on the three contracts? What does this figure mean?

Problems

1. From Table 12.4, write a brief report on the pattern of car changes, suggesting possible reasons for the pattern. Also give in your report estimates of median car sizes last year and this year for those surveyed, giving your assumptions in estimating these.

2. Over the last two years the Market Research department of Tied washing powder company has been conducting an objective longitudinal study of 200 households chosen at random. In their field there are only three major brands: Tied, Daze and Scurf. Two years ago, of the 200 households, 50 used Daze, 80 used Scurf and 70 used Tied. Each of these households were recently contacted, with the following results. Of the

50 Daze users 12 had changed to Scurf and 8 to Tied. Of the 80 Scurf users 5 had changed to Daze and 10 to Tied. Of the 70 Tied users 5 had changed to Daze and 18 to Scurf.

Produce a report for the management of Tied including information on current and past market shares, and probabilities of brand switching from and to Tied from each of the other brands.

13
The Binomial and Poisson Distributions

13.1 The binomial distribution*

Consider a situation where the following three conditions apply:

1. We have a series of 'trials', 'tests', 'experiments' or 'observations', each of which has two and only two possible outcomes, for example a test of n components, each of which is either good or faulty

and 2. Each test is 'independent'; that is the result of any given test tells you nothing about the result of the preceding or succeeding test.

and 3. If we call one result 'A' and the other result 'B' then:

Pr(Result A) = p (a constant)
Pr(Result B) = $1-p$ (also a constant)

For example: Probability (faulty component) = 0.08 (8 per cent)
Probability (good component) = 0.92 (92 per cent)

This is a *binomial situation*: and

If n = number of tests
p = Pr(A-Result)
r = number of A-Results obtained in the n tests
then r is distributed as a binomial distribution.

To illustrate this further let us look at the outcomes (in terms of numbers good and faulty) which could result from taking various sizes of samples of components, if the probability of any individual component being faulty were 0.08 (8 per cent).

Sample n = 2 (we take two components at random). There are *three* possible results: none faulty, exactly one faulty, and two faulty.

$Pr(r = 0) = Pr$(none faulty)
$= Pr$(first good *and* second good)
$= Pr$(first good) $\times Pr$(second good)
$= 0.92 \times 0.92 = 0.92^2$

$Pr(r = 1) = Pr$(exactly one faulty)
$= Pr$(first faulty *and* second good) $+ Pr$(first good *and* second faulty)
$= [Pr$(first faulty) $\times Pr$(second good)$] + [Pr$(first good) $\times Pr$(second faulty)$]$
$= [0.08 \times 0.92] + [0.92 \times 0.08]$

If we consider this last line we shall see that there are two orders in which we might obtain one good and one faulty component in our sample of two: faulty then good or good then faulty. The probability of getting any given *one* of these is 0.08×0.92, so we add this once for each order.

$$Pr(r = 1) = \begin{bmatrix} \text{number of possible orders} \\ \text{of one good and one faulty} \end{bmatrix} \times \begin{bmatrix} \text{probability of getting} \\ \text{any given order} \end{bmatrix}$$
$$= 2 \times 0.08 \times 0.92$$

Finally, $Pr(r = 2) = Pr$[first faulty *and* second faulty] $= (0.08)^2$ (analogous to $r = 0$).

Table 13.1 summarizes this.

Sample n = 3 (we take three components at random). Table 13.2 shows the *four* possibilities for r, and the probability of each. To find the probability that $r = 1$, we say:

$Pr(r = 1) = $ (number of orders) $\times$ (probability of any given order)
$= 3 \times 0.08 \times (0.92)^2$

there being three orders in which we could find one faulty and two good components.

Sample n = 5 (we take *five* components at random). Obviously we can set out a similar distribution, as shown in Table 13.3. If we look at this table it should be clear that, for example:

$Pr(r = 2) = $ (number of orders) $\times$ (probability of any given order)
$= 10 \times (0.08)^2 (0.92)^3$

But how since there are ten orders in which we could find two faulty and one good component: do we know that there are exactly ten orders in which we can arrange two faulty and three good components? As some readers may

Table 13.1

Number faulty (when $n = 2$) x	Probability of this number being faulty $Pr(r = x)$	
0	$(0.92)^2$	$= 0.8464$
1	$2(0.08)(0.92)$	$= 0.1472$
2	$(0.08)^2$	$= 0.0064$
Total		1.0000

Table 13.2

Number faulty (when $n = 3$) x	Probability of this number being faulty $Pr(r = x)$	
0	$(0.92)^3$	$= 0.778688$
1	$3(0.08)(0.92)^2$	$= 0.203136$
2	$3(0.08)^2(0.92)$	$= 0.017664$
3	$(0.08)^3$	$= 0.000512$
Total		1.000000

Table 13.3

Number faulty (when $n = 5$) x	Probability of this number being faulty $Pr(r = x)$	
0	$(0.92)^5$	$= 0.6590815232$
1	$5(0.08)(0.92)^4$	$= 0.2865571840$
2	$10(0.08)^2(0.92)^3$	$= 0.0498360320$
3	$10(0.08)^3(0.92)^2$	$= 0.0043335680$
4	$5(0.08)^4(0.92)$	$= 0.0001884160$
5	$(0.08)^5$	$= 0.0000032768$
Total		1.0000000000

remember from studying 'permutations and combinations' the formula for the number of orders is:

$$\frac{n!}{x!\,(n-x)!}$$

where $n!$ (pronounced 'n factorial') means

$$n \times (n-1) \times (n-2) \times (n-3) \times \ldots \times 3 \times 2 \times 1$$

Since the probability of getting x faulty and $(n-x)$ good in some *given* order will be $p^x(1-p)^{n-x}$, the whole formula is:

$$Pr(r = x, \text{ in sample of } n) = \frac{n!}{x!\,(n-x)!} \times p^x(1-p)^{(n-x)}$$

where n = size of sample
 p = probability any component taken at random from the population is faulty
 r = number faulty in the sample

Thus, if $n = 5, p = 0.08$

$$Pr(2 \text{ faulty}) = \frac{5!}{2!\,3!} \times (0.08)^2 (0.92)^3$$

$$= \frac{5 \times 4 \times 3 \times 2 \times 1}{(2 \times 1) \times (3 \times 2 \times 1)} \times (0.08)^2 (0.92)^3$$

$$= 10(0.08)^2 (0.92)^3 = 0.0498360320$$

Further example

If a sample of eight is taken, what is the probability of exactly three faulty?

$$Pr(r = 3) = \frac{8!}{3!\,5!} (0.08)^3 (0.92)^5$$

$$= 0.01889718543$$

(A complete distribution for $n = 8$ could, of course, be compiled)

Diagram of binomial distribution

It is possible to draw a diagram to show the probability distribution given in Table 13.3; this is given in Fig. 13.1.

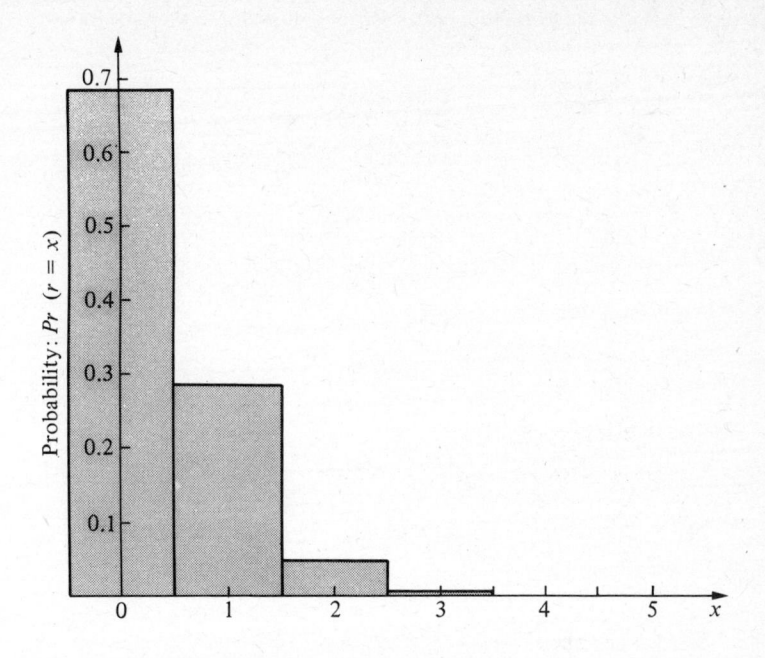

Figure 13.1 *Binomial distribution for n = 5, p = 0.08*

Cumulative probability and tables

It may be as well to begin with a reminder of some basic notation:

⩽ means 'less than or equal to' e.g. '3 ⩽ 5'
< means 'less than' e.g. '4 < 5'
⩾ means 'greater than or equal to' e.g. '6 ⩾ 6'
> means 'greater than' e.g. '6 > 4'.

Table 13.4 shows probabilities ($Pr(r = x)$) and cumulative probabilities ($Pr(r \leqslant x)$). Cumulative probability is a similar idea to that of cumulative frequency, which was considered in sections 8.3 and 9.2.

On any line: $Pr(r \leqslant x)$ = sum of all $Pr(r = x)$ values up to and including that line, using the addition rule with mutually exclusive events, for example the third entry = $0.9954747392 = Pr(r \leqslant 2) = Pr(r = 0) + Pr(r = 1) + Pr(r = 2)$.

Table 13.4 Cumulative probability for binomial distribution, $n = 5$

Number faulty (when $n = 5$) x	Probability $Pr(r = x)$	Cumulative Probability $Pr(r \leq x)$
0	0.6590815232	0.6590815232
1	0.2865571840	0.9456387072
2	0.0498360320	0.9954747392
3	0.0043335680	0.9998083072
4	0.0001884160	0.9999967232
5	0.0000032768	1.0000000000

Because of the way it is calculated we can always use the cumulative to find again the original probabilities:

(a) $Pr(r = 3) = Pr(r \leq 3) - Pr(r \leq 2)$
                $= 0.9998083072 - 0.9954747392 = 0.004333568$

(b) $Pr(r \geq 3) = 1 - Pr(r \leq 2) = 1 - 0.9954747392$
                $= 0.0045252608$

Using tables. The speed with which one can calculate binomial probabilities with a calculator depends on the facilities it has. At present at least, few have a special binomial facility. Calculators with factorial and power functions may be useful, although there is a danger of overflowing. But with any more basic machine, it is much easier to use tables. Binomial tables usually give the cumulative probabilities, as shown in the Table 1 in Appendix A. It is, of course, unnecessary to give the number of decimal places given in Table 13.4, and figures correct to four decimal places are given. This means that beyond a certain point the cumulative probability rounds to 1.0000 for all values of x; for $n = 5$ and $p = 0.08$ this point is $x = 4$. The tables are used as we have already used Table 13.4.

Further example

In a sampling scheme for a flowline production, ten components are taken and tested every hour.

(a) If $p = 0.15$ what is probability of one or less defective?
     answer: $Pr(r \leq 1) = 0.5443$

(b) If $p = 0.15$ what is the probability of more than one defective?
     answer: $Pr(r > 1) = 1 - Pr(r \leq 1) = 1 - 0.5443 = 0.4557$

(c) If $p = 0.02$ what is probability of under two defectives?
answer: $Pr(r < 2) = Pr(r \leq 1) = 0.5443$

(d) If p rose to 0.6 what would be probability of two or fewer defectives? To do this we have to turn the problem around and ask: If
$$p = Pr(\text{good}) = 0.4,$$
what is the probability of getting eight or more good?
$$= 1 - Pr(\text{less than eight good})$$
$$= 1 - Pr(\text{seven or less good})$$
$$= 1 - 0.9877 = 0.0123$$

Notes (a) The binomial tables given in this textbook are not intended to be comprehensive but are for demonstration purposes. More comprehensive tables are available.

(b) Provided n is reasonably large, if binomial tables are unavailable the binomial distribution can be approximated by a Normal or a Poisson distribution as appropriate (see note on tables). These distributions are dealt with next.

13.2 The Poisson distribution*

This section gives a very brief outline of another distribution sometimes linked with the binomial, called the 'Poisson distribution'.

Suppose that at a very large telephone exchange calls arrive purely at random. This means that the number of calls arriving in any given time interval tells us nothing about the number which are likely to arrive in the preceding or succeeding intervals. All that we know is that *in the long run* there is a mean of, say, three per second at a particular time of day. If we have this pattern, then we can find a probability distribution for the number of calls arriving in any given second; this is shown in Table 13.5. To calculate these we use the formula:

$$Pr(R = x) = e^{-m} \cdot \frac{m^x}{x!}$$

where R = number of arrivals during one time unit (in this example, one second); m = mean number which arrive during the time unit; e is a natural constant (like π), having a value of 2.7183 to four decimal places. For anyone likely to need to calculate Poisson probabilities (say for an examination) it

Table 13.5 *Poisson distribution: calls arriving at telephone exchange in one-second period where mean = 3 per second*

Number of calls x	Probability of this number $Pr(R = x)$
0	0.0498
1	0.1494
2	0.2240
3	0.2240
4	0.1680
5	0.1008
$\geqslant 6$	0.0839

may be worth getting a calculator which can give $x!$, m^x and e^{-m} painlessly. There are tables available of e^{-m}, but rather than this I prefer to give a sample Poisson table (Table 2, Appendix A). Like the binomial table, Table 1, it is given in terms of the cumulative probability: it gives $Pr(R \leqslant x)$ for various values of x. Beyond a certain value of x this is 1.0000 to four decimal places.

Example

At a large telephone exchange three calls on average arrive per second. Find the probability that:

(a) in one second exactly five calls arrive:
$$Pr(R = 5) = Pr(R \leqslant 5) - Pr(R \leqslant 4)$$
$$= 0.9161 - 0.8153 = 0.1008$$

(b) in two seconds exactly ten calls arrive:

To do this we need to note that the mean number, m, in *two* seconds will be $2 \times 3 = 6$ calls.
$$Pr(R = 10) = Pr(R \leqslant 10) - Pr(R \leqslant 9)$$
$$= 0.9574 - 0.9161 = 0.0413$$

Note that these two probabilities are not the same.

13.3 Summary

Binomial distribution

If we have a situation where there are a number (n) of trials, tests or experiments, each of which produces one of two results (A or B) with constant

The binomial and poisson distributions–205

probabilities (p and $1-p$ respectively), then the number of A results (r) is distributed as a binomial distribution.

$$\text{Probability that } (r = x) = \frac{n!}{r!\,(n-r)!}\, p^r\,(1-p)^{n-r}$$

Poisson distribution

If there are no set 'trials', but a situation in which events (such as telephone calls or faults) can happen at any moment, and if those events are unconnected and purely at random, then the number (R) of events during any given time period is distributed as a Poisson distribution:

$$\text{Probability that } (R = x) = \frac{e^{-m}\,m^x}{x!}$$

Recall and exercise questions

1. Define the situation in which the binomial distribution applies.
2. Define a situation in which a Poisson distribution might arise.
3. A sample of five components are randomly selected and tested from a flowline which produces an average of 4 per cent defective. Calculate and check using tables the probability that:
   (a) none is faulty;
   (b) exactly one is faulty;
   (c) exactly three are faulty;
   (d) the number faulty is less than or equal to two;
   (e) less than three are faulty;
   (f) three or more are faulty;
   (g) either exactly two, exactly three, or exactly four are faulty;
   (h) between two and four inclusive are faulty.
4. In transcribing digits there is a 1 per cent chance that any given digit will be wrongly transcribed. If 20 digits are transcribed, what is the probability of:
   (a) no transcription errors?
   (b) exactly one error?
   (c) at least one error?
5. In a workshop there are five machines. At any given time each one has an independent chance of 96 per cent that it is working. What is the probability that at a randomly chosen moment:
   (a) at least four are working?
   (b) all are working?

6. The incoming calls at a telephone exchange arrive purely at random. During a weekday afternoon they average five per minute.

   What is the probability of:
   (a) exactly five in a given minute?
   (b) more than five in a given minute?
   (c) between two and five inclusive in a given minute?

   What is the probability of:
   (d) more than ten in any given two minute period?
   (e) exactly three in a 30-second period?

7. Tilkinpons Ltd produce float glass by a new continuous process in strips four metres wide. Periodically faults occur in the glass, and the company would like to find out whether or not these occur purely at random.

   Over some days the glass has been cut into sheets six metres long. The number of faults in each of 100 sheets has been counted, and found to be as follows:

Number of faults	Number of sheets with this many faults
0	34
1	32
2	13
3	14
4	5
5	2

   (a) Check that the mean for these figures is 1.3 faults per sheet.
   (b) Briefly explain the implications if the faults do occur purely at random.
   (c) If the number of faults on a sheet has a Poisson distribution with mean of 1.3, find the probability that there are (i) no faults on a sheet and (ii) two faults on a sheet.

Problems

1. A company is considering the introduction of a quality-control scheme, and at present has several alternatives in mind:

   (a) From the continuous flowline production a sample of five will be taken at hourly intervals. If any of these five are faulty then the machine will be overhauled.

(b) From the continuous flowline production a sample of 15 will be taken at hourly intervals. If at least two of the 15 are faulty, then the machine will be overhauled.

(c) From the continuous flowline production a sample of fifteen will be taken at hourly intervals. If any of the fifteen are faulty then a further sample of 20 will be taken immediately, and the machine overhauled if and only if at least two of these 20 are faulty.

If the process is operating as usual then the company expect that about one per cent of components coming off the line will be faulty; this is acceptable. If the defective rate is only 1 per cent then the company would prefer not to stop the machine for overhaul. If, on the other hand, the machine *does* need to be overhauled, then the defective rate may rise to around 10 per cent defectives. This would be unacceptable, and if a sampling scheme missed its occurrence this would be undesirable.

Write a brief report indicating the effects of the different schemes suggested.

14
The Normal Distribution

14.1 Continuous probability distributions

In section 8.1 we looked at the construction of histograms using percentage frequency densities. One example was given in Table 8.6 and Figure 8.6 showing the heights in centimetres of a group of 200 men. Figure 14.1 repeats this example. We remember that the area of each block is equal to the percentage of

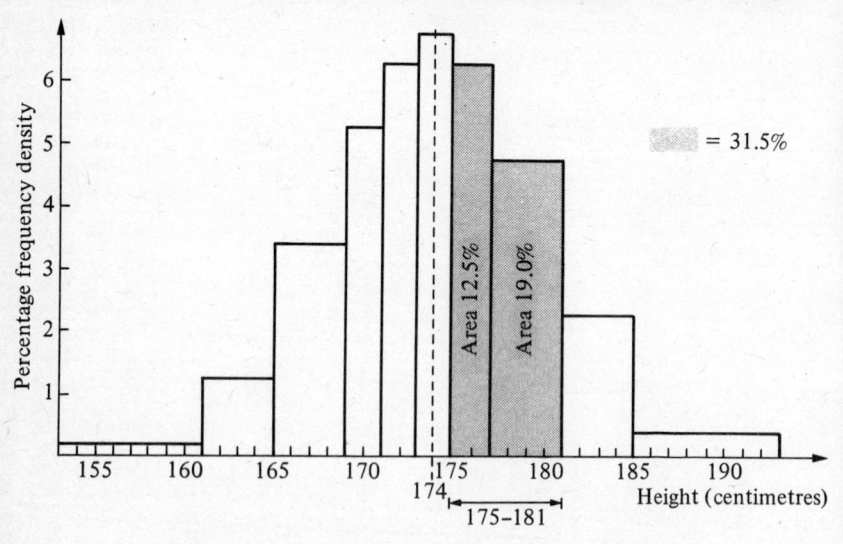

Figure 14.1 *Heights of a group of 200 men (UK)*

observations which fall into the interval upon which that block is based. The percentage between 175 and 177 is 12.5, the area of that block, and the total area 31.5 of the two blocks based on 175-177 and 177-181 is the total percentage of observations in the interval 175-181.

Percentage between 175 and 181 = Total area of blocks in this interval
= 31.5 per cent.

Suppose, then, that we select one man at random from the 200 in the group. If 31.5 per cent of the heights fall in the range 175-181, then there will be a 31.5 per cent chance that our randomly selected height will come within that range.

Pr(randomly selected height is between 175 and 181) = 31.5 per cent (0.315).

For any interval, therefore, the total area of the histogram blocks which have their bases falling within that interval represents both:

(a) the percentage or proportion of the whole group falling within that interval, and
(b) the percentage chance (i.e. the probability expressed as a percentage) that any observation selected randomly from the 200 will fall within that interval.

Now suppose that instead of 200 male adults from the UK we have a group of 20 000 such adults. With a larger number of observations we can break the figures down into more class intervals, each of which can be shorter in length. Figure 14.2 shows how a histogram using percentage frequency densities might

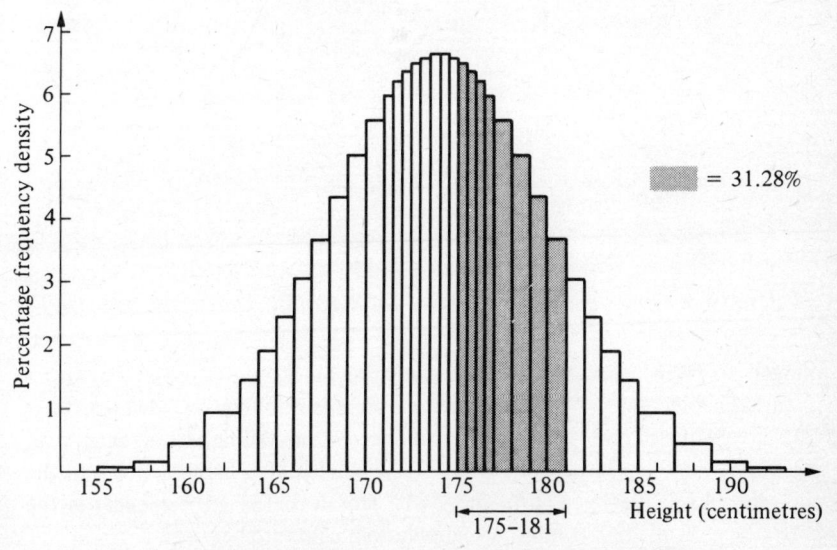

Figure 14.2 *Heights of a group of 20 000 men (UK)*

look for these 20 000. As the scales are the same as on the histogram for 200 men the total area is still the same (100 per cent). But it has a much 'smoother' appearance, due to the larger number of class intervals. As before, the total area of any set of adjacent blocks will be the probability that an observation selected at random from the group falls within the interval on which the blocks are based. If we take the same interval as before, we may find that approximately the same percentage of heights fall within it:

Pr(randomly selected height is between 175 and 181) = 31.28 per cent (0.3128)

Finally, suppose that instead of a group of 200 or 20 000 we have a group of 20 000 000 adult males from the UK. We can now make the class intervals very narrow indeed. A histogram using percentage frequency densities has such narrow class intervals that they effectively merge into a continuous block, as in Fig. 14.3. The tops of the histogram blocks have effectively

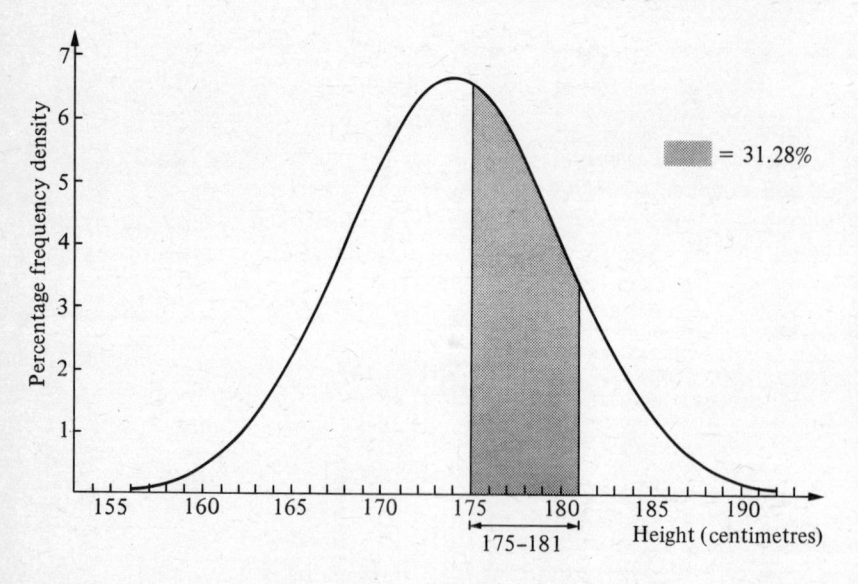

Figure 14.3 *Frequency density curve for heights of 20 million men (UK)*

merged to form a continuous *curve* and the blocks themselves form an indistinguishable area. If we take any set of adjacent blocks, however, their total area will still give the percentage of observations falling within the interval on which they are based. In Fig. 14.3 the shaded area is the area of all the adjacent blocks between 175 and 181. This area (31.28 per cent) is the percentage of all the 20 million men whose heights are between 175 and 181 centimetres. This also implies that there is a 31.28 per cent chance that the height of any randomly selected man will fall between 175 and 181. The area bounded by an interval on the horizontal axis, the vertical lines at each end of that interval, and the curve, is called the *area under the curve* for that interval.

For such a percentage frequency density curve we may therefore conclude:

1. The area under the curve for any interval is the percentage of observations which fall within that interval.
2. This area is also, therefore, the percentage chance that a random observation falls in that interval.
3. The total area under the curve (i.e. between the curve and the horizontal axis) must be 100 per cent.

Probability density

One of the things we have noted about our 'percentage frequency density curves' is that an area under the curve for any interval is the 'percentage chance' that a randomly selected observation will fall within that interval. Now, in Chapter 12 we noted that a probability can be expressed either as a percentage (that is, as a proportion of 100), or as a number between 0 and 1 (that is, as a proportion of one). In a density curve context it is more usual to use the 0-1 notation, and Fig. 14.4 shows this. As may be seen, this curve is identical

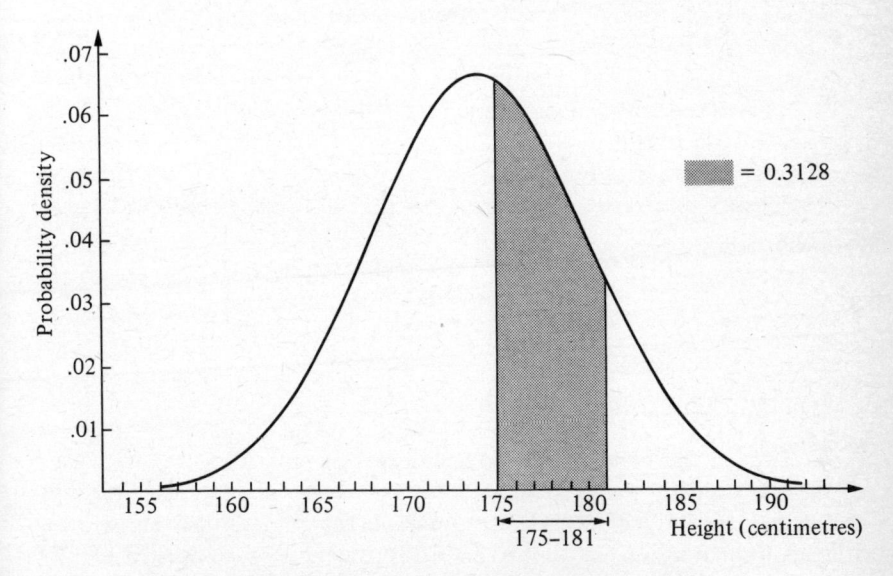

Figure 14.4 *Probability density curve for men's heights (UK)*

to that in Fig. 14.3, but the axis is now shown as a 'probability density'. This means each figure on it is (1/100) of the 'percentage frequency density'.

Summary

A *probability density curve* relates to situations where we randomly select an observation of a continuous variable (like a man's height) from a large number (for instance 20 million) of possible ones. It can be any shape, but from the way we derived it we know:

1. It can never 'double back': it has only one value for any given value along the horizontal scale.
2. The total area under the curve (bounded by the curve and the horizontal axis) is 1.00 (or 100 per cent).
3. For any interval on the horizontal axis, the area under the curve gives the proportion of the population which fall in that interval.
4. This area therefore also gives the probability that a randomly selected observation will fall in that interval.

Further example

Fig. 14.5 is a probability density curve for ages of people from the Netherlands. Like all probability density curves, the total area under the curve in Fig. 14.5 is 1.00, but unlike our previous example it is markedly positively skewed (for the meaning of this see section 9.5). The area under the curve for the interval 50-60 (shaded) is 0.1012. This means that the proportion of Dutch people who are between 50 and 60 years old is 0.1012 (10.12 per cent). The chance of a randomly selected Dutch person being between 50 and 60 years old is 0.1012.

14.2 The normal distribution

While there can be a great variety of shapes of probability density curves, there is one particular shape which arises in a number of important situations; this is a shape called the *normal distribution*. The word 'normal' here is being used in a technical sense - it does not imply that there is something 'abnormal' about other shapes! Fig. 14.6 shows a normal probability density curve shape.

The normal distribution–213

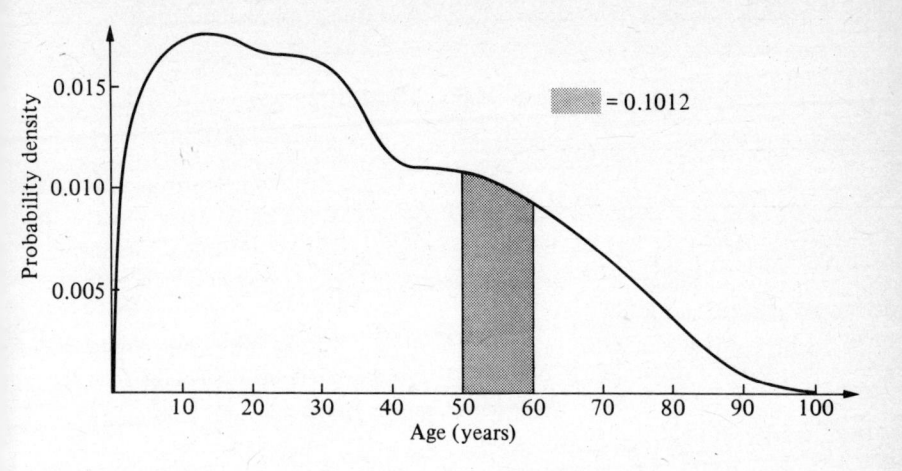

Figure 14.5 *Probability density curve for age in years of randomly selected people in the Netherlands*

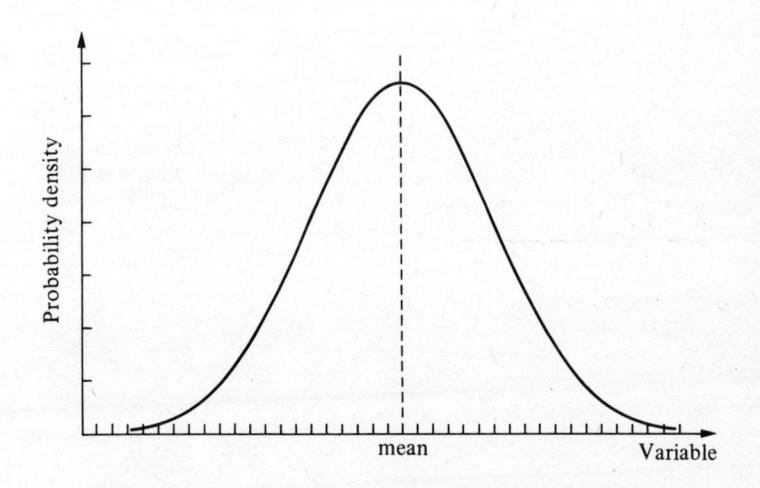

Figure 14.6 *A normal probability density curve*

A normal distribution curve is shaped like a bell. It is symmetrical, tailing away in each direction from a central mode or high point; because of its symmetry this central point is the mean and median as well as the only mode. Its 'normality', the fact that it *is* a normal distribution, determines very precisely its *shape*, but does not determine its location or dispersion.

We need only two numbers to specify a normal distribution curve completely:

(a) a measure of its average (usually given as its mean), and
(b) a measure of its dispersion (usually the standard deviation)

From these two numbers it is possible to draw up the appropriate normal probability density curve. For example, the probability density curve for men's heights in Fig. 14.3 is a normal curve. Using the fact that its mean is 174 cm and standard deviation is 6 cm, the whole curve could be drawn without any further information.

In Fig. 14.7 is shown a comparison of the probability density curve for the UK men, with the similar normal curve for heights of men in Ying Tong (an imaginary expanding third world market). Their average height is 167 cm and the standard deviation is only 4 cm. The total area of the YT curve is

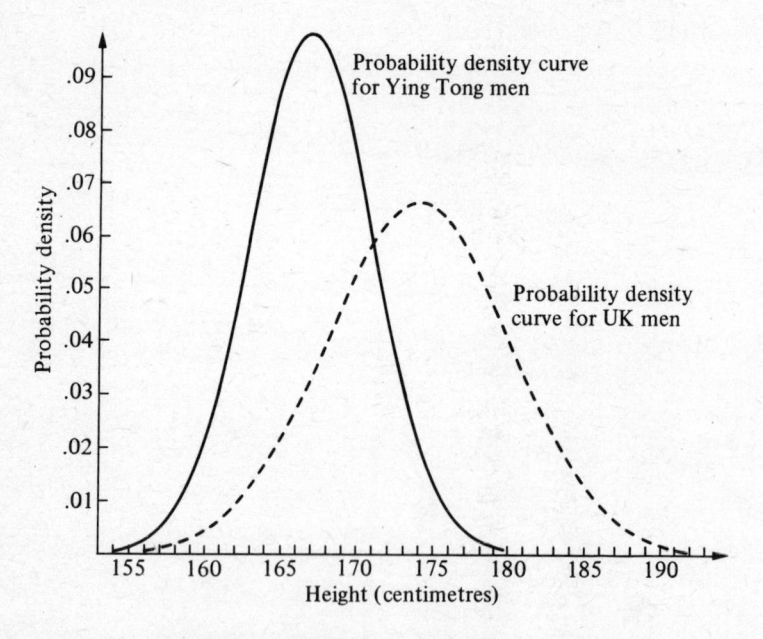

Figure 14.7 *Probability density curve for heights of men in Ying Tong compared with that for UK men*

100 per cent, the same as the UK curve. But the mean for the YT curve is less so the peak comes to the left of that for the UK, and the standard deviation for the YT curve is less so it is taller and narrower than that for the UK.

Normal areas

A very basic fact about areas under normal curves is that *for any normal distribution, a vertical line drawn a given number of standard deviations away from the mean will divide the total area in the same proportions.*

For example, for *any* normal distribution, if we draw a vertical line $1\frac{1}{2}$ standard deviations above the mean, it will divide the total area into one area of 0.933 (93.3 per cent) and another area of 0.067 (6.7 per cent).
This particular example is illustrated in Fig. 14.8 and 14.9 for the two distributions of heights we have already considered. In both diagrams we have drawn a vertical line (labelled X) at $1\frac{1}{2}$ standard deviations above the mean, which, remember, comes at the bump in the middle. For the UK distribution the mean is 174, and the standard deviation is 6. A line $1\frac{1}{2}$ standard deviations above the mean is at $1\frac{1}{2} \times 6$ above 174. That is:

$$X = 174 + 1\frac{1}{2} \times 6 = 183$$

Similarly, for the YT distribution the mean is 167 and the standard deviation is 4, so a line $1\frac{1}{2}$ standard deviations above the mean will be at:

$$X = 167 + 1\frac{1}{2} \times 4 = 173$$

In both diagrams, the line drawn $1\frac{1}{2}$ standard deviations above the mean divides the area into two parts, one of 0.933 and one of 0.067. This tells us that 6.7 per cent of UK men are 183 cm or more, and 6.7 per cent of YT men are 173 cm or more.
 Now, we remember that the normal distribution is *symmetrical* about its mean; so what would happen if we drew a vertical line $1\frac{1}{2}$ standard deviations *below* the mean? Fig. 14.10 and 14.11 show this. In both cases the line again divides the total area into a larger and a smaller area. But now the larger area (0.933) is to the right of the line, while the smaller (0.067 is to its left.

Note: The letter z is commonly used to indicate the number of standard deviations a line (X) is above the mean. If it is actually below (to the left of) the mean, then z will be negative.

Summary

In any normal distribution a vertical line drawn z standard deviations away from the mean will divide the total area in the same proportions. A line to the left of the mean (z negative) will have the larger part of the area to its right, while one drawn to the right of the mean (z positive) will have the larger part

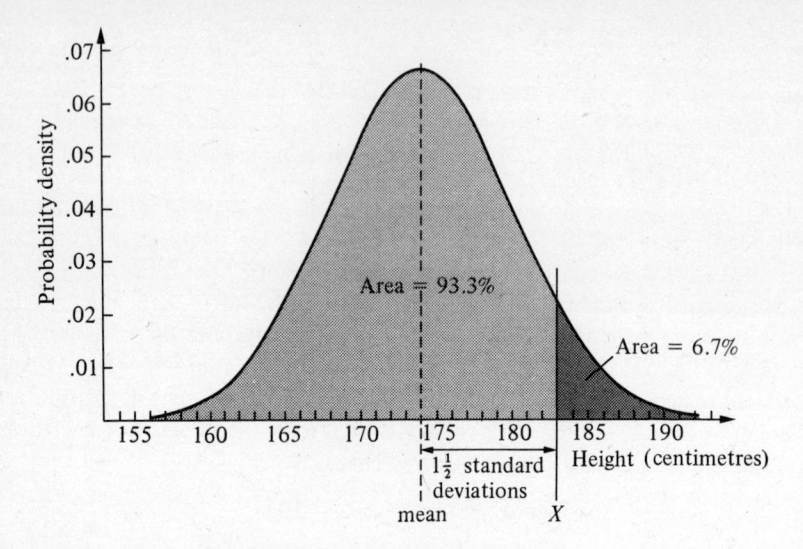

Figure 14.8 *Heights of men (UK)*

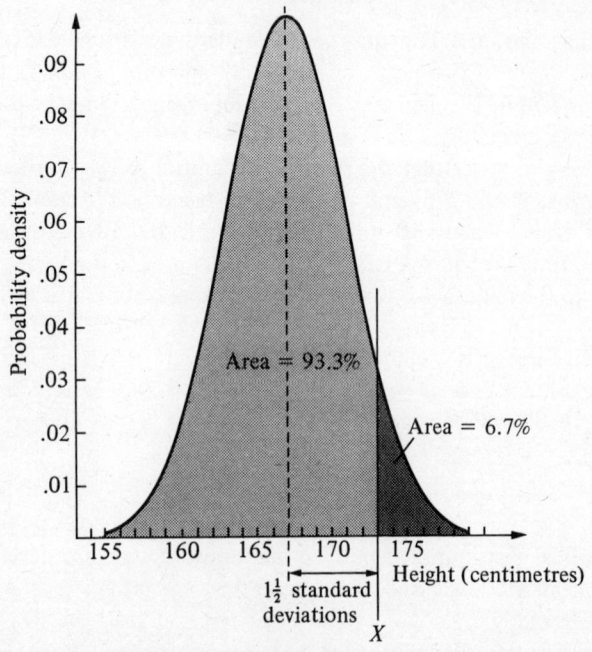

Figure 14.9 *Heights of men in Ying Tong*

The normal distribution—217

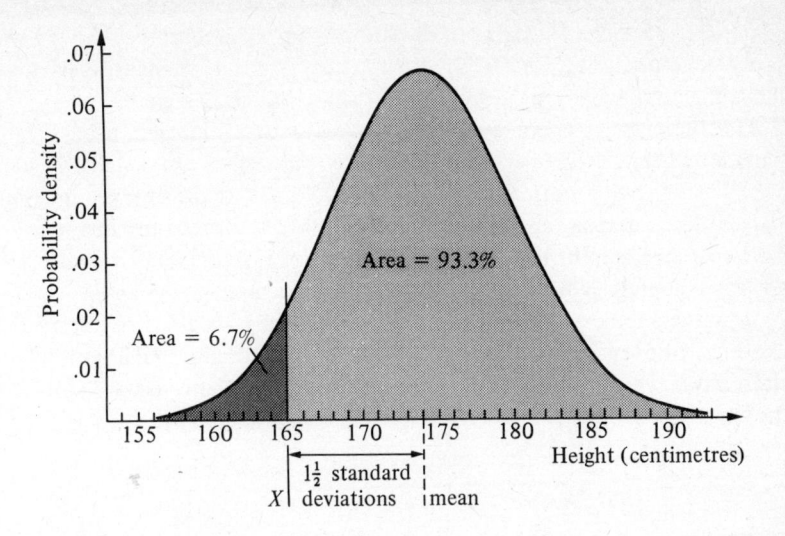

Figure 14.10 *Heights of men (UK)*

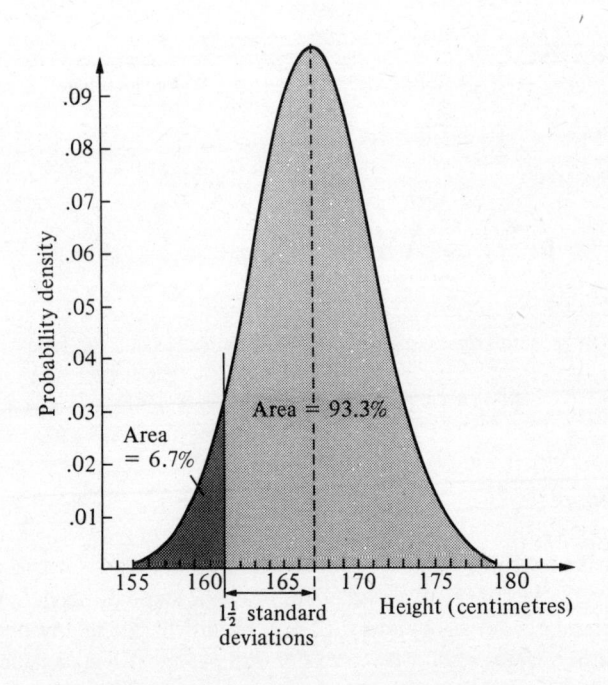

Figure 14.11 *Heights of men in Ying Tong*

to its left. So a line $1\frac{1}{2}$ standard deviations below the mean ($z = -1\frac{1}{2}$) will have 0.067 to its left and 0.933 to its right. A line $1\frac{1}{2}$ standard deviations above the mean will have 0.933 to its left and 0.067 to its right.

The business of larger and smaller areas to right and left need not be learned by heart. I would strongly advise the reader always to sketch a diagram when dealing with normal distributions. On this can be marked relevant magnitudes and areas, and it is obvious from the position of any line whether the larger area is to its left or its right. You require, please note, a sketch, and not an accurately plotted curve. It is, however, advisable to practise to make your sketches look reasonably normal in shape. The normal distribution looks like Figs. 14.6-14.11, which are all accurately plotted. Fig. 14.12 shows some alternatives which I sometimes receive. (I use sketches on pages 222-6, but they are reasonable.)

A normal distribution

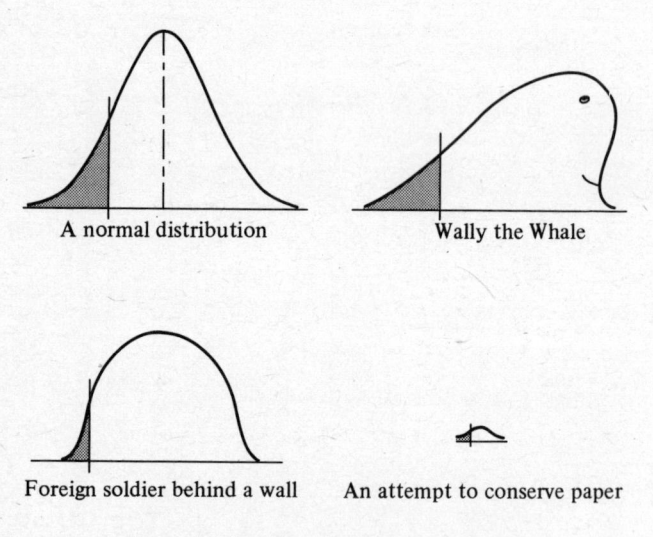

Wally the Whale

Foreign soldier behind a wall

An attempt to conserve paper

Figure 14.12 *Some alternative diagrams*

Further example

We took as an example lines $1\frac{1}{2}$ standard deviations away from the mean. As a second example consider vertical lines drawn 2 standard deviations away. A line 2 standard deviations away from a mean will divide any normal distribution into a larger area of 0.97725 (97.725 per cent) and a smaller area of 0.02275 (2.275 per cent). Fig. 14.13 illustrates this for the heights of UK men. A vertical line at 186 is two standard deviations *above* (or to the right of) the mean, and it cuts off a small 'right-hand tail' area of 0.02275 (2.275 per cent). Likewise, a line at 162 is two standard deviations *below* the mean, and cuts off a small 'left-hand tail' area of 0.02275 (2.275 per cent).

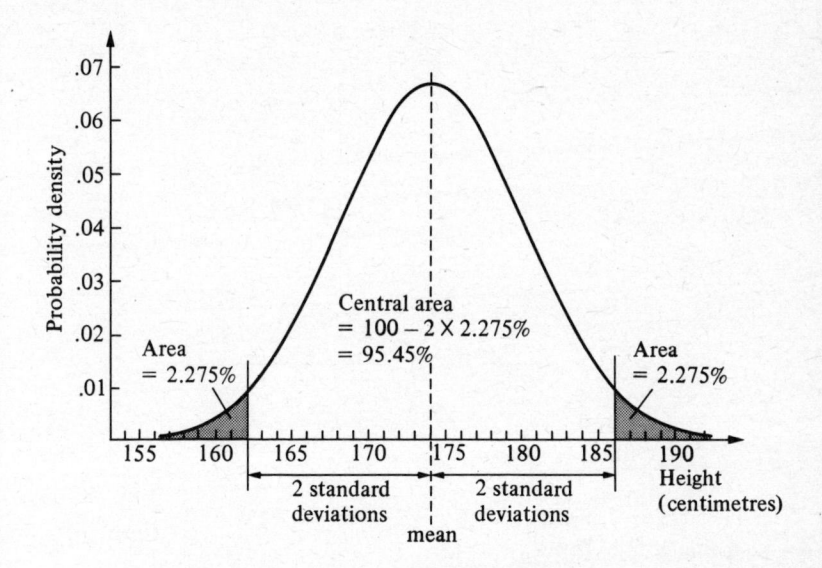

Figure 14.13 *Heights of men (UK)*

This leads us to the further interesting point that, since the total area is 1.00 (or 100 per cent), the area left in the middle *between* the line at 162 and the line at 186 is:

$$1 - 2 \times 0.02275 = 0.9545$$
or as a percentage: $100 - 2 \times 2.275 = 95.45$ per cent

For the UK, then, 95.45 per cent of men have heights between 162 cm and 186 cm, and a man drawn at random will have a 95.45 per cent chance (or 0.9545 probability) of having a height between 162 cm and 186 cm.

Figure 14.14 shows the percentages of the area under any normal curve enclosed by lines at one, two, and three standard deviations either side of the mean. In this form this is useful information. It enables us to say much about where the observations of a normal distribution will lie, given only its mean and its standard deviation. However, we may need more precise figures for areas under a normal curve, and these are provided by tables.

220–*Applied business statistics*

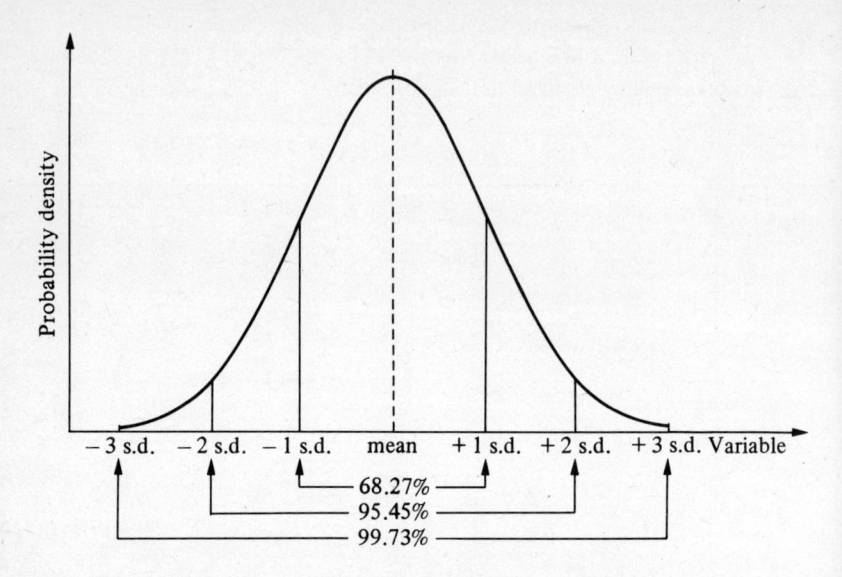

```
                          |
                         /|\
                        / | \
                       /  |  \
                      /   |   \
                     /    |    \
                    /     |     \
                   /      |      \
                  /       |       \
─────────────────/        |        \─────────────────→
  −3 s.d.  −2 s.d.  −1 s.d.  mean  +1 s.d.  +2 s.d.  +3 s.d.  Variable
                         68.27%
                         95.45%
                         99.73%
```

Figure 14.14 *Percentages of normal curve areas*

14.3 Using normal tables

We need to develop the idea of using normal distribution tables to find various areas under a normal curve. The normal curve will be taken to refer to a very large population (remembering the technical use of 'population' to mean a group of something under study. It may be useful first to recap on the key terms and symbols:

Mean of population = μ (pronounced 'mu'). $\bar{x}$ will be used in distinction for the mean of a *sample*.

Standard deviation of population = s.d.(x). Many books use the Greek σ (*sigma*) for this, so I will give it in brackets for key formulae.

Suppose that, as above, we are told that the heights of UK men are normally distributed, and the distribution has:

$$\text{Mean } \mu = 174 \text{ cm}$$
$$\text{Standard deviation s.d.}(x) = 6 \text{ cm}$$

Then suppose that we are asked what percentage of men have heights of 181 cm or more.

The first thing to do is to sketch a diagram as in Fig. 14.15. We wish to know the area to the right of a line at 181 cm as this will give us the percentage of height 181 cm or more. To use the tables we will need to find the number of standard deviations between 181 and the mean 174. The distance is 181 − 174 = 7 cm and the standard deviation is 6 cm, so to find the number of standard deviations in this distance we have to ask how many sixes there are in seven. Since there are 7/6 sixes in seven there are 7/6 (about 1.17) standard deviations between 181 and the mean.

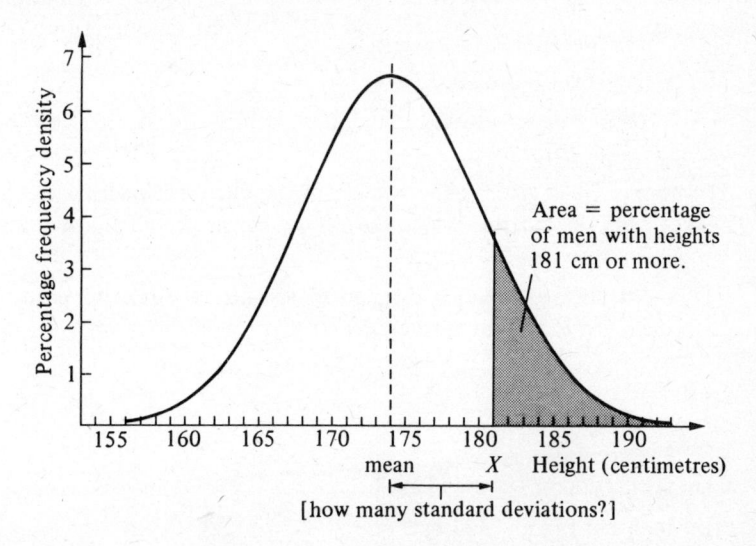

Figure 14.15 *Heights of men (UK)*

This operation, finding how many standard deviations there are between some value X and the mean, is fairly common in dealing with normal distributions. The number of standard deviations is called by the specific term 'z value'. Before continuing, let us be clear how we found it:

$$z = \frac{X - \text{mean}}{\text{standard deviation}} = \frac{X - \mu}{\text{s.d.}(x)} = \frac{181 - 174}{6} = 1.17$$

$$z = \frac{X - \mu}{\text{s.d.}(x)} \text{ is sometimes written } \frac{X - \mu}{\sigma}$$

This formula will frequently be used to calculate the number of standard deviations between a line at X and the mean. Having found our z-value, normal tables will enable us to find the areas either side of X.

The tables give, for any z value, the size of the *smaller* of the two areas into which the line divides the total area. They give the area as a proportion of one rather than of 100 per cent, so we need to multiply the figure by 100 to get a percentage. The first decimal place of z is given in the left-hand column, and the second decimal place across the top. Fig. 14.16 shows how to use the tables for $z = 1.17$. From the tables, 12.1 per cent of UK men are 181 cm or more tall: the probability that a man chosen at random is 181 cm or more is 0.121 or 12.1 per cent.

We can, in fact, use the tables to find any area of a normal distribution if we know the mean and standard deviation. We need only remember that:

1. The total area is 1.00 (or 100 per cent).
2. The curve is symmetrical about the mean, which implies that each half has an area of 0.5 or 50 per cent.
3. The normal tables give for any z value (the number of standard deviations between line X and the mean) the size of the smaller of the two areas into which X divides the whole.
4. We should always sketch a diagram to visualize the relative positions of X, the mean μ, and the various areas.

Further examples

1. What proportion of UK men have heights:
    (a) between 172 and 181 cm?
    (b) between 176 and 181 cm?
    (c) between 170 and 172 cm?
    (d) over 170 cm?

    (a)

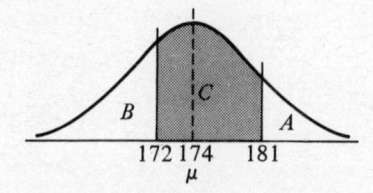

    Remember $z = \dfrac{X - \mu}{\text{s.d.}(x)}$

    To find the area C we need to find A and B and subtract both from 1.00.

    $$z_A = \frac{181 - 174}{6} = 1.17 \quad \begin{pmatrix}\text{taken correct to two places of decimals}\\ \text{since this is all that the tables provide}\end{pmatrix}$$

    hence from tables area A is 0.1210 (as above)

$z = 1.1\ 7$

z	.00	.01	.02	.03	.04	.05	.06	.07	.08	.09
0.0	0.50000	0.49601	0.49202	0.48803	0.48405	0.48096	0.47608	0.47210	0.46812	0.46414
0.1	0.46017	0.45620	0.45224	0.44828	0.44433	0.44038	0.43644	0.43251	0.42858	0.42465
0.2	0.42074	0.41683	0.41294	0.40905	0.40517	0.40129	0.39743	0.39358	0.38976	0.38591
0.3	0.38209	0.37828	0.37448	0.37070	0.36693	0.36317	0.35942	0.35569	0.35197	0.34827
0.4	0.34458	0.34090	0.33724	0.33360	0.32997	0.32636	0.32276	0.31918	0.31561	0.31207
0.5	0.30854	0.30503	0.30153	0.29806	0.29460	0.29116	0.25774	0.28434	0.28096	0.27760
0.6	0.27425	0.27093	0.26763	0.26435	0.26109	0.25745	0.25463	0.25143	0.24825	0.24550
0.7	0.24196	0.23885	0.23576	0.23270	0.22965	0.22663	0.22363	0.22065	0.21770	0.21476
0.8	0.21186	0.20897	0.20611	0.20327	0.20045	0.19766	0.19489	0.19215	0.18943	0.18673
0.9	0.18406	0.18141	0.17879	0.17619	0.17361	0.17106	0.16835	0.16602	0.16354	0.16109
1.0	0.15866	0.15625	0.15386	0.15151	0.14917	0.14686	0.14457	0.14231	0.14007	0.13786
1.1	0.13567	0.13350	0.13136	0.12924	0.12714	0.12507	0.12302	0.12100	0.11900	0.11702
1.2	0.11507	0.11314	0.11123	0.10935	0.10749	0.10565	0.10383	0.10204	0.10027	0.09853
1.3	0.09680	0.09510	0.09342	0.09176	0.09012	0.08851	0.08691	0.08534	0.08379	0.08226
1.4	0.08076	0.07927	0.07780	0.07636	0.07493	0.07353	0.07215	0.07678	0.06944	0.06811
1.5	0.06681	0.06552	0.06426	0.06301	0.06178	0.06057	0.05938	0.05821	0.05705	0.05592
1.6	0.05480	0.05370	0.05262	0.05155	0.05050	0.04947	0.04846	0.04746	0.04648	0.04551
1.7	0.04457	0.04363	0.04272	0.04182	0.04093	0.04006	0.03920	0.03836	0.03754	0.03673
1.8	0.03593	0.03515	0.03438	0.03362	0.03288	0.03216	0.03144	0.03074	0.03005	0.02938
1.9	[illegible]	[illegible]	0.02743	0.02680	0.02619	0.02559	0.02500	0.02442	0.02385	0.02330
2.0				[illegible]	0.02068	0.02018	[illegible]	0.01923	0.01876	0.01834

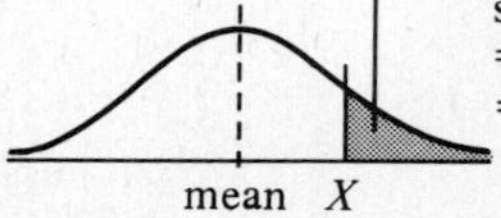

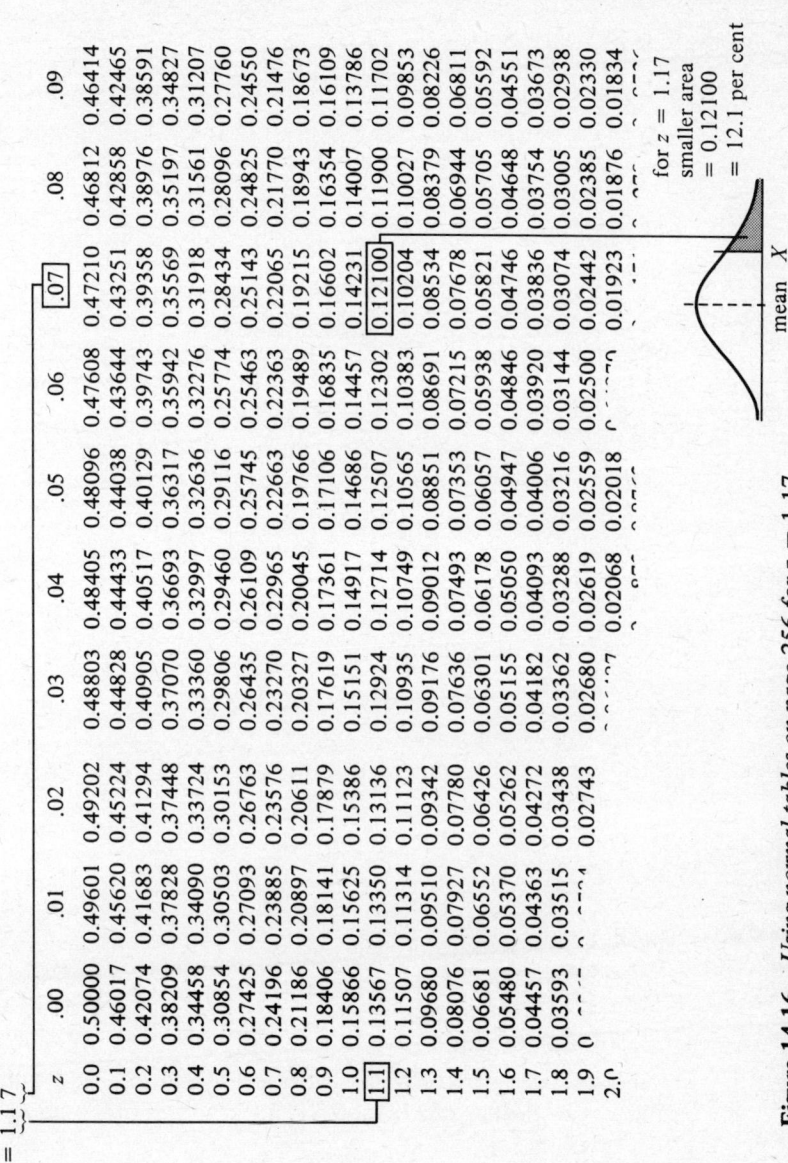

for $z = 1.17$
smaller area
$= 0.12100$
$= 12.1$ per cent

Figure 14.16 *Using normal tables on page 256 for $z = 1.17$*

$$z_B = \frac{172-174}{6} = -0.33 \text{ (the } - \text{ indicates } X \text{ to left of mean)}$$

hence from tables area B is 0.3707 (B is the smaller area here)
Thus: area $C = 1 - 0.1210 - 0.3707 = 0.5083$
so 50.83 per cent of heights are between 172 and 181 centimetres.

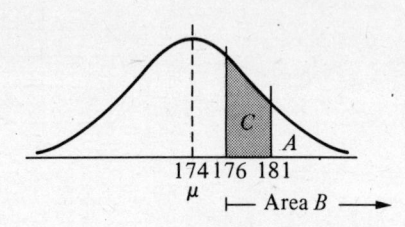

(b)

174 176 181
μ
⊢— Area B ⟶

To find the area C here we need to find the area to the right of 176 (area B) and subtract from it the area to the right of 181 (area A).

$$z_A = 1.17 \text{ (as above) so area } A \text{ is } 0.121.$$

$$z_B = \frac{176-174}{6} = 0.33$$

hence from tables area B is 0.3707

$$\text{area } C = \text{area } B - \text{area } A = 0.3707 - 0.1210$$
$$= 0.2497$$

so 24.97 per cent of heights are between 176 and 181 centimetres.

(c)

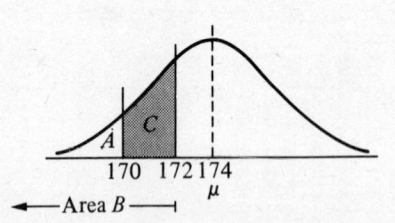

170 172 174
μ
⟵Area B ⊣

The required area here is C, that between 170 and 172. This will be the area to the left of 172 (area B) minus the area to the left of 170 (area A).

$$z_A = \frac{170-174}{6} = -0.67$$

hence from tables area A is 0.25143

$$z_B = \frac{172-174}{6} = -0.33$$

hence from tables area B is 0.3707

Thus area C = area B − area A = 0.3707 − 0.25143
= 0.11927

so 11.927 per cent of heights are between 170 and 172 centimetres.

(d)

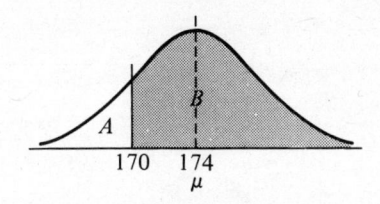

The required area here is area B, which is 1 − area A.

$$z_A = \frac{170-174}{6} = -0.67$$

hence from tables area A is 0.25143

Thus area B = 1 − area A = 1 − 0.25143 = 0.74857

so 74.857 per cent of men are over 170 centimetres tall.

2. What height will only 5 per cent of men exceed?

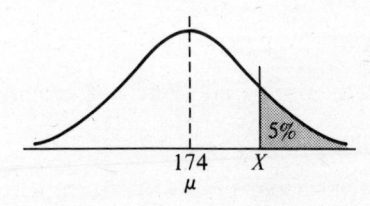

This necessitates using the tables the other way around, to find a z value corresponding to a given area. The area in the right-hand tail is 5 per cent or 0.05. In the tables the areas decrease as z increases, so we can run down the areas until we find:

for $z = +1.64$ area in tail is 0.05050

$z = +1.65$ area in tail is 0.04947

so for a tail (i.e. the smaller area) of 0.05 we must take z somewhere between the two - say 1.645.

But what is z? Remember that it is the number of standard deviations between the mean (μ) and a line at X.

$$z = \frac{X-\mu}{\text{s.d.}(x)} \text{ so, rearranging } X = \mu + z \times \text{s.d.}(x).$$

Here $X = 174 + 1.645 \times 6 = 183.87$ cm
so 5 per cent of men exceed 183.87 cm in height.

3. What interval symmetrical about the mean will contain 90 per cent of men's heights?

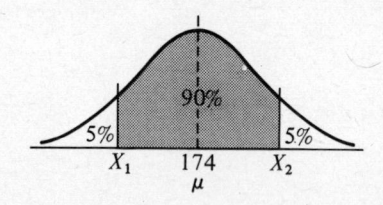

This follows on from the previous example. To have 90 per cent in the centre area, between X_1 and X_2, requires 5 per cent in each tail. The z value which gives us a 5 per cent tail (0.05 in the smaller of the two areas into which X divides the total) is 1.645. Thus:

$$X_2 = \mu + 1.645 \times \text{s.d.}(x) = 174 + 1.645 \times 6 = 183.87 \text{ cm}$$

Similarly:

$$X_1 = \mu - 1.645 \times \text{s.d.}(x) = 174 - 1.645 \times 6 = 164.13 \text{ cm}.$$

So 90 per cent of heights are contained in the interval 164.13 to 183.87 cm.

Note. Such an interval is sometimes written:

$$\mu \pm 1.645 \times \text{s.d.}(x)$$

i.e. $174 \pm 1.645 \times 6$

The symbol ± means 'plus or minus', and written thus it indicates the *interval* from $(174 - 1.645 \times 6)$ to $(174 + 1.645 \times 6)$.

14.4 Summary

Probability density curve

The area bounded by an interval on the horizontal axis, the vertical lines at each end of the interval, and a probability density curve, gives us:

(a) the proportion of the whole group or population which fall within that interval;

(b) the probability that a single observation selected at random falls within that interval.

Features. The area between a probability density curve and the horizontal axis is always 100 per cent or 1.00.

Such a curve can be any shape but it can never 'double back' on itself.

Normal distribution

The normal distribution is a particular kind of probability density curve. It can have any mean and standard deviation, but it is always shaped like a bell.

Features. The normal curve:

(a) is symmetrical about a central mode, which is also its median and mean;

(b) has its total area divided in given proportions by a vertical line X at a given number (z) of standard deviations away from the mean (μ).

Normal tables. Normal tables list the z values (i.e. numbers of standard deviations away from the mean) against the corresponding areas under the curve.

Normal problems. It is advisable always to sketch a diagram, to relate the appropriate z values and areas visually. The formula below will often be useful (but should be understood!):

$$z = \frac{X-\mu}{\text{s.d.}(x)} \quad \text{or} \quad \frac{X-\mu}{\sigma}$$

Recall and exercise questions

1. Explain, with a sketch, the meaning of the 'area under a section of a probability curve'.
2*. Table 14.1 shows the age breakdown for the population of France.

Table 14.1 *Age distribution of the population of France*

Exact age[a] (years)	Number of people (thousands)
0- 5	3914
5-10	4185
10-15	4288
15-20	4249
20-25	4226
25-30	4396
30-35	3361
35-40	2931
40-45	3206
45-50	3301
50-55	3168
55-60	2329
60-65	2195
65-70	2376
70-75	2034
75-80	1463
over 80	1351
Total	52973

[a] Most published tables give age at last birthday (which is a discrete variable). But tables of exact ages may be derived from these.

(a) Draw a histogram utilizing percentage frequency densities.
(b) Superimpose on your histogram a sketch of what you would expect the probability density curve to look like for the age of a randomly selected French person. (Hint: Imagine what it would look like if the class intervals were very much narrower but the heights of the curve were the same).

3. The heights of men in the Republic of Ying Tong are normally distributed with a mean of 167 cm and a standard deviation of 4 cm.
   (i) What proportion of them have heights:
       (a) over 172 cm?
       (b) between 172-173 cm?
       (c) between 165-173 cm?
       (d) between 167-173 cm?
   (ii) Within what range would 90 per cent of the heights lie?

4. A machine fills 500-gram jars of coffee automatically. In fact it is found that the weight of coffee inserted differs from jar to jar, and it falls into a normal distribution with a standard deviation of 0.6 grams. The machine can be set so that the mean weight inserted is a given value between 499 and 502 grams.
   (i) At present it is set to a mean of 501 grams. What proportion of jars contain:
      (a) under 500 grams?
      (b) over 503 grams?
      (c) between 499.5 and 503 grams?
   (ii) If the mean is set to 501.5 grams what difference would this make to the proportions of (i)?
   (iii) To what should the mean be set so that only one per cent of jars will contain under the stated weight of 500 grams?

5. The national mean for IQ is 100, and the standard deviation is 13. If IQs are distributed normally:
   (i) What proportion of people have IQs:
      (a) over 130?
      (b) under 80?
      (c) between 80 and 120?
      (d) between 90 and 130?
   (ii) Within what range would the IQs of 95 per cent of people lie?
   (iii) If further education is to be available for the top 20 per cent of IQs, within what range should applicants have their IQ?

6. Observed demand for petrol at a filling station has been seen to be normal with a mean of 4950 litres per weekday, and a standard deviation of 320 litres.
   (i) On what percentage of weekdays would demand be:
      (a) under 4500 litres?
      (b) under 4950 litres?
      (c) between 4400 and 5000 litres?
   (ii) How much petrol should be available for a weekday so that there is only 1 per cent chance of running out?

7. A certain type of resistor is mass produced, with normally distributed resistances, having mean 110 ohms and standard deviation 3 ohms. They are tested, one by one, to see if they lie within the specification limits 110 ± 4 ohms.
   (a) What proportion of resistors are expected to lie outside the specification limits?

(b) A profit of 6p is made on resistors within the specification limits, and a loss of 3p on those outside the limits. What is the average profit expected on a resistor?

8. Duplicop Services Ltd rent out their own office copying machine. As part of their standard contract their service engineer will call after every 50 000 copies. The part which nearly always needs replacing soonest is the master roller, which is always replaced on the service visit as a routine. If this should go wrong between normal services (i.e. before the 50 000 copies have gone through, then Duplicop have to pay for their engineer to make an extra visit. They therefore want as many master rollers as possible to last beyond 50 000 copies.

There are two available makes of master roller which will fit their machine. Droppit and Dratt Ltd offer one for which they claim a mean of 60 000 copies and standard deviation of 6000. The other company, Masterplan Rollers, claim a mean of 55 000 and a standard deviation of 2500 copies.

(a) Advise Duplicop as to which would be the best supplier.
(b) If Duplicop decided to increase the interval to 52 000 copies between regular service visits, how would this affect the situation?

Problems

1. An electronic unit has two main components. Components of type A have been found to have lives which fall into a normal distribution with a mean of 340 days and a standard deviation of 20 days. Components of type B have lives normally distributed with a mean of 315 days and standard deviation of 9 days.

If either of the two components in a unit fail, then the whole unit will malfunction. What percentage of units will last:
(a) under 300 days?
(b) over 330 days?

An alternative to component B called component B2 is available; components of this type have lives distributed normally with a mean of 310 days and a standard deviation of 5 days. The company are considering replacing the electronic units routinely in a service after 300 days use to avoid the inconvenience of a failure during use. Someone has suggested that if this policy is adopted using the alternative B2 component will be better than using B, but this seems odd since the average life of B2 is lower. Write a memo to the production manager to clarify the situation.

2. A clothing manufacturer intends to commence export to the Republic of Ruritania. A sample survey has shown that the heights of Ruritanian men are approximately normally distributed, with a mean of 172 cm and a standard deviation of 5 cm.

He intends to offer five lengths: 'Very Long' 'Long' 'Medium' 'Short' and 'Very Short'. Each of these lengths would cover a different band of sizes (for instance, 'Long' might cover 174-179 cm), but at present he is not sure how to apportion his five bands of sizes.

Suggestion A is that he find the range within which 99 per cent of heights lie, divide it equally into five size bands, and produce different numbers of garments in each band (and so produce more of the most popular sizes).

Suggestion B is that he divide up his five bands of sizes so that there will be equal numbers of customers in each band (the bands will therefore be of unequal length).

Produce a report making clear the implications of each suggestion in numerical terms, and comment on them.

3. A particular mechanical unit requires shafts of diameter 300 mm. A shaft of over 300.11 mm will not fit into the unit; this will be detected at the assembly stage and it can be discarded at a cost of £1.20. If, however, the shaft is under 299.7 mm, then it will get through assembly but will malfunction when the unit is tested. In this case it will cost £2.30 to replace and discard the shaft.

The lathe which produces the shafts can be set so that the mean diameter of shafts produced is any value, but the shafts will fall into a normal distribution, with a standard deviation of 0.06 mm.

(a) Produce a brief report for the production manager, incorporating a graph showing costs against different settings for the average, and making a recommendation for optimal setting.

(b) Outline how your answer might change if he had the option of measuring all the shafts produced before assembly, at a cost of £y each. In this way shafts which were too small, as well as those which were too large, could be discarded before assembly.

15
The Distribution of the Sampling Mean

15.1 The concept of sampling mean distribution

The concept of the sampling mean distribution is basic to understanding statistical sampling. It will be developed here in terms of a particular example.

English eggs are now graded into seven numbered grades by weight. Grade 3 eggs are supposed to be between 60 and 65 grams. Table 15.1 shows the weights actually recorded for every egg in 48 boxes of grade 3 eggs, each containing six eggs. The table also shows the average weight of eggs (i.e. arithmetic mean) in each box.

The weights of the 288 individual eggs can be put into a frequency table, as shown in Table 15.2. From this table the mean may be calculated for these individual eggs using the formula in section 9.1. The *dispersion* of the individual weights may be measured by finding the variance and the standard deviation, which we remember was defined as the square root of the variance (see section 10.4 for definitions, explanations and formulae). From the table:

Mean $\mu = 62.44$
Variance $= 1.95$
Standard deviation $=$ s.d.$(x) = \sqrt{1.95} = 1.40$

(Note: We will call the mean for a single box of eggs $\bar{x}$, and the mean for the whole 288 eggs μ). Table 15.2 can also be used to construct a probability density curve, as shown in Fig. 15.1.

Table 15.2 and Figure 15.1 refer to the 288 weights of individual eggs. But in our original Table, 15.1, these individuals were randomly divided into 48 boxes, each containing six eggs. The mean weight for each box was listed, giving 48 *sampling mean* figures, each the mean of a sample of six eggs. These 48 sampling means can themselves be grouped into a frequency table (Table

The distribution of the sampling mean

Table 15.1 Weights (grams) of eggs in boxes of six

Box number	1st egg	2nd egg	3rd egg	4th egg	5th egg	6th egg	Mean weight in box
1	61.53	61.14	62.28	60.33	62.97	63.12	61.895
2	64.73	61.65	62.81	62.38	60.63	60.72	62.153
3	63.91	63.97	61.67	62.56	62.53	63.71	63.058
4	63.29	60.55	61.59	63.68	61.27	61.23	61.935
5	63.05	62.18	61.06	61.68	62.14	60.32	61.139
6	60.71	62.13	61.71	64.76	62.34	60.57	62.037
7	60.87	60.68	63.65	61.83	61.22	60.51	61.460
8	61.78	62.47	63.22	63.96	61.99	60.66	62.347
9	61.01	64.10	63.25	60.98	62.05	64.88	62.711
10	63.02	63.83	62.23	64.34	63.61	60.20	62.872
11	64.21	63.41	62.30	60.44	64.40	62.97	62.955
12	61.61	62.75	62.46	64.81	63.65	60.21	62.582
13	64.10	62.65	61.54	64.45	62.91	63.25	63.150
14	61.63	61.19	62.35	63.20	60.21	61.69	61.712
15	64.09	62.39	60.00	63.99	64.66	63.78	63.118
16	64.67	61.54	61.61	60.76	61.29	63.11	62.163
17	62.93	64.82	62.55	60.56	62.56	63.06	62.747
18	64.09	60.20	60.73	62.54	64.91	60.51	62.163
19	64.76	61.39	60.61	61.72	61.39	60.82	61.782
20	63.59	60.72	62.59	64.66	60.39	62.40	62.392
21	62.03	62.54	60.89	62.62	64.77	61.60	62.408
22	65.06	64.29	61.57	60.92	62.19	61.95	62.663
23	60.43	59.87	60.33	64.58	64.37	64.40	62.330
24	63.66	62.56	62.13	62.89	63.84	61.93	62.835
25	63.15	60.11	63.53	60.72	60.67	60.80	61.455
26	63.05	63.04	63.31	63.15	63.49	62.13	63.027
27	64.17	60.93	64.56	64.44	60.65	61.29	62.673
28	64.63	64.36	62.59	63.44	61.09	63.93	63.340
29	61.51	62.96	62.32	62.54	62.95	62.98	62.543
30	61.33	60.32	64.70	63.14	62.19	61.18	62.143
31	63.38	60.43	64.28	60.25	64.39	64.06	62.798
32	64.35	61.35	63.00	60.63	62.25	61.45	62.505
33	60.47	63.64	62.96	60.13	60.10	62.21	61.568
34	62.15	64.44	63.94	63.66	64.71	62.45	63.558
35	64.04	64.49	62.12	63.58	61.16	62.20	62.932
36	61.11	60.08	61.57	61.54	63.02	60.40	61.225
37	60.55	62.86	61.41	63.84	64.50	61.11	62.378
38	63.03	63.56	62.93	61.19	60.61	63.03	62.308
39	63.67	61.24	63.96	62.32	64.71	61.02	62.820
40	63.32	62.68	60.33	61.08	61.25	63.28	61.990
41	63.10	61.38	60.40	64.55	61.21	63.18	62.305
42	63.79	63.44	64.81	64.75	61.81	61.80	63.400
43	60.82	61.41	61.10	63.52	64.08	62.13	62.177
44	64.08	64.78	60.07	62.13	62.60	61.61	62.545
45	64.22	60.02	62.61	60.49	61.00	63.47	61.968
46	64.33	61.52	62.56	60.78	61.03	63.69	62.318
47	63.97	60.41	63.85	61.20	62.99	61.93	62.392
48	61.91	62.50	60.01	62.29	64.59	60.73	62.005

234—Applied business statistics

Table 15.2 *Table showing weight in grams of 288 grade 3 eggs*

Weight (class-interval)	No. of eggs (f)
−60.49	28
60.5−60.99	30
61.0−61.49	31
61.5−61.99	29
62.0−62.49	30
62.5−62.99	31
63.0−63.49	31
63.5−63.99	29
64.0−64.49	25
64.5−	24
Total	288

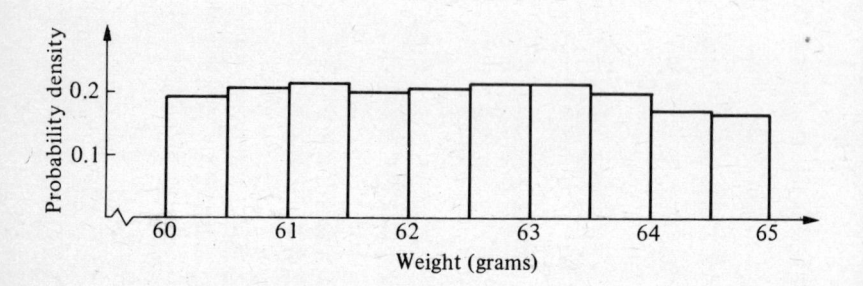

Figure 15.1 *Probability density for the weights of grade 3 eggs*

15.3). This table shows a *distribution of sampling means* (based on the 48 available values), and from the table a mean and standard deviation can be calculated:

Mean = 62.41 (we can call this $\mu_{\bar{x}}$)

Variance of sampling means = 0.306

Standard deviation of sampling means = s.d.$(\bar{x}) = \sqrt{0.306} = 0.553$

From Table 15.3 we can also draw up a probability density curve. This is shown in Fig. 15.2, superimposed for comparison on the probability density curve for individuals shown on the same scale. Table 15.3 and Fig. 15.2, then, represent the distribution of sampling means - the distribution of the 48 means of samples of size six.

Table 15.3 Table showing mean weight in grams in each of 48 boxes of six eggs

Weight (class-interval)	No. of boxes (f)
61.00–61.25	1
61.25–61.50	2
61.50–61.75	3
61.75–62.00	5
62.00–62.25	7
62.25–62.50	9
62.50–62.75	8
62.75–63.00	6
63.00–63.25	4
63.25–63.50	2
63.50–63.75	1
Total	48

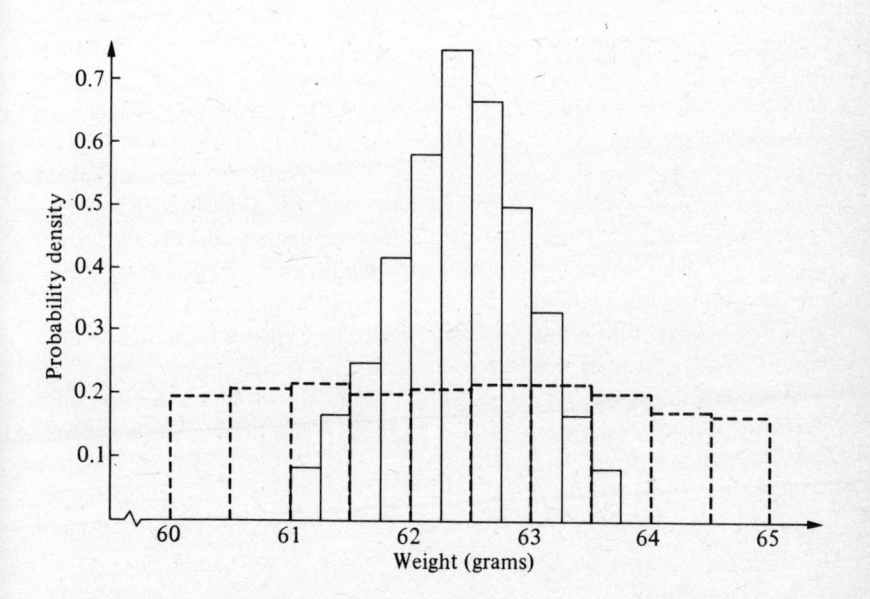

Figure 15.2 Probability density for mean weights in boxes of eggs (that for individuals is shown as a dotted line)

15.2 The central limit theorem

Suppose that we now compare the shape of the distribution of individuals (Table 15.2 and Fig. 15.1) with the distribution of sampling means (Table 15.3 and Fig. 15.2). As Fig. 15.2 makes clear, their shapes are quite different: the distribution of individuals (based on the 288 available) is more or less flat and the observations are spread more or less evenly between the limits of 60 and 65 grams (apart from a slight 'tailing off' to the right). But the distribution of sampling means (based on the 48 available) is shaped like a bell, tailing off symmetrically either side of a central mode. In short, it looks like a normal distribution.

This is no coincidence or chance happening. There is a standard mathematical result known as the *central limit theorem*, which may be stated:

> Whatever the shape of a distribution of individuals in a population, a distribution of sampling means tends to be normal in shape (for samples of sufficient size).

The proof of this theorem is beyond our present scope, but it has certainly been demonstrated by the above example.

How big do the samples have to be before the distribution of sampling means can be treated as normal? That depends. If the distribution of *individuals* was near normal, than *any* size of samples will give sampling means which are normally distributed. As we have seen, for a flat-topped distribution of individuals a sample size of six gives reasonable normality to the distribution of sampling means, while one of, say, 30 will be quite adequate. If we are dealing with random samples of, say, 100, then practically any distribution encountered in business practice can be assumed to be normal. All this, of course, assumes that the individuals in a sample are randomly and independently selected. Obviously if we purposely *pick out* all the large ones, for example, or use a sampling method which produces a consistent bias then these comments may be invalid.

15.3 The standard error of the mean

Fig. 15.2 shows the different shapes of the distributions of individuals and of sampling means. But we may also see that the individual values are more spread out (or *dispersed*) than the sampling means. How do the two distributions compare numerically in their averages and dispersions? Table 15.4 summarizes the values of the mean, variance and standard deviation (square root of the variance) for the two distributions.

Table 15.4 Comparison of the distributions of individual weights and sampling means

	Mean	Standard deviation	Variance
Distribution of 288 individual weights	$\mu = 62.4$	s.d.$(x) = 1.4$	Variance $(x) = 1.9$
Distribution of 48 sampling means of samples of size six	$\mu_{\bar{x}} = 62.4$	s.d.$(\bar{x}) = 0.55$	Variance $(\bar{x}) = 0.31$

First, regarding the averages. We see immediately that the means for the two distributions (μ and $\mu_{\bar{x}}$) are the same.

Second, regarding dispersion. The relationship here is not so obvious, but if we look we will see that the variance for the sampling mean distribution is approximately one-sixth of the variance for the individuals (i.e. 0.31 is about one-sixth of 1.9). This is, in fact, because six was the number in each of the samples from which we formed our sampling mean distribution, that is six was the *sample size*.

This example, then, has illustrated what is a standard result:

$$\text{Variance}(\bar{x}) = \frac{\text{Variance}(x)}{6} = \frac{\text{Variance}(x)}{\text{Sample size}}$$

From this the relationship between the *standard deviations* can be derived if we remember that the standard deviation is defined as the square root of the variance.

$$\sqrt{\text{Variance}(\bar{x})} = \sqrt{\frac{\text{Variance}(x)}{\text{Sample size}}} = \frac{\sqrt{\text{Variance}(x)}}{\sqrt{\text{Sample size}}}$$

$$\text{s.d.}(\bar{x}) = \frac{\text{s.d.}(x)}{\sqrt{\text{Sample size}}} = \frac{\text{s.d.}(x)}{\sqrt{n}}$$

The standard deviation of the distribution of sampling means, s.d.$(\bar{x})$, is sometimes called the *standard error of the mean* and so I shall write it as s.e.$(\bar{x})$. It is important to note that this term, strictly speaking, applies to the distribution of sampling means, and not to an individual sampling mean as such.

Summary

Sampling mean distributions tend to be normal in shape, and that the following numerical relationships apply between the averages and dispersions of the distribution of individuals and the distribution of sampling means:

$$\text{Mean } \mu_{\bar{x}} = \text{Mean } \mu$$
(for distribution of sampling means) (for distribution of individuals)

$$\frac{\text{Standard deviation s.d.}(\bar{x})}{\text{(for distribution of sampling means)}} = \frac{\text{Standard deviation s.d.}(\bar{x}) \text{ (For distribution of individuals)}}{\sqrt{\text{Sample size } n}}$$

$$= \text{Standard error of the mean s.e.}(\bar{x})$$

Note: Some books write s.e.$(\bar{x})$ as $\sigma_{\bar{x}} = \dfrac{\sigma}{\sqrt{n}}$

15.4 Applications

If the distribution of sampling means may be taken as normal in shape, and is of a known mean and standard deviation (i.e. s.e.$(\bar{x})$), we may use the normal tables (Appendix A, Table 3) to find the probability that a random sampling mean falls within any given range.

Example

What is the probability that the mean weight of six random grade 3 eggs will be 62.9 grams or more?

In this example we know that the sampling mean distribution for samples of six grade 3 eggs has:

$$\text{Mean } \mu = 62.41$$
$$\text{s.e.}(\bar{x}) = 0.553$$

We can therefore use these values in the usual formula:

$$z = \frac{X - \text{mean}}{\text{Standard deviation}}$$

which here is: $z = \dfrac{X - \mu}{\text{s.e.}(\bar{x})} = \dfrac{62.9 - 62.41}{0.553} = 0.886$

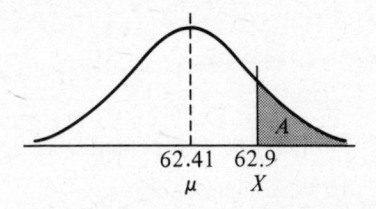

From normal tables this gives area $A = 0.187$ so in about 18.7 per cent of cases the sample mean would be 62.9 or more.

Now let us consider how we would get s.e.$(\bar{x})$, that is the standard deviation of a sampling mean distribution. For the grade 3 eggs, we listed the weights of 48 samples of six eggs, and then estimated s.e.$(\bar{x})$ from these 48 results. But this was really for purposes of illustration. More usually, s.e.$(\bar{x})$ would be estimated using a known (or estimated) standard deviation for individuals:

$$\text{s.e.}(\bar{x}) = \text{s.d.}(\bar{x}) = \frac{\text{s.d.}(x)}{\sqrt{n}}$$

Usually I will calculate s.e.$(\bar{x})$ in this way as a separate step before substituting in a z-value formula:

$$\frac{X - \mu}{\text{s.e.}(\bar{x})}$$

But many texts combine the two formulae to get:

$$z = \frac{X - \mu}{(\text{s.d.}(x)/\sqrt{n})} \quad \text{or} \quad \frac{X - \mu}{(\sigma/\sqrt{n})}$$

Some readers may wish to learn this as a standard formula, but it is implied in combining the standard error and z-value formulae.

Further example

Grade 4 eggs usually have a distribution of weights with a mean of 57.6 grams and a standard deviation of 1.45 grams.
What is the probability that:

(a) the mean weight of eggs in a box of six will be under 56.4 grams?
(b) the mean weight of eggs in a sample of 48 eggs will be between 57.3 and 57.8 grams?
(c) the total weight of a box of six eggs will exceed 350 grams?

(a) The distribution of mean weights of samples of six will have:

Mean $\mu_{\bar{x}} = \mu = 57.6$ grams

$$\text{s.e.}(\bar{x}) = \frac{\text{s.d.}(x)}{\sqrt{n}} = \frac{1.45}{\sqrt{6}} = 0.592 \text{ grams}$$

$$z = \frac{X - \mu}{\text{s.e.}(\bar{x})} = \frac{56.4 - 57.6}{0.592} = -2.027$$

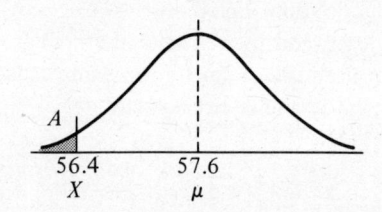

The minus sign indicates that the smaller area is to the left of the line, as the diagram shows. Looking up $z = 2.03$ in normal tables we get: area $A = 0.02118$ (2.118 per cent). So the probability that the mean weight of eggs in a box of six is under 56.4 grams will be 0.02118 (or 2.118 per cent).

(b) The distribution of means of samples of 48 will have:

Mean $\mu_{\bar{x}} = \mu = 57.6$ grams

$$\text{s.e.}(\bar{x}) = \frac{\text{s.d.}(x)}{\sqrt{n}} = \frac{1.45}{\sqrt{48}} = 0.2093 \text{ grams}$$

$$z_A = \frac{X_A - \mu}{\text{s.e.}(\bar{x})} = \frac{57.3 - 57.6}{0.2093} = -1.43$$

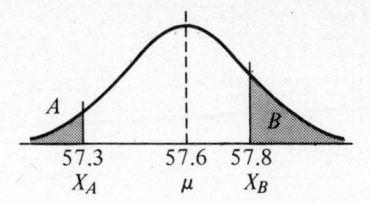

Hence from tables $A = 0.07636$

$$z_B = \frac{X_B - \mu}{\text{s.e.}(\bar{x})} = \frac{57.8 - 57.6}{0.2093} = 0.96$$

Hence from tables $B = 0.16853$

The proportion between 57.3 and 57.8 will be:

$1.00 - 0.07636 - 0.16853 = 0.75511$ (about 75.5 per cent)

(c) To find the probability that the total weight of six random eggs will exceed 350 grams, we need to note that if the total exceeds 350 grams then the *mean* weight for one egg in the sample of six will exceed $(350/6) = 58.33$ grams, since the mean is the total divided by the number in the sample.

The problem then reduces to finding the probability that the mean of a sample of six eggs is greater than 58.33 grams.

Using the same approach as in part (a) the answer to this may be found to be 0.10935 (10.935 per cent; $z = 1.23$).

There remains the wider question of why we might wish to calculate probabilities about sampling means anyway. This really, is dealt with in the next chapter, but a very brief indication of one reason may be given here.

Suppose that we wished to determine whether a large batch of grade 4 eggs really did have an overall mean of around 57.6 grams. We might find the mean weight of a sample. Obviously this could not be expected to be *exactly* 57.6 grams: sometimes we might happen to pick more larger eggs, and sometimes more smaller. But the question would be *how large* a difference we would be prepared to ascribe to chance in the egg selection, and what size of difference would lead us to conclude that an overall mean figure (μ) of 57.6 grams was unlikely to be correct.

15.5 Summary

Sampling mean distribution

As well as a distribution of individuals there is a distribution of means of samples each of size n, drawn randomly from the population. This is termed the sampling mean distribution.

Central limit theorem

Whatever the shape of the distribution of individuals, the sampling mean distribution tends to be normal in shape if the sample size is large enough.

Average and dispersion

The average of the sampling mean distribution is given by:

Mean $\mu_{\bar{x}} = \mu$ (The mean of the distribution of the individuals)

The standard deviation of the sampling mean distribution (also called the *standard error of the mean* is given by:

$$\text{s.e.}(\bar{x}) = \text{s.d.}(\bar{x})$$

$$= \frac{\text{Standard deviation of individuals}}{\sqrt{\text{Sample size}}}$$

$$= \frac{\text{s.d.}(x)}{\sqrt{n}} \quad \text{or} \quad \frac{\sigma}{\sqrt{n}}$$

z-value

For the sampling mean distribution a z value is given by:

$$z = \frac{X - \mu}{\text{s.e.}(\bar{x})} \quad \text{or} \quad \frac{X - \mu}{(\text{s.d.}(x)/\sqrt{n})} \quad \text{or} \quad \frac{X - \mu}{(\sigma/\sqrt{n})}$$

Recall and exercise questions

1. State the central limit theorem.
2. What is the standard deviation of the sampling mean distribution in terms of that for the individuals?
3. What is the meaning of the phrase: 'standard error of the mean'?
4. A particular type of component has been found to give on average 1250 hours of use, with a standard deviation of 180 hours. If a random sample of 80 components are used:
   (a) what is the probability that their average life is:
      (i) less than 1240 hours?
      (ii) more than 1290 hours?
      (iii) between 1220 and 1270 hours?
   (b) within what range is it 95 per cent probable that the mean of the 80 will lie?

5. Butter is packed in tubs marked '250 g' (250 grams), four of which make up a pack marked '1000 g'. In fact the machine which fills the tubs dispenses weights in a normal distribution with a mean of 250.8 grams and a standard deviation of 0.6 grams.
   (a) What percentage of *tubs* contain less than the 250 grams marked on them?
   (b) In what percentage of packs of four tubs will the average weight of the four be less than 250 grams? (Assume tubs are randomly allocated into packs.)

6. A random sample of 50 female manual workers is taken. What is the probability (using the data in Table 8.12) that the mean weekly hours of the 50 is:
   (a) over 45 hours?
   (b) under 40 hours?
   (c) between 30 and 40 hours?

Problem

1. Even Dale Yoghurt Company currently produce yoghurt cartons labelled 'approx. weight 160 grams'. In fact the machine now in use to fill the cartons fills with an average of 163 grams and a standard deviation of 1.7 grams. The mean of these distributed weights could actually be reset to any desired level, though the standard deviation cannot be reduced or altered.

    The marketing department would like to print 'minimum contents 160 grams' on each carton. This, however, would require that no more than one in a hundred cartons had less than this stated contents. Some directors have asked whether to ensure this would mean significantly increasing the mean setting on the machine (so giving away more yoghurt).

    An alternative suggestion has come from someone who pointed out that the cartons are currently packed in sets of four cartons. Suppose that nothing was printed on individual cartons, but each pack of four had 'minimum average contents of cartons 160 grams'. He wonders whether this would make any difference to the mean to which the machine could be set so that only one in a hundred packs would contain four cartons with average weight below that stated.

    Write a brief report indicating the percentage of cartons currently being filled to less than 160 grams, and the mean settings which would be necessary for the two alternative new schemes.

16
Significance Testing

16.1 The significance test concept

Oldenlay Ltd produce a large number of eggs (with a little help from their hens) for distribution each week. They have a machine which automatically cleans the eggs, weighs them, sorts them into the different numbered grades by weight, and boxes them up in boxes of six. Periodically eggs break while they are being weighed, leaving some egg deposited on the weighing pan. When this happens, the recorded weights of eggs will include the extra deposit, and the eggs assigned to any grade will be lighter than they should be. Whenever this situation arises, Oldenlay would like to discover it as soon as possible, to avoid any possible action from the Weights and Measures Inspectors.

They are proposing the following method of quality control. Periodically, a strip of eight plastic boxes each containing six eggs will be taken, and weighed by hand. The weight of the strip of eight plastic boxes is known exactly, and can be deducted, to leave the total weight of the 48 eggs. This can be divided by 48 to get a precise mean weight for the sample of 48 eggs. From this they hope to be able to discover whether or not the machine is weighing correctly.

Let us consider how their quality control measures would be applied to grade 4 eggs. If the machine is weighing and grading correctly, then for eggs marked 'grade 4' the distribution of individual weights will have a mean μ of 57.6 grams and a standard deviation s.d.(x) of 1.45 grams. Sample means for random samples of 48 eggs will fall into a sampling mean distribution as we saw in the previous chapter; this will be normal in shape and will have:

Mean of the sampling mean distribution $\mu_x = \mu = 57.6$ grams

Standard error of the sampling means s.e.$(\bar{x}) = \dfrac{\text{s.d.}(x)}{\sqrt{n}}$

$$= (1.45/\sqrt{48}) = 0.2093 \text{ grams}$$

Obviously, therefore, even when the machine is weighing and grading correctly, *some* sample means will be lower than others just by chance. Only if we get a sample mean which is *very* low will we tend to question the machine's accuracy. But just *how* low would a sample mean have to be before we decided that it would be unreasonable to ascribe it to chance, and concluded that the machine had gone wrong? To answer this we need to remember the general facts about areas under the normal curve which are illustrated in Fig. 16.1 (and compare Fig. 14.14).

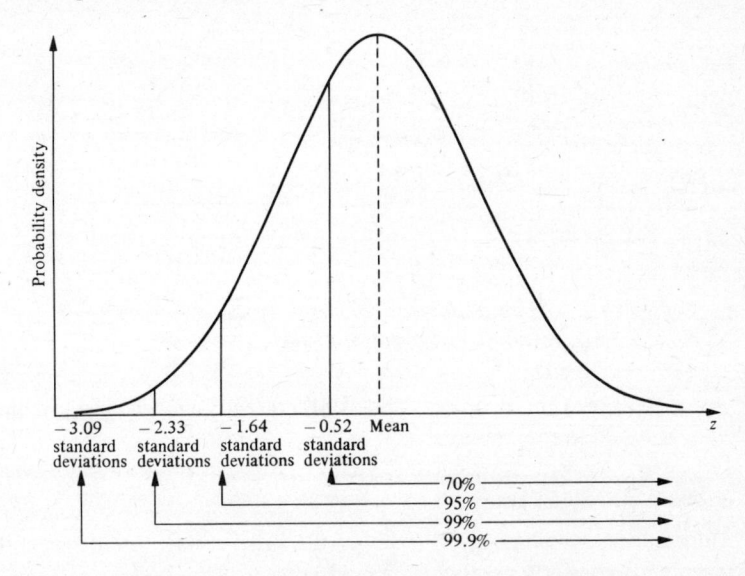

Figure 16.1 *Standard areas of any normal curve*

Normal tables show us, for example, that in 99 per cent of instances a random observation is not more than 2.33 standard deviations below the mean: that is, z is greater than or equal to -2.33 ($z \geqslant -2.33$). Applying these same areas to the present example of samples means of 48 correctly assigned grade 4 eggs, we get the distribution shown in Fig. 16.2. If, therefore, the machine is weighing and grading accurately then, for example, in 99 per cent of samples the sample mean will have a z value greater than -2.33. This implies that in 99 per cent of instances the sample mean itself will be greater than $(\mu_{\bar{x}} - 2.33 \times \text{s.e.}(\bar{x})) = 57.112$ grams. Suppose, then, that we get a sample mean *below* this figure of 57.112 grams? Either a one in a hundred (1 per cent) chance has come up, or else the machine is no longer weighing and grading correctly. The second of these possibilities seems the more likely. This kind of reasoning can be illustrated further if we consider the implications of four different possible values for the sample mean. All of these are below the

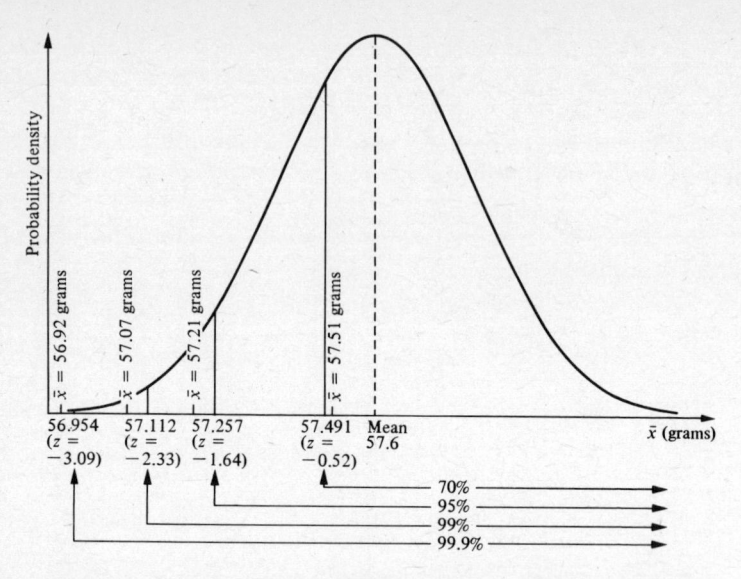

Figure 16.2 *Areas under normal curve for sampling means of weights of 48 eggs*

average (57.6 grams) for sample means of 48 correctly graded eggs, but they are below it by different amounts:

1. *Suppose $\bar{x} = 57.51$ grams:*

    Though this sampling mean is lower than the average, even when the machine is correctly grading we would expect a sample mean this low in over 30 per cent of cases. We may see this in either of two ways:

    (a) by comparing the actual sample mean $\bar{x} = 57.51$ grams with the 30 per cent value (57.491) shown in Fig. 16.1. Here the sample mean is larger (57.51 > 57.491) so *more than* 30 per cent of sample means would be at least as low as the sampling mean obtained

    (b) by comparing the z-value: $z = \dfrac{\bar{x} - \mu}{\text{s.e.}(\bar{x})} = \dfrac{57.51 - 57.6}{0.2093} = -0.43$

    and seeing that it is larger than the z-value (−0.52) for a 30 per cent tail.

    Note that for negative numbers −0.43 is *larger* than −0.52.

    Since, therefore, we should expect a sample mean this low more than 30 per cent of the time when the machine is grading correctly, there is no reason to doubt its accuracy. There is no evidence from this sample mean that the machine has gone wrong.

2. *Suppose $\bar{x} = 57.21$ grams:*
   If the machine is grading correctly then in fewer than 5 per cent of instances will we get a sample mean below 57.257 grams (or, equivalently, a z-value less than -1.64), as may be seen from Fig. 16.2. A sample mean of 57.21 grams is under 57.257 (or, equivalently, its z-value -1.86 is less than -1.64). This means that if the machine is grading correctly then in less than 5 per cent of instances will we get a sample mean as low as the value 57.21. We might well, therefore, feel that this offered some evidence that the machine had gone wrong; otherwise we would have to assume that an unluckily low sample mean had chanced to come up.

3. *Suppose $\bar{x} = 57.07$ grams:*
   When the machine is grading correctly, in only 1 per cent of instances will we get a sample mean below 57.112 grams (or, equivalently, a z-value less than -2.33). The sample mean $\bar{x} = 57.07$ is less than 57.112 (and its z-value is -2.53, which is less than -2.33). This implies that if the machine is grading correctly then in *fewer* than 1 per cent of instances would the sample mean be as low as the one we have obtained. Unless we are prepared to accept that this unlucky one in a hundred chance has come up, we will see this as evidence that the machine has gone wrong.

4. *Suppose $\bar{x} = 56.92$ grams:*
   Fig. 16.2 shows that if the machine is correctly grading then in only 0.1 per cent of instances will we get sample mean less than 56.954 grams (or a z-value below -3.09.) If we obtain $\bar{x} = 56.92$ then we know that since it is less than 56.954 (and its z value is -3.25, which is less than -3.09) a value this low would come up less than 0.1 per cent of the time when the machine is grading correctly. Unless, therefore, we are prepared to accept that the present sample is that one in a thousand (or 0.1 per cent) chance, we will consider that this is strong evidence that the machine has gone wrong.

In essence, what we are doing is this: we recognize that sampling means below 57.6 grams *will* sometimes occur by chance, depending on which eggs happen to go into the boxes selected. The question is: *how far below 57.6 grams would a value of the sample mean ($\bar{x}$) have to be before we concluded that it was implausible to ascribe a value so low to chance, and decided that the machine had gone wrong?*

We are basically looking at the probability:

Pr[(sample mean is this low)   *if*   (machine is still accurate)]

Or this could be expressed in the more usual language of conditional probability (see section 12.2) as:

Pr[(sample mean is this low)   *given*   (machine is still accurate)]

Whichever expression we choose to use, should this probability turn out to be very low, rather than believe that a very unlikely combination of low-weight eggs has happened to come up by chance, we will feel it more sensible to conclude that the machine is not weighing accurately. We cannot actually find a figure for Pr[(machine is still accurate) *if* (sample mean is this low)], for (as shown in section 12.2) this would be quite different from Pr[(sample mean is this low) *if* (machine is still accurate)]. It is clear, however, that if the latter expression (which we *can* find) is low, then it is reasonable to doubt the machine's accuracy.

But how low is 'low'? That is, how low would this probability have to be before we decided that the machine was inaccurate? We have to decide on some kind of 'lower boundary'. If Pr[(sample mean is this low) *if* (machine is still accurate)] falls below this lower boundary, then we shall conclude that the machine is no longer accurate. The value of this lower boundary is called the *significance level*, and the appropriate significance level to apply to a given situation may depend on various factors, such as our strength of belief in the machine's accuracy before taking our sample, and the consequences of a wrong decision. Commonly, however, statisticians use the round numbers of 5%, 1% and 0.1%. That is, it is common to adopt a significance level of 5% (or 0.05), 1% (or 0.01), or 0.1% (0.001). Like all probability figures these may be expressed either as percentages (e.g. 5%) or as fractions of one (e.g. 0.05). I shall usually refer to them as percentages, but this is just a question of personal preference.

Suppose, then, that we decide that, say, a 5% significance level is appropriate; the implications of this decision are clarified in Fig. 16.3. If the machine is still working correctly, then there would be only a 5 per cent chance of getting a sample mean result below 57.257 grams (implying a z-value less than -1.64). If the actual sample mean turns out to be below this, we would conclude that there was evidence at the 5% level of significance that the machine was not weighing accurately. The value 57.257 grams (which has $z = -1.64$) is called the *critical value* for the sample mean, and the region below it is called the *critical region*. A sample mean within the critical region constitutes evidence at that level of significance that the machine is not weighing accurately.

16.2 Formalizing testing

In order to generalize easily, it is useful to set up our testing procedure more formally. We begin with a *null hypothesis H_0*:

H_0 The eggs marked 'grade 4' come from a distribution with mean $\mu = 57.6$ gram and standard deviation s.d.$(x) = 1.45$ grams.

[Figure: Bell curve showing 95% area with critical region for 5% significance level shaded on the left tail, with Probability density on y-axis and $\bar{x}$ (grams) on x-axis. Key values marked: 57.257 ($z = -1.64$) and Mean 57.6.]

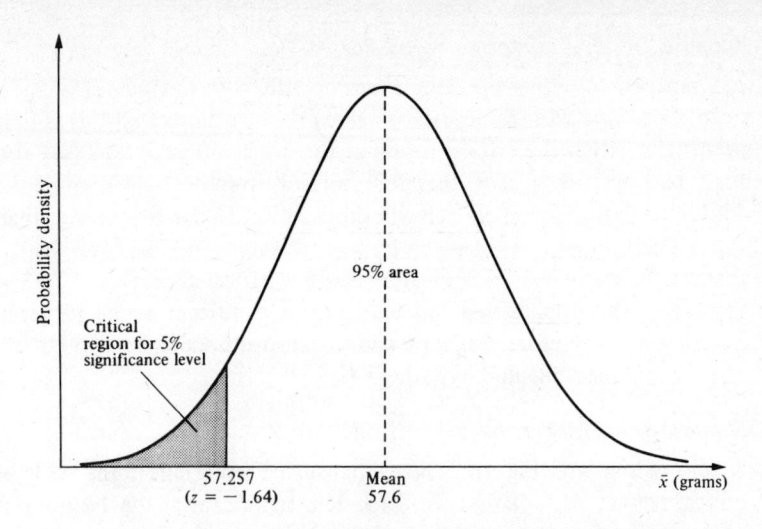

Figure 16.3 *5% critical region to test for a possible reduction in population mean*

The sample is taken to put this hypothesis to the test, but bearing in mind that it is not the only possibility. In this instance we are assuming that if the mean weight of eggs marked 'grade 4' does differ from 57.6 grams then there are good reasons to expect that it will be less. Thus we have an *alternative hypothesis* H_1:

H_1: The eggs marked 'grade 4' come from a distribution with a mean μ less than 57.6 grams and a standard deviation s.d.$(x) = 1.45$ grams.

We then have a *test result R*:

R: In a random sample of $n = 48$ eggs classed as grade 4 we had a sample mean of $\bar{x}$.

We decide on a *significance level* which gives us a *critical region*; for example for the 5% significance level $\bar{x}$ is significant if it is below the critical value of 57.257 grams (that is if its z-value is less than -1.64).

If the test result is in the critical region, then it is significant at that level, and we would say that it constitutes evidence to reject the null hypothesis at that level of significance.

We may now consider the implications of the values for $\bar{x}$ given as 1. to 4. earlier in this section:

1. Suppose $\bar{x} = 57.51$ grams $(z = -0.43)$

   This is not within the critical region even at the 5% level of significance, so there is no evidence from the sample to reject the null hypothesis H_0.

2. Suppose $\bar{x} = 57.21$ grams $(z = -1.86)$

   A comparison of either the sample mean $\bar{x}$ with the critical value (57.257), or of its z-value with the critical z-value (-1.64) shows that this sample mean falls within the critical region at the 5% significance level (see also Figs. 16.3 and 16.4). It is, therefore, a result which is significant at the 5% level. A further comparison will show, however, that $\bar{x}$ is greater than the critical value (57.112) at the 1% level of significance or, equivalently, that its z-value of -1.86 is greater than the critical z-value of -2.33 at this level. This implies that this result is *not* significant at the 1% level. Such a result, therefore, might be said to constitute some, but not strong, evidence to reject the null hypothesis H_0.

3. Suppose $\bar{x} = 57.07$ grams $(z = -2.53)$

   A comparison with Fig. 16.2 shows that such a result falls in the 1% level critical region, but not that for 0.1%. It is significant at the 1% but not at the 0.1% level. It is, of course, also significant at the 5% level, but this need not be mentioned since a result significant at the 1% level is always significant at the 5% level too. For a test result significant at the 1% but not at the 0.1% level one might say that there was strong evidence to reject the null hypothesis H_0.

4. Suppose $\bar{x} = 56.91$ grams $(z = -3.25)$

   This is below 56.954 (and its z-value is less than -3.09), so it is in the critical region at the 0.1% level, and is significant at that level. This implies that the test result constitutes very strong evidence to reject H_0.

The mechanics of a significance test

As we saw in the previous section, a significance test works by considering the probability Pr[(sample mean is this low) *if* (machine is still accurate)]. Its underlying logic is that if this probability is low then we will doubt the machine's accuracy (that is, we will doubt H_0). In practice, however, it proved convenient to identify critical regions for 5%, 1% and 0.1% levels of significance, and see whether the sample mean obtained fell within any of those regions. But we also saw that we could discover whether the sample result was in the critical region by considering either its actual value or its z-value. Thus at the 5% level of significance Fig. 16.3 showed that we would know the sample mean to be in the critical region if we knew either that its actual value was less than 57.257 grams or that its z-value was less than -1.64. In practice we find that it is easier to work with the z-values, and test the significance of a sample mean by comparing its z-values with the critical ones.

Significance testing

Let us suppose that we have taken a sample of 48 eggs, and found the mean to be 57.21 grams. We wish to test the hypothesis that the population mean is 57.6 grams, against the alternative hypothesis that it is less. The test would be set out as follows:

Null H_0: The eggs marked 'grade 4' come from a distribution with mean $\mu = 57.6$ grams and a standard deviation s.d.$(x) = 1.45$ grams.

Alternative H_1: The eggs marked 'grade 4' come from a distribution with mean $\mu < 57.6$ grams, and a standard deviation s.d.$(x) = 1.45$ grams.

Test result R: In a random sample of $n = 48$ eggs classed as grade 4, there was a sample mean of $\bar{x} = 57.21$ grams.

We may compute a test statistic: $z = \dfrac{\bar{x} - \mu}{\text{s.e.}(x)} = \dfrac{57.21 - 57.6}{(1.45/\sqrt{48})} = -1.86$

We then compare this with a standard table of critical z values:
For a 5% significance level z is critical if less than -1.64
For a 1% significance level z is critical if less than -2.33
For a 0.1% significance level z is critical if less than -3.09

In this instance the result is significant at the 5% but not at the 1% level, so we conclude that the sample offers some, but not strong, evidence to reject H_0.

16.2 Levels, errors and conclusions

Any significance test is applied to make a decision about the truth or falsity of a null hypothesis. There are therefore obviously two kinds of 'error' we can make; in this context the word 'error' does mean a kind of mistake, not a numerical value:

A *type 1 error* would be to reject wrongly a true null hypothesis.
A *type 2 error* would be to accept wrongly a false null hypothesis.

Our choice of significance level affects the chances of either of these errors occurring: Thus:

(a) A *1% significance level* gives us a 1 per cent chance of wrongly rejecting any true null hypothesis tested (type 1 error).

(b) A *5% significance level* gives us five times as much chance of wrongly rejecting any true null hypothesis tested (type 1 error) *but* it gives us *less* chance of wrongly accepting a false null hypothesis (type 2 error).

A 1% level requires a more 'way-out' value before rejecting the null hypothesis. It will therefore cause us to reject wrongly fewer true null hypotheses, but to accept wrongly more false null hypotheses, than will a 5% level.

I have mentioned 5%, 1% and 0.1% significance levels. These are convenient round numbers, but their general use in preference to, say, 3% or 6% is purely conventional.

The choice of which significance level to adopt will really depend on the circumstances. It may depend on how strongly we believed in the null hypothesis before the test; a well supported null hypothesis will require more evidence (say a 0.1% level) to overthrow it. It may depend on the practical implications of making a type 1 or type 2 error, especially if, for example, the effect of one would be merely irritating while the other would be a disaster!

Conclusions of tests

The point about to be made is implicit in the previous section, but is of sufficient importance to be explicitly emphasized here. It concerns the possible kinds of conclusion one can draw from a significance test.

Let us recap on our conclusions from the four alternative supposed $\bar{x}$ results in the previous section:

1. $\bar{x} = 57.51$: not significant at the 5% level.
    Conclude: the sample result does not constitute any real evidence to reject H_0 and conclude that the machine is wrong.
2. $\bar{x} = 57.21$: significant at the 5% but not at the 1% level.
    Conclude: the sample result constitutes some but not strong evidence to reject H_0 and conclude that the machine is wrong.
3. $\bar{x} = 57.07$: significant at the 1% but not at the 0.1% level.
    Conclude: the sample result constitutes strong evidence to reject H_0 and conclude that the machine is wrong.
4. $\bar{x} = 56.92$: significant at the 0.1% level.
    Conclude: the sample result constitutes very strong evidence to reject H_0 and conclude that the machine is wrong.

Results 3 and 4 should be clear, but there is one mistake sometimes made. This is to say that if, for example, the result is 57.09 then 'there is less than a 1 per cent chance that the null hypothesis H_0 is true'. It would be nice to be able to make statements like this about the probabilities of hypotheses being true, but unfortunately, in conventional statistics at least, it is not possible to do so. A result which is 'significant at 1%' does not tell us the probability of the hypothesis being true; it tells us only the chance of getting a test result such as we have obtained *if* the hypothesis is true. We have already seen in section 16.1 that we are basically dealing with $Pr[$(result this low) given (H_0 is true)] which is not the same as $Pr[(H_0$ is true) *given* (result this low)]. We can only ever obtain the former on which to make decisions, not the latter.

More confusion, however, often arises over a result such as result 1 above, where the result $\bar{x} = 57.51$ grams is *not* significant. What such a result means is that we have *failed to find significant evidence to reject H_0*. But this does *not* mean that we have proven H_0 to be correct: it means only that H_0 remains *plausible* in the light of the sample evidence.

To further illustrate this point we might note that had we begun with some other null hypothesis (that the mean μ was 54.2 grams, for example) then we could have found that exactly the same sample test result was not significant for this hypothesized value either. That is, *either* hypothesis would remain plausible had we begun with it as our null hypothesis.

It is true that the correct conclusion to draw from a result like $\bar{x} = 57.51$ grams is often given simply as 'the result is not significant, so we must accept the null hypothesis'. But this should not mislead us. We accept it *only* in the sense that our prior disposition to believe in its truth, based, presumably, on some sensible grounds, is undisturbed by the sample result. The more strictly correct conclusion based on it is that 'there is insufficient evidence to reject the null hypothesis'. While such an insufficiency of evidence *could* result from the null hypothesis being at least approximately true, it could on the other hand simply be that we need to take a larger sample before a proper decision can be made.

16.4 One-tailed and two-tailed tests

In the significance test example developed in the previous two sections it has been assumed that our possible basis for rejecting H_0 and concluding that the machine was faulty would be a very *low* sampling mean. We did not consider that a very *high* sampling mean might throw doubt on the accuracy of the weighing process.

We made this assumption, of course, because of a particular property of the machine: that when it begins to weigh wrongly it is usually because of broken egg sticking to the weighing pan, which causes an overestimate of the weight of each egg. But the assumption led us to adopt a critical region, at whatever significance level we chose, which consisted of just the left-hand tail of the lowest values. For obvious reasons a test based on such assumptions is called a *one-tailed test*, and its use was made formally explicit by the use of an alternative hypothesis H_1 that the mean was *less than* 57.6 grams.

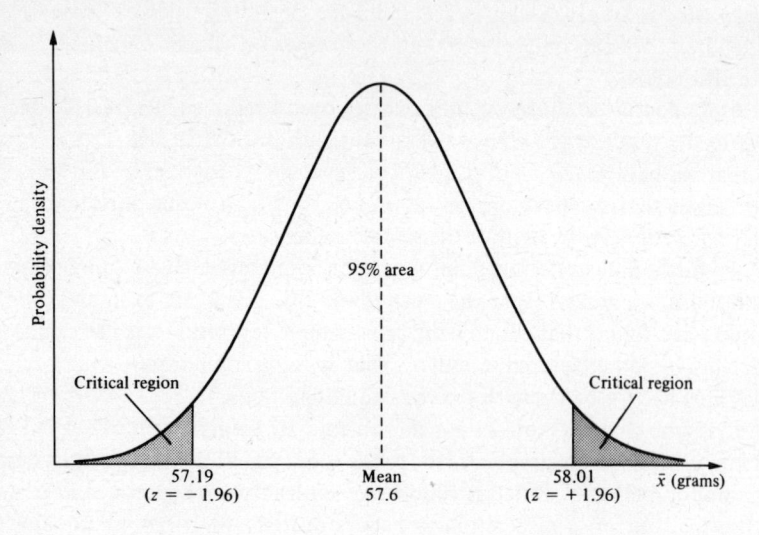

Figure 16.4 *5% critical region for a two-tailed test*

Suppose, however, that it was discovered that, periodically, because of a technical fault, the machine might begin to underestimate the weights instead of overestimating them. We might then decide that we had better adopt a critical region which led us to reject the null hypothesis if a sample value was *either* too low *or* too high. Fig. 16.4 illustrates this for a 5 per cent level of significance. Formally this can be set up as a hypothesis test:

Null hypothesis H_0: Mean $\mu = 57.6$ grams s.d.$(x) = 1.45$ grams

Alternative hypothesis H_0: Mean $\mu \neq 57.6$ grams s.d.$(x) = 1.45$ grams

Test result R: In a sample of $n = 48$, the sample mean $\bar{x}$ was obtained.

The use of the 'not equal to' sign ($\neq$) indicates a *two-tailed test*: the critical region has a left-hand and a right-hand tail as shown in Fig. 16.4. The value of $\bar{x}$ will be significant at the 5% level if it is more than 1.96 standard errors *either* below *or* above the hypothesized population mean of 57.6 grams. As before, we can approach this by calculating the actual critical values for $\bar{x}$ itself: it will be significant at the 5% level if it is either below 57.19 grams or above 58.01 grams. It will be simpler, however, to calculate a z-value for the sample mean, which will be a test statistic. For a two-tailed test z will show that the result is significant at the 5% level if it is either below -1.96 or above $+1.96$.

One important point to note may be illustrated if we consider one of the results, $\bar{x} = 57.21$ grams, looked at in sections 16.2 and 16.3. The z-value for this is -1.86. This is below -1.64 and so is significant at the 5% level for a one-tailed test, but is not below -1.96, so it is *not* significant at the 5% level for a two-tailed test. The choice of a one-tailed or two-tailed test can, therefore, make a difference to the conclusion.

Choosing a test

The *kinds* of conclusions possible from one-tailed and two-tailed tests are, of course, the same. But since their critical regions are differently positioned they could sometimes give different conclusions. Our choice is between (a) a one-tailed test, against an alternative of a lower population mean (which has a left-hand-tail critical region), (b) a one-tailed test, against an alternative of a higher population mean (which has a right-hand-tail critical region), and (c) a two-tailed test (which has a two-tailed critical region). How should we decide between them?

The key question is: Would we wish to doubt the null hypothesis for a sufficiently far out sample mean $\bar{x}$, whether that $\bar{x}$ were above or below the hypothesized population mean?

Sometimes this may be obvious, for example:

1. If we are testing the resistance to fracture of a bolt used to attach an aeroplane wing we will not mind much if the sample has a high mean, but we will be distinctly worried if it has too low a mean! A one-tailed test will therefore be suitable.
2. If we are testing the weights of our food products in containers labelled with the supposed weight, what then? We may be concerned that a low sampling mean might indicate the machine is giving short measure (which could result in a fine), but we will also be concerned that a high sampling mean might indicate that the machine is giving too much produce (which would reduce profits). Therefore a two-tailed test might be appropriate.

Quite often, however, it may be by no means clear whether a one-tailed or a two-tailed test would be appropriate. Moreover the decision is not always trivial, for as we saw a result like $\bar{x} = 57.21$ for the eggs was significant at the 5 per cent level for a one-tailed test but not for a two-tailed test. In practice, all that one can hope for is that either the choice between a one-tailed and a two-tailed test seems obvious, or else that the result is so far significant or non-significant that the choice makes no difference.

In examinations, it is usual for the question to be worded so as to indicate whether a one- or two-tailed test is expected thus:

(a) 'determine whether this sample result is significantly *lower than* the ...' (this indicates a one-tailed test);

(b) 'determine whether this sample result is significantly *different from* the ...' (this indicates a two-tailed test).

The critical z-values

We have seen that there are three possibilities for the alternative hypothesis in a hypothesis test on the mean. For any of them the simplest and usual way to conduct the test will be to compute the z-value corresponding to the sample mean obtained and compare it with the critical z-values. These may be listed as follows:

1. For a *one-tailed test*, testing against the alternative that the population mean is *lower* than that in the null hypothesis:

    Result is significant at the 5% level if z is below -1.64
    Result is significant at the 1% level if z is below -2.33
    Result is significant at the 0.1% level if z is below -3.09

2. For a *one-tailed test*, testing against the alternative that the population mean is greater than that in the null hypothesis:

    Result is significant at the 5% level if z is above $+1.64$
    Result is significant at the 1% level if z is above $+2.33$
    Result is significant at the 0.1% level if z is above $+3.09$

3. For a *two-tailed test*, testing against the alternative that the population mean is *different from* that in the null hypothesis:

    Result if significant at the 5% level if z is above 1.96 or below -1.96
    Result is significant at the 1% level if z is above 2.58 or below -2.58
    Result is significant at the 0.1% level if z is above 3.29 or below -3.29

16.5 Summary

Null hypothesis

We begin with a 'null hypothesis' which we wish to test against a stated alternative on the basis of a sample result.

Significance concept

If belief in the truth of a particular null hypothesis requires us to believe that a very unlikely sample event has occurred, then it seems more reasonable to doubt the truth of that hypothesis.

Significance level

We decide on a probability value below which, given that a particular hypothesis were true, the probability of a sample result occurring by chance would be low enough to make us conclude the hypothesis false.

This value is a significance level. Conventionally 5%, 1% and 0.1% are popular.

Critical region

From the significance level and the shape of the sample mean distribution we can find what values of the sample mean are 'significant', that is, constitute evidence that the tested hypothesis is false. These values form a critical region, and its boundaries with the non-significant values are termed the *critical values*.

One- and two-tails

If a low *or* a high sample mean would make us doubt the null hypothesis then a two-tailed critical region and two-tailed test are applicable. If a mean only in one particular direction would make us doubt the null hypothesis then a one-tailed critical region and one-tailed test are applicable.

Mechanics

In practice the simplest approach is to find the critical z-values for the various significance levels, and then see if the z-value for the sampling mean actually obtained is inside or outside these critical values. The z-value obtained may be called the *test statistic*.

Conclusions

If the result is not significant then there is 'insufficient evidence to reject the null hypothesis'. If it is significant then there is 'some', 'strong' or 'very strong' evidence to reject the null hypothesis, depending on the level at which it is significant.

Recall and exercise questions

1. What is a 'null hypothesis'?
2. What is the implication, if a significance test is:
   (a) significant at the 1% level?
   (b) not significant at the 5% level?
3. If a test result is significant at the 1% level, which of the following is true:
   (a) it is also significant at the 5% level?
   (b) it is also significant at the 0.1% level?
4. A machine is set to produce giggle spindles with a mean diameter of 30 m and a standard deviation of 0.1 mm. To test the setting a sample of 20 are carefully measured, and the mean of the 20 is found to be 30.04 m.
   (a) Do a one-tailed test of significance at the 5% level to see if there is evidence that the mean setting is actually above 30 mm.
   (b) Do a two-tailed test of significance at the 5% level to see if there is evidence that the mean setting differs from 30 mm.
   In both answers state the meaning of your answer clearly.
5. A machine automatically fills '250 gram' packs of coffee. Its usual setting is such that the distribution of weights in packs has a mean of 253 gram and a standard deviation of 1.5 gram. To test the setting a sample of 30 packs is taken and their contents weighed. The mean of the sample is 252.45 grams. Is there any evidence that the setting is different from usual?

6. An engineering company is considering the purchase of a 24 volt turn-hazard interrupter unit. They require this to last for 120 000 cycles on average before failure. The manufacturers, Flashit Ltd, have claimed that their unit does have an average life at least this high, with deviation of about 13 000 cycles. To test the claim the company obtains six units from Flashit, and tests the number of cycles until the contacts weld and the light unit stays permanently on. The figures they obtain are 96 889, 99 949, 101 139, 116 088, 121 017 and 121 237 cycles. Do these test results show anything about the manufacturer's claims?

7. Published statistics of sizes of manufacturing units in the timber and furniture trades show an average size of 47 employees per unit, with a standard deviation of 76.3. A questionnaire is sent out to 500 randomly selected units, and only 100 reply. The survey organizers would like to know whether the average size of unit among those who replied is significantly less than the average for the industry as a whole. The mean number of employees amongst the companies who replied was 35.7. Advise the survey organizers concerning their question.

Problems

1. There is increasing concern about the incidence of shoplifting, and shop management are seeking to understand more about its causes and the kinds of people involved. A security company recently compiled figures for the ages of those arrested for shoplifting different kinds of merchandise. These are shown in Table 16.1.

Table 16.1 *Ages of shoplifters: percentage of arrested in each age group*

Age	Large group of all kinds of goods	Sample of 59 thefts of books
0–15	14.1	15.0
15–21	9.3	25.6
21–31	18.5	36.9
31–41	14.4	3.2
41–51	12.3	5.2
51–61	12.1	1.9
61–71	12.3	7.0
71–	7.0	3.2
	100.0	100.0

Prepare an article for a Retailer's journal, comparing the distribution of ages in the sample of those arrested for book thefts with the distribution of ages in those arrested for thefts of all kinds. Your article should include, with explanation, a significance test to see if the mean of the sample of 59 is significantly different from that of the whole population of arrested shoplifters.

2. ABC Engineering Company are at present involved in a bitter dispute over pay and conditions. The union spokesman claims that too much overtime is needed for his members to 'make ends meet', and that a typical member of his union will work an average of one extra hour per day as overtime, making six per week in a six-day week. Management have denied that average overtime worked is as high as this.

Table 16.2 shows the figures for the 1220 workers (this is in a 'closed shop', so it includes all workers in the sector).

Table 16.2 *Average hours overtime worked, in last quarter*

Average hours overtime per week	Number of workers doing these hours
0-1	180
1-2	80
2-4	160
4-6	180
6-8	300
8-10	180
10-12	100
12-16	40

Further controversy has now arisen in that 30 workers were randomly selected by management to meet an independent inquiry. On checking the records of these 30 the union claim that the average overtime they worked seems 'significantly less than the average for the whole group'. They suspect that the 30 were selected because they worked less than average – in short a 'fix'. The management deny this, and insist that the workers were randomly chosen as requested by the inquiry. Their mean number of hours of overtime was actually 4.75.

Write a report for the management:

(a) present the distribution of overtime worked.
(b) Find appropriate measures of average, and explaining how the apparent discrepancy between claims of union and management could be based on genuinely different interests leading to a different average being considered appropriate.

(c) comment on whether there is statistical evidence that the 30 may have been unfairly (i.e. not randomly) selected. Also comment on how, if possible the management might have fulfilled the inquiry request without any chance of appearing to 'fix' the results. (Assume they didn't!)

17
Estimation

17.1 Bias, error and precision

The three terms 'bias', 'error' and 'precision' were introduced in section 4.3. To recap:

1. *Error* is the difference between an estimate and the 'true' figure for the numerical value it is estimating.
2. *Bias* is a term which applies not to an individual estimate as such, but to the whole set of estimates which could be obtained by repeatedly applying the same procedure. If *on average* the estimates tend to consistently over- or under-estimate the value, then there is a bias.
3. *Precision* is a term which also applies to the whole set of possible estimates rather than to just one individual estimate. It refers to how close *on average* the estimates get to the actual value: that is, whether, on the whole, the errors are large or small.

It will be useful to reconsider these concepts in the light of our work on the distribution of sampling means. To do this we shall consider an example.

Suppose that we wished to do a survey to find the average monthly household expenditure on hairdressing and beauty treatment for the county of Mumbleshire. The actual distribution of individual expenditures might have, say, a mean of £3.76 and a standard deviation of £4.50 (and be highly positively skewed). We would not know this, of course, and might consider several alternative methods to estimate the mean figure:

1. Pick out 900 names and addresses at random from the Mumbleshire telephone directory and obtain information from all of them.
2. Pick out 400 names at random from the Mumbleshire Register of Electors and visit them to obtain information.

3. Pick out 100 names at random from the Mumbleshire Register of Electors and visit them to obtain information.

Any one of these three sampling procedures would give us a sample from which a sample mean could be calculated, and this mean could be used to estimate the population mean. But each one of the procedures would give a different sample mean each time it was repeated, and, as we saw in section 15.1, these different sampling means would fall into a sampling mean distribution. Each of the three procedures will produce sampling means from different sampling mean distributions, and these three distributions are shown in Fig. 17.1. We remember that we are intending to use our sample mean to estimate the unknown population mean, which has also been marked on Fig. 17.1. So what kind of accuracy will each sampling procedure give us in this?

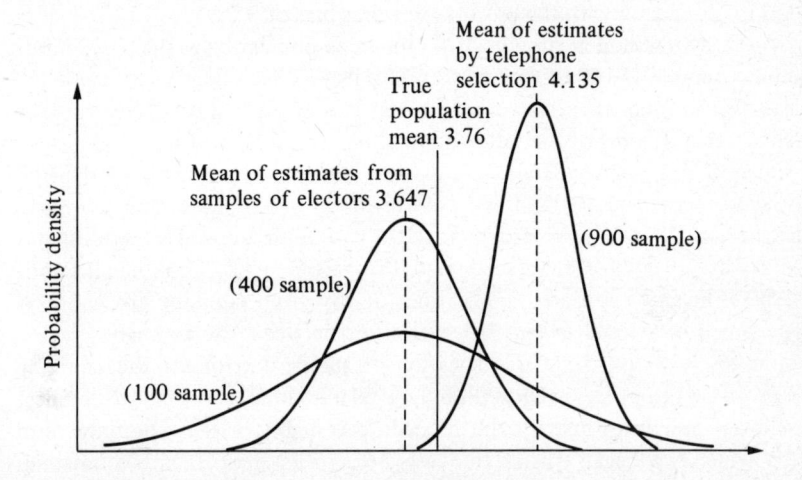

Figure 17.1 *Distributions of sampling means for three different sampling procedures*

Not everyone can afford to be on the telephone, and in any case surveys show it to be a middle-class rather than a working-class priority. Thus the distribution of expenditures claimed by telephone subscribers might have much the same standard deviation (4.5) as that for all households, but its average would be considerably higher. In fact, it proves to be £4.135, compared with a population mean for all households of only £3.76. This implies that the sampling mean distribution of samples of 900 random telephone subscribers has a mean of £4.135, and a standard error of

$$(\text{s.d.}(x)/\sqrt{n}) = (4.5/\sqrt{900}) = 0.15$$

Most of the distribution, as shown in Fig. 17.1, lies *above* the population mean £3.76: that is, we will usually get an *overestimate* of the population mean for all households. The difference between the true population mean (£3.76) and the mean of the sampling mean distribution (£4.135) is called the *bias* of the procedure. Here it is (4.135 − 3.76) which is 0.375.

The second method, taking a random sample of 400 from the Register of Electors, would produce a sampling mean distribution as shown. Its standard error is $(s.d.(x)/\sqrt{n}) = (4.5/\sqrt{400}) = 0.225$. This is a larger standard error than that for the previous procedure because the sample is smaller. But the average of this sampling mean distribution is much nearer the true household population mean, for there are fewer types of household excluded from the sampling frame. It does produce, on average, a slight underestimate of the true expenditure, because people tend to understate their expenditure on such items. But this downwards bias is only (3.647−3.76), which comes to −0.113, compared with the previous upwards bias of +0.375.

The third method is sampling from the same population as the second, but taking a sample of 100 instead of 400. This leaves the average of the sampling mean distribution as the same, so the bias is unaltered. But it gives a rather higher standard error of the distribution at $(4.5/\sqrt{100}) = 0.45$.

So in the diagrams of Fig. 17.1, the *bias* is the difference between the true population mean (3.76) and the average of the appropriate sampling mean distribution. It is the amount by which the procedure over- or underestimates on average. The *precision* relates to the variation between the different estimates given by repeated applications of any given sampling procedure. A high precision would mean little variation between the estimates, which would imply a low standard deviation or standard error for the sampling mean distribution. Here the 900 random telephone subscribers give the best precision, but unfortunately the procedure is highly biased. The third term with which we began this section was *error*. The error of any estimate is simply the amount it differs from the true population mean. Thus the term 'error' differs from bias and precision in that it applies to an individual estimate rather than to features of the distribution of estimates. Any one of our three procedures could, in an individual case, give an error better or worse than the others. We can say, however, that the second procedure (400 electors) has no worse bias than the third (100 electors), but has a higher precision. Thus, though it is not certain, it is *likely* that it will give an estimate with a smaller error.

All three of our procedures described above were random samples, though one used a different sampling frame from the other two. But what of a quota sample? Provided that a suitable sampling frame is available, a random sample is usually less *biased* than a quota sample. But a quota sample is usually cheaper per unit sampled than a random sample. This means that a larger quota sample can be taken for the same cost, which increases the *precision* of the estimate. In many practical situations the past experience of a survey organization would make them confident that the possible bias of the quota sample was more than offset by its better precision for the same cost.

17.2   Confidence intervals

Let us now return to consider a further aspect of the example developed in Chapter 16. This concerned the accurate weighing of a sample of 48 eggs to determine whether or not the population mean differed from 57.6 grams. Suppose that we obtain a sample mean of 56.91 grams. This (see Chapter 16) is significantly lower than 57.6 grams at the 0.1% level, and so we will conclude that the population mean (the setting of the weighing machine) has changed from 57.6 grams.

But suppose that we want to know what the actual new mean setting is? Our only source of information is the sample, so the best single estimate is 56.91 grams. But how accurate is this figure? To answer this question we may remember that according to the rules we know for the distribution of the sampling mean, sample mean values are distributed around the population mean as in Fig. 17.2. From our normal tables we know that 95 per cent of

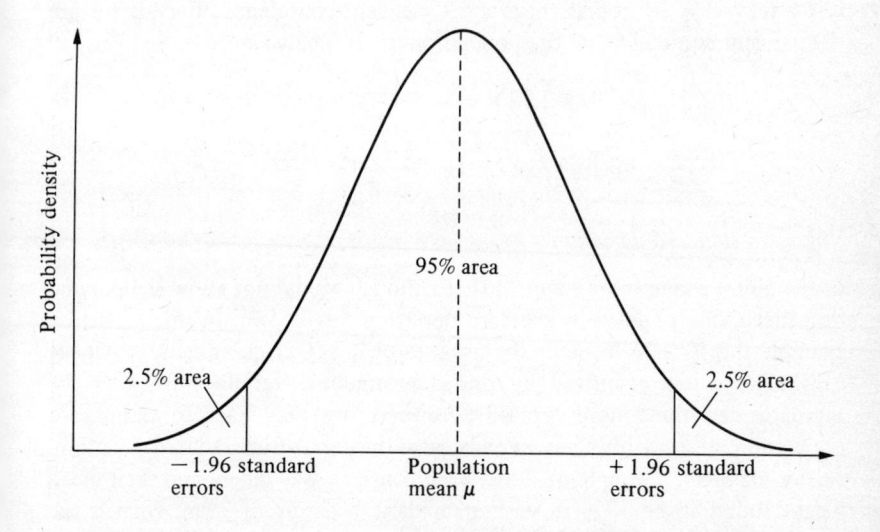

Figure 17.2   *Distribution of sampling means around the population mean*

the time the sample mean will be within plus or minus (±) 1.96 standard errors of the population mean μ. That is:

95 per cent of the time $\bar{x}$ is within $\pm 1.96 \times \dfrac{\text{s.d.}(x)}{\sqrt{n}}$ of μ;

logically, therefore, we may say:

95 per cent of the time μ is within $\pm 1.96 \times \dfrac{\text{s.d.}(x)}{\sqrt{n}}$ of $\bar{x}$.

This gives us a way of stating an interval within which we are 95 per cent sure that the population mean μ will lie. We simply take the interval of

$$1.96 \dfrac{\text{s.d.}(x)}{\sqrt{n}} \text{ either side of } \bar{x}:$$

that is:

$$56.91 \pm 1.96 \times \dfrac{1.45}{\sqrt{48}}$$
$$= \text{from } (56.91 - 0.41) \text{ to } (56.91 + 0.41)$$
$$= \text{from } 56.50 \text{ to } 57.32$$

We are, then, 95 per cent sure on the basis of the sample that the population mean μ is between 56.50 and 57.32 grams. This interval is known as a *confidence interval*. We say that we are '95 per cent confident that μ is between 56.50 and 57.32 grams'.

We may also, of course, have a 99 per cent confidence interval: we are '99 per cent confident that the population mean is between:

$$56.91 \pm 2.58 \times \dfrac{1.45}{\sqrt{48}}$$

i.e. between 56.37 and 57.45 grams'.

Unknown standard deviation

In the above example we assumed that although we did not know the population mean, the population standard deviation *was* known. In this particular example this is realistic, since the usual error in egg weighing causes each egg weight to be overestimated by the same amount, but does not alter the standard deviation. More typically, however, we may have to estimate a population mean without any knowledge of the population standard deviation.

For example, a sample of 200 Ruritanian men was taken, and their mean height found to be 172 cm with a standard deviation of 5 cm. What is the best estimate of the mean for the whole population? What we can do here is use the sample itself to estimate the value of s.d.(x). For a sample as large as 200 the s.d.(x) for the whole population will be given, to a good approximation, by the standard deviation *within the sample*: that is s.d.(x) is approximated by s.d.(within the sample).

We may therefore use our sample standard deviation of 5 cm to form a confidence interval: we are 95 per cent confident that the population mean μ is between

$$172 \pm 1.96 \times \frac{5}{\sqrt{200}}$$

i.e. between 171.31 and 172.69 cm

This is only valid for large samples. If the sample were small (say size 10), then one could not assume that the standard deviation within the sample was necessarily a good estimate of that within the population as a whole. On the other hand, it also assumes that the population is very much larger than the sample. If the population is relatively small various adjustments should be made, but they are beyond our present scope.

Notation

It may be useful to note some common terms and notation:
1. A single figure estimate (e.g. 172) is sometimes called a *point estimate* and a confidence interval an *interval estimate* of a population mean.
2. Some books use the Greek letter σ (sigma) to indicate what I have called s.d.(x), and the letter s to indicate the standard deviation *within* the sample.

*Confidence intervals for proportions**

There are situations where we are not dealing with measurable values such as heights, but simply with *attributes* which each sample member either has or has not. This is sometimes called *attribute sampling*. Often, we wish to estimate the proportion P of a large population having some attribute, based on the proportion p of a sample of size n which have the attribute.

Example

A council has a large number of ratepayers, and wishes to estimate the proportion of these who paid their rates by the required date. A sample of 100 are randomly selected, and only 45 of these paid up on time. What can be concluded?

The proportion p of the sample who paid on time was $(45/100) = 0.45$. This is the best available single figure estimate, or *point estimate* for the population proportion P. But a confidence interval for P at 95 per cent confidence can be given using the formula:

$$p \pm 1.96 \sqrt{\frac{p \times (1-p)}{n}} = 0.45 \pm 1.96 \times \sqrt{\frac{0.45 \times 0.55}{100}}$$

Thus we are 95 per cent confident that the population proportion P lies between 0.352 and 0.548, or between 35.2 per cent and 54.8 per cent if we prefer to state it as a percentage. This particular example was first raised on page 30 in connection with survey methods.

The formula for a 95 per cent confidence interval (a 99 per cent one would simply use 2.58 instead of 1.96) can be used for attribute sampling whenever the sample is large and the population is very large compared with the sample.

Confidence, random and quota samples

The concept of confidence intervals is connected with that of *precision* and was looked at first in Chapter 4. Strictly speaking, confidence limits can be given only if the sample is random, but some survey organizations have found in practice that 'limits' can be given for certain kinds of quota samples. These 'limits', however, cannot be based on formulae given in this section, and experience indicates wider intervals than those calculated for random samples.

17.3 Summary

Bias, error and precision

Error is the individual estimate's difference from the true value. Bias is the difference between the mean of the distribution of estimates obtained by the particular sampling procedure used and the true value of the population mean. Precision may be measured by the standard deviation of the distribution of estimates.

Confidence

On the basis of a sample we may assign, using normal tables, an interval within which we are a particular percentage certain that the population mean will lie;

thus, for example, we can be 95 per cent confident that μ is within 1.96 standard errors of $\bar{x}$, that is, between:

$$\bar{x} \pm 1.96 \times \frac{\text{s.d.}(x)}{\sqrt{n}} \quad \left[\bar{x} \pm 1.96 \times \frac{\sigma}{\sqrt{n}} \right]$$

For reasonably large samples this may be approximated by:

$$\bar{x} \pm 1.96 \times \frac{\text{s.d.(within sample)}}{\sqrt{n}} \quad \left[\bar{x} \pm 1.96 \times \frac{s}{\sqrt{n}} \right]$$

*Confidence for proportions**

A confidence interval, at the 95 per cent level of confidence, for the population proportion P may be approximated for large samples by:

$$p \pm 1.96 \times \sqrt{\frac{p \times (1-p)}{n}}$$

where p is the sample proportion.

Recall and exercise questions

1. Briefly explain, using a diagram, the meaning of the terms *error*, *bias* and *precision* in the context of estimating a population mean.
2. Briefly explain the concept of a *confidence interval*.
3. A tyre manufacturer has developed a new 'Supertyre 186', which has been tested for endurance on 150 trucks in commercial use. The mean endurance was 15 000 miles, with a standard deviation within the sample of 2200 miles. What is the 95 per cent confidence interval for the mean endurance to be expected if the tyres go into full production? What is the 99 per cent confidence interval?
4. A survey has been carried out to estimate the average weekly household expenditure in Mumbleshire on life assurance, contributions to pension funds, etc. A random sample of 200 was selected, and re-visited until contacted. The mean figure obtained for the 200 was £5.12, with a standard deviation within the sample of £4.30. Give a 95 per cent confidence interval for the true mean of all Mumbleshire residents (assume no bias).
5.* A random sample of 200 were taken from the Register of Electors and asked whether they could remember the brand name in a particular television advertisement which was described to them. Only 78 could

remember it. Give a 95 per cent confidence interval for the proportion of the electors in the region who would remember the brand name from this advertisement.

Problems

1. A company has a number of spurling machines, and has recently kept records of the times these will run before needing service. Table 17.1 shows records of 110 such times.

    Table 17.1 *Times between service of spurling machines*

Time between service (tens of hours)	Frequency
40–42	7
42–44	19
44–46	30
46–48	21
48–50	25
50–56	8
Total	110

    Write a report based on these recorded times:
    (a) showing the pattern of distribution of times in the sample;
    (b) giving 95 and 99 per cent confidence intervals for the mean time between services to be expected for all such machines, over a long period.

2. Bugem pest control have recently been conducting agricultural tests to see if the yield of potatoes is improved by the use of *Carbohiran 1* in the control of potato cyst nematode. Trials have been conducted to find the yield in kilograms per plot, where each plot consists of four four-metre rows, one metre apart. For untreated plots the yield has been found to have a mean of 27.6 kg, and a standard deviation of 4.2 kg within the test area. Six plots treated with *Carbohiran 1* have given the following yields (in kilograms): 44.7, 52.0, 50.8, 41.7, 49.9, 47.7.

(a) Do these treated plots yield a significantly greater amount than the untreated ones?

(b) On the basis of the sample result, within what range would you be 99 per cent certain that the mean yield of treated plots would lie?

(c) From a figure for plot yield, an estimate can be made of the equivalent yield in tonnes per hectare by multiplying by a factor of 0.867; thus the mean for untreated planting would be

$$27.6 \times 0.867 = 23.9 \text{ tonnes per hectare}$$

Their tests were done in England, but Bugem are now considering marketing *Carbohiran 1* in Wales, where the published yield for the same year was 35.5 tonnes per hectare (source: *Welsh Agricultural Statistics*). Draft a document, including any appropriate figures, setting out the evidence for the effectiveness of *Carbohiran 1* and its potential for use in Wales.

(*Note*. The 0.867 conversion factor may be used to convert average figures, and also confidence limits if you wish.)

18
Further Significance Testing*

18.1 Basic elements of significance tests

In Chapter 16 we looked in detail at one particular kind of significance test. This was of the form shown in this example:

Null hypothesis H_0: The eggs marked 'grade 4' come from a distribution with mean $\mu = 57.6$ grams and standard deviation s.d.$(x) = 1.45$ grams.

Alternative hypothesis H_1: The eggs marked 'grade 4' come from a distribution with μ less than 57.6 grams, and a standard deviation s.d.$(x) = 1.45$ grams.

Test result R: In a random sample of $n = 48$ eggs classed as grade 4 we had a mean of $\bar{x} = 57.21$ grams.

Test statistic: From the test result we calculated a test statistic z:

$$z = \frac{\bar{x} - \mu}{\text{s.e.}(\bar{x})} = \frac{\bar{x} - \mu}{(\text{s.d.}(x)/\sqrt{n})}$$

$$= \frac{57.21 - 57.6}{(1.45/\sqrt{48})} = -1.86$$

This test statistic was compared with the critical values for a one-tailed test, as given in section 16.4:
  Result is significant at the 5% level if z is below -1.64
  Result is significant at the 1% level if z is below -2.33
  Result is significant at the 0.1% level if z is below -3.09

In this instance it is below −1.64 but not below −2.33, so the result is significant at the 5% but not the 1% level.

There are various other kinds of significance test, all of which contain elements similar to this one:

Null hypothesis. This must be a *specific* hypothesis. For example $\mu = 57.6$ grams is specific enough to have calculable implications for probabilities of various sample results. A hypothesis like $\mu < 57.6$ grams would not be sufficiently specific.

Alternative hypothesis. This may be specific or vague. Most usually it will be vague, though it may indicate whether a test is a one- or a two-tailed test.

Test result R. This will be the result of some specific survey or test, giving a sample value (for example a sample mean $\bar{x}$).

Test statistic. From the test result a test statistic may be calculated in a more convenient form for actually doing the test (for example the z-value for a sample mean).

Critical values. There will be critical values for the test statistic, usually available in the form of a table. If the test statistic falls in the critical region bounded by these critical values, then the result is significant and constitutes evidence against the null hypothesis. If the test statistic does not fall in the critical region, then the result is not significant, and there is insufficient evidence from that result to reject the null hypothesis.

The rest of this chapter contains intentionally brief outlines of some of the more common significance tests in use. For each type just one example will be given.

18.2 Tests of population means − unknown standard deviation

In the examples we have looked at the value of s.d.(x) has been a part of the assumptions of the hypothesis under test. In some instances, however, we know nothing of the value of s.d.(x) other than what can be estimated from the sample itself.

Example. The national published figures show that average household expenditure on postage, telephones and telegrams is £1.08 per week. A sample of 150 residents of Mumbleshire were interviewed, and their mean expenditure was found to be £1.21 per week with a standard deviation within this sample of 82p per week. Does this indicate a mean expenditure for Mumbleshire which differs from the national average?

In this example we have a *large sample* (150). This means that we can use the standard deviation *within* the sample as a good approximation to the standard deviation for the population as a whole; that is, s.d.(within the sample) will give a good approximation to s.d.(x). Our significance test will therefore proceed as follows:

Null hypothesis H_0: The population mean μ for Mumbleshire is £1.08.

Alternative hypothesis H_1: The population mean for Mumbleshire is not £1.08 (this alternative indicates a two-tailed test).

Test result R: A sample is taken from which the mean $\bar{x}$ and s.d.(within sample) can be found. Here $\bar{x} = 1.21$, s.d.(within sample) $= 0.82$ and $n = 150$.

Test statistic:
$$z = \frac{\bar{x} - \mu}{\text{s.d.(within sample)}/\sqrt{n}} \quad \left[\frac{\bar{x} - \mu}{s/\sqrt{n}}\right]$$

Here $z = \dfrac{1.21 - 1.08}{0.82/\sqrt{150}} = 1.94$

Critical values: These will be the same as those from the normal tables appropriate to a two-tailed test where the standard deviation is known. Thus, for example, the result is significant at the 5% level if z is above 1.96 or below -1.96 (see section 16.4).

Conclusion: The z-value obtained (1.94) does not fall in the critical region, so the result is not significant at the 5% level. There is insufficient evidence from this result to reject the null hypothesis that £1.08 is the mean expenditure for Mumbleshire.

Small samples

If we have a similar situation but have a *small* sample, say 10, then this approach will not do. The standard deviation of observations within a small sample cannot be guaranteed to give a good approximation to the population standard deviation s.d.(x). Instead we use, therefore, 'Student's t-test'.

Example. The manufacturers of a particular engine, the TX56, claim that it should run an average of 120 000 miles before needing replacement or major overhaul. A haulage company, Ulent Carriers, bought six trucks fitted with the TX56, which have been subjected to average use. It has been found that the mileage before breakdown on these six were: 95 000, 90 000, 100 000 110 000, 115 000, and 125 000 miles. Do these results differ significantly from the company's claims?

Null hypothesis H_0: The population mean μ = 120 000 miles.

Alternative hypothesis H_1: Population mean μ not equal to 120 000 miles (this requires a two-tailed test).

Test Result: Six results (listed as above), $n = 6$.

Test Statistic:
$$t = \frac{\bar{x} - \mu}{\sqrt{\frac{\frac{1}{n-1}\Sigma(\bar{x}-x)^2}{n}}}$$

Here $t = -2.63$

This t-statistic is very similar to the z-value used where the population standard deviation is known, but instead it estimates that population standard deviation using the standard deviation between the observations in a small sample. This estimation adds to the uncertainty, and so a t-statistic has to be larger than the z-value for a similar size sample before the result is significant. The smaller the sample on which the standard deviation is estimated, the larger the t-statistic value must be before the result is significant at a given level. The critical values for this t-statistic, and their relationship with the sample size, are shown in Appendix A, Table 4.

Critical values: For a sample of size $n = 6$, and a two tailed test, the critical values are:

t is significant at the 5% level if above 2.57 or below −2.57

t is significant at the 1% level if above 4.03 or below −4.03

t is significant at the 0.1% level if above 6.87 or below −6.87

Conclusion: $t = -2.63$ is significant at the 5% but not at the 1% level. There is some, but not strong, evidence that the manufacturer's claim of $\mu = 120\,000$ miles is incorrect.

Note: As well as sample size (which relates only to the situation I have just been describing), the t tables are labelled with *degrees of freedom*. In the situation presently discussed here, the number of degrees of freedom is always the sample size minus one (in this case $6-1 = 5$). But the concept of degrees of freedom is more general, and allows the tables to be used in other contexts as well, so most t tables carry this label rather than 'sample size'.

18.3 Difference between two means

In certain situations we wish to test whether there is a difference between the means of two different populations, without knowing what either of those population means is. Usually in such instances we do not know what the standard deviations are either, and have no particular reason to assume that they are the same for the two populations. I am going to look only at situations where the sample sizes are large, and therefore the standard deviations of both populations can be estimated accurately from the samples. For small samples the procedure is both complex and controversial.

Example

To test which of two cellulose paints is the quicker to dry, a sample of 100 panels was painted with 'Gollycover' and a second sample of 150 with 'Redisplash'. The results were:

100 Gollycover: Mean drying time 3.41 hours, s.d.(within sample) 0.66 hours

150 Redisplash: Mean drying time 3.57 hours, s.d.(within sample) 0.54 hours.

Is the difference between them significant?

Null hypothesis H_0:   Mean (Redisplash) μ_1 = Mean (Gollycover) μ_2

Alternative hypothesis H_1: Mean (Gollycover) $\neq$ Mean (Redisplash)

Test result R: A large sample from each population (here 100 from Gollycover and 150 from Redisplash).

Test statistic:
$$z = \frac{(\bar{x} \text{ for 1st sample}) - (\bar{x} \text{ for 2nd sample})}{\sqrt{\dfrac{[\text{s.d.(sample 1)}]^2}{(\text{Size of sample 1})} + \dfrac{[\text{s.d.(sample 2)}]^2}{(\text{Size of sample 2})}}} \left[\frac{\bar{x}_1 - \bar{x}_2}{\sqrt{\dfrac{s_1^2}{n_1} + \dfrac{s_2^2}{n_2}}} \right]$$

here $z = -2.02$

Critical values: Our alternative hypothesis indicates that either of the two population means might be greater, so a two-tailed test is needed. Since z here is distributed approximately normally, the same critical values may be used as those given at the end of section 16.4

Conclusion: The observed value $z = -2.02$ is significant at the 5% level (since it is below -1.96) but not at the 1% level (since it is not below -2.58). There is some, but not strong, evidence for a difference between the mean drying time for Gollycover and Redisplash.

18.4 Tests on proportions

Sometimes we are interested not in a measured value, but in the proportion or percentage of a population falling into one of two alternative categories.

Example

The published figures for the North West Region show that of those taking driving tests the proportion who passed was 0.48 (48 per cent). One particular examiner has passed only 19 out of the 50 he has examined (which is 0.38 or 38 per cent). Is there evidence that he is passing a percentage different from the regional average, or could his apparently different pass rate on this particular sample be ascribed to chance?

Null hypothesis H_0: Mean proportion being passed by examiner $P = 0.48$

Alternative hypothesis H_1: $P \neq 0.48$

Test result R: In a sample of $n = 50$, proportion 0.38 passed.

Test statistic:

$$z = \frac{p - P}{\sqrt{\frac{P(1-P)}{n}}}$$

where P is hypothesized proportion
p is sample proportion
n is sample size

which here is $z = \dfrac{0.38 - 0.48}{\sqrt{\dfrac{0.48 \times 0.52}{50}}} = -1.42$

Critical values: A two-tailed test is indicated by the alternative hypothesis, which would allow the proportion to be either above or below 0.48. Since for a reasonably large sample z is approximately normally distributed, the critical values will be the usual ones for a two-tailed normal distribution.

Conclusion: The observed $z = -1.42$ is not significant at the 5% level, and there is no evidence from the sample to conclude that the examiner is passing a different proportion from the regional average.

Note. Some books call the population proportion π, but note that this has nothing to do with circles. I have used a capital P to avoid confusion.

Difference between two proportions

Suppose that, instead of testing a stated value for a single population proportion we wish to test whether there is a difference between the proportions in two different populations.

Example. A driving test centre has two examiners: Mr Clunk and Mr Click. To see whether one is being stricter than the other, the records of recent tests are examined. Mr Clunk passed 24 of his last 60 applicants, and Mr Click passed 36 of his last 70.

Null hypothesis H_0:   There is no difference between the pass rates for the two examiners ($P_1 = P_2$)

Alternative hypothesis H_1:   There is a difference ($P_1 \neq P_2$)

Test result R: We have samples (size n_1 and n_2) from each of the two populations, and proportions (p_1 and p_2) for each sample.

Test statistic:

$$z = \frac{p_1 - p_2}{\sqrt{\left(\dfrac{n_1 p_1 + n_2 p_2}{n_1 + n_2}\right)\left(1 - \dfrac{n_1 p_1 + n_2 p_2}{n_1 + n_2}\right)\left(\dfrac{1}{n_1} + \dfrac{1}{n_2}\right)}}$$

which is sometimes approximated by:

$$z = \frac{p_1 - p_2}{\sqrt{\dfrac{p_1(1-p_1)}{n_1} + \dfrac{p_2(1-p_2)}{n_2}}}$$

where p_1 = proportion in 1st sample (24/60) = 0.4
         p_2 = proportion in 2nd sample (36/70) = 0.51
         n_1 = sample size of 1st sample (60)
         n_2 = sample size of second sample (70)

so here, using the full formula, $z = -1.25$.

Critical values: The alternative hypothesis indicates a two-tailed test, and z is approximately normally distributed. The critical values are therefore the usual ones for any z-value.

Conclusion: An observed value of $z = -1.25$ is not significant at the 5% level, so there is no evidence to reject the null hypothesis that the examiners are passing equal proportions.

18.5 Chi-squared contingency tests

These tests have a wide area of application in situations where we have a table containing cross-categories and wish to know whether one category is affected by another.

Example

Suppose that a company does a consumer test to find preferences for shampoo and obtains the results given in Table 18.1. As may be seen, a total of 80 men

Table 18.1 *Shampoo preferences*

	Prefer Blowit	Prefer BO6	Don't know	Total
Men	30	40	10	80
Women	25	20	15	60
Total	55	60	25	**140**

and 60 women participated in the tests. What we want to know is whether there is any evidence of a difference in attitude between the sexes. Since the computation procedures are longer than for the previous tests, it may be useful to summarize the steps first before performing them:

Null hypothesis H_0: There is no association between sex and shampoo preference, that is, men and women have the same preferences in shampoo.

Alternative hypothesis H_1: There is an association between sex and shampoo preference.

Test result R: A *contingency table* like Table 18.1.

Test statistic: This is chi-square (χ^2), which is computed as shown below.

Critical values: Usually only one-sided tests are used, and the critical values depend on the *degrees of freedom* for the statistic as explained below. Table 5, Appendix A, lists the critical values.

Further significance testing*—281

We begin by considering the overall proportions of preferences.

$\frac{55}{140}$ prefer Blowit $\frac{60}{140}$ prefer BO6 $\frac{25}{140}$ don't know

If there is no difference between men's and women's preferences, then we would expect these proportions, averaged over a number of samples like this, to apply both to men and to women; for example:

the average number of women preferring Blowit will be $\frac{55}{140} \times 60 = 23.57$

the average number of women preferring BO6 will be $\frac{60}{140} \times 60 = 25.71$

In this way, for each category in which we have an *observed frequency O*, we can also calculate an *expected frequency E* – the long-run average expected in that category if the null hypothesis is true. This is shown in Table 18.2. The

Table 18.2 *Chi-square table*

	Prefer Blowit O	E	Prefer BO6 O	E	Don't know O	E
Men	30	31.43	40	34.29	10	14.29
Women	25	23.57	20	25.71	15	10.71

formula to obtain E in each case is:

$$E = \frac{\text{row total} \times \text{column total}}{\text{overall total}}$$

The test statistic is called *chi-square value* (Greek letter χ^2), and it is calculated from the values of O and E using the formula:

$$\chi^2 = \Sigma \frac{(O-E)^2}{E}$$

This is shown in Table 18.3. To find the critical values for this test statistic

Table 18.3 *Chi-square calculations*

Observed (O)	Expected (E)	$\frac{(O-E)^2}{E}$
30	31.43	0.0651
40	34.29	0.9508
10	14.29	1.2879
25	23.57	0.0868
20	25.71	1.2681
15	10.71	1.7184
		$\chi^2 = 5.3771$

we have to use chi-square tables. We need first to calculate the number of *degrees of freedom*:

$$\begin{aligned}\text{Degrees of freedom} &= (\text{rows}-1)\times(\text{columns}-1)\\ &= (2-1)\times(3-1)\\ &= 2\end{aligned}$$

With the chi-square test there is no choice between one-tail or two-tail test, and the tables (Appendix A, Table 5) list critical values for 2 degrees of freedom as:

χ^2 is significant at 5% level if greater than 5.99
χ^2 is significant at the 1% level if greater than 9.21
χ^2 is significant at the 0.1% level if greater than 13.82

This particular result of $\chi^2 = 5.38$ is therefore not significant at the 5% level, and there is insufficient evidence to show a difference between the preferences of men and women for different types of shampoo.

Recall and exercise questions

1. State the major elements of any hypothesis test, and the possible conclusions which it can give.

2. The manufacturers of a parts kit for the TX34 engine claim that the whole engine can be assembled in a 'standard time' (in work study jargon) of 63 minutes; this assumes that the worker is working at standard rate and allowances have been made for relaxation, and so on. The Bridge Motor Company have been assembling these units for several weeks. A work study is then done, and finds that on 100 observations of assembly, the mean standard time (after multiplying by rating, and adding allowances) was 70 minutes, with a standard deviation within the sample of 24 minutes. Is there any evidence that the manufacturers' claim is inaccurate?

3. A key component in Ace Co.'s Flimflam machines eventually fails and has to be replaced. Past experience has shown the average life of units to be 645 hours' use. Ace have recently bought ten of these components from a different supplier from their usual one, and would like to know whether the new units are a significant improvement in terms of endurance. The ten new units lasted 630, 640, 651, 666, 652, 643, 652, 660, 650 and 657 hours respectively. What does this show? Explain your answer and its implications for Ace Co. policy.

4. A company has introduced a trial scheme of paying a bonus to employees who have an anti-flu vaccination. This was tried in certain sections, while other sections (matched for type of work) were left without this incentive. During the winter period the 150 workers in the sections given the bonus showed an average period of incapacity due to illness of 9.6 working days, with a standard deviation of 3 working days. The 180 workers in sections not given the bonus showed an average of 10.5 working days' incapacity, and a standard deviation of 3.8 working days. Does this show that the bonus scheme has had any effect on incapacity?

5. A market research survey has been carried out in the South West region into opinions on carpeting. One of the questions concerned whether people were owner-occupiers or rented their homes. Out of 400 people interviewed, 150 owned their homes and 250 rented. Is this significantly different from the published figures for the South West: 45 per cent owner-occupiers?

6. A mail-order company runs two catalogue schemes. A recent area of interest to the company is whether a higher proportion of their 'Homefund' catalogue customers than their 'Regency Fair' catalogue customers are in car-owning households. A survey with follow-up of 150 'Homefund' customers showed 82 to be in households owning cars, while among 160 'Regency Fair' customers only 81 owned cars. Does this prove anything?

7. A hi-fi manufacturer makes music centres with three alternative finishes: silver, gold and wood veneer. They have recently conducted a survey to find the appeal of the three finishes, and have compiled the figures in Table 18.4. Is there any evidence that there is a difference between the preferences of different social classes?

Table 18.4

Social class	Gold	Silver	Wood	Total
AB	20	40	10	70
C1	22	35	8	65
C2	20	30	15	65
DE	10	30	10	50
Total	72	135	43	**250**

Problem

Until recently a local authority had a policy of purchasing older terraced houses to let to tenants. A recent local election saw a change of control in the council, and the new controlling group are reversing that policy. They would like to have a good estimate of the market value of the properties in their ownership, and a local estate agent (who happens to be an uncle of Mr Fred Beare, a leading councillor) has estimated the average value of such property to be £15 800. A second councillor, Mrs Hope Best, has obtained figures for the actual sale prices of 30 similar properties sold in the last couple of months (prices are not rising steeply at present). Table 18.5 shows these. Mrs Best

Table 18.5

Price of houses	Number
14 800–15 099	5
15 100–15 399	8
15 400–15 699	8
15 700–15 999	6
16 000–16 300	3
Total	30

thinks that this indicates a lower average price than £15 800, but Mr Beare dismisses this: 'I back Uncle Harry's experience, and anyway you can't tell anything from a mere sample . . .'

Use the figures in Table 18.5 to present a report to the Council, indicating the pattern of distribution of prices, and whether the sample does indicate that the average market value of such houses is indeed less than £15 800.

19
Regression

19.1 The need for regression

Suppose that I wish to have a quick statistical way to estimate the market value of any given detached house in a particular locality, for example the area of Fulwood in Preston. I take a random sample of eleven houses from estate agents' details, and find the average (mean) price to be £47 000. But, rather than simply use this average figure for all such detached houses, I may feel that the rateable value of any particular house (obtainable from the rating records) will enable me better to judge its market value. So I tabulate the rateable values of my eleven sample houses against their market values, as in Table 19.1. As shown in section 7.1, these values may be represented

Table 19.1 *Market price and rateable value of houses in Fulwood*

House number	Market price (£000)	Rateable value (£)
1	51.0	360
2	59.0	302
3	72.5	298
4	72.5	352
5	45.5	290
6	47.5	333
7	38.8	274
8	39.5	289
9	36.0	242
10	30.0	272
11	27.0	215

graphically by means of a scatter diagram of rateable value against price. Fig. 19.1 shows this (compare with Figs 7.1 and 7.2). The scatter diagram (or scattergraph) Fig. 19.1 confirms our belief that a higher rateable value tends to indicate a higher market value (I am taking market price and market value as the same).

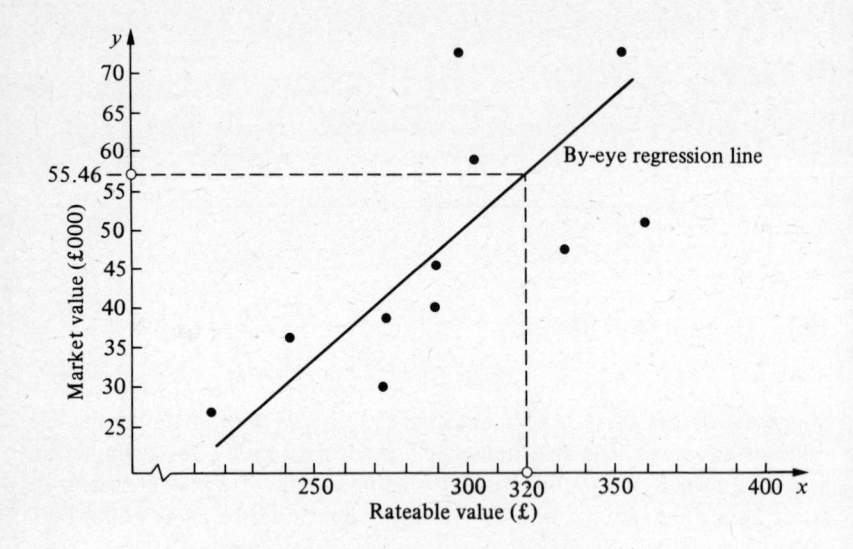

Figure 19.1 *'By-eye' regression line*

But we need more than this. We need a line plotted on the scattergraph which will enable us to predict the market value of a house from its rateable value. Such a line is called a *regression line*; the simplest form is obviously a straight line. By guessing where I think a line goes best through the existing points, I have plotted such a straight line regression line 'by eye' on Fig. 19.1. The method of prediction has been illustrated for a rateable value of £320. Reading up to the line, and then across to the y-axis, we get a prediction of a market value of 55.46 (£000).

However, a guess at a good line 'by eye' is not the best way to get a line for such predictions. There are two common ways of improving on a simple visual guess: the *three-point method* and *least-squares regression*. The reader should be aware of the basic approach of both methods, but need not necessarily be practised in the calculation of both.

19.2 A simple three-point method*

A simple method of obtaining a line from which to predict y figures from x figures consists of the following steps:

1. Rearrange the pairs of values so that the x-values are in ascending order of magnitude.
2. Find the mean $\bar{x}$ of the x-values, and the mean $\bar{y}$ of the y-values.
3. Take all the pairs for which the x-value exceeds $\bar{x}$, and find the mean for the x-values and the y-values in this group (it should contain about half the observations).
4. Take all the pairs for which the x value is less than $\bar{x}$, and find the means for the x-values and the y-values in this group.
5. Plot the three pairs of means obtained in steps 2, 3 and 4 on the scattergraph. Usually, they should fall reasonably close to a straight line, which can then be plotted by eye.

In our example these steps can be done as follows:

Market price (y)	Rateable value (x)
36.0	215
27.0	242
30.3	272
38.8	274
39.5	289
45.5	290
72.5	298
59.0	302
47.5	333
72.5	352
51.0	360
519.3	3227

for the six values with x below $\bar{x}$:
    mean of x-values = 1582/6 = 263.7
    mean of y-values = 217.1/6 = 36.2

overall means $\bar{x}$ = 3227/11 = 293.4 ($\bar{y}$ = 47.2)

for the five values with x above $\bar{x}$:
    mean of x-values = 1645/5 = 329
    mean of y-values = 302.5/5 = 60.5

The three points to plot are therefore: (263.7, 36.2) (x = 263.7, y = 36.2)
(293.4, 47.2) (x = 293.4, y = 47.2)
(329, 60.5) (x = 329, y = 60.5)

Fig. 19.2 shows these points plotted on the scattergraph, with a line drawn through them from which predictions of y-values from x-values can be made.

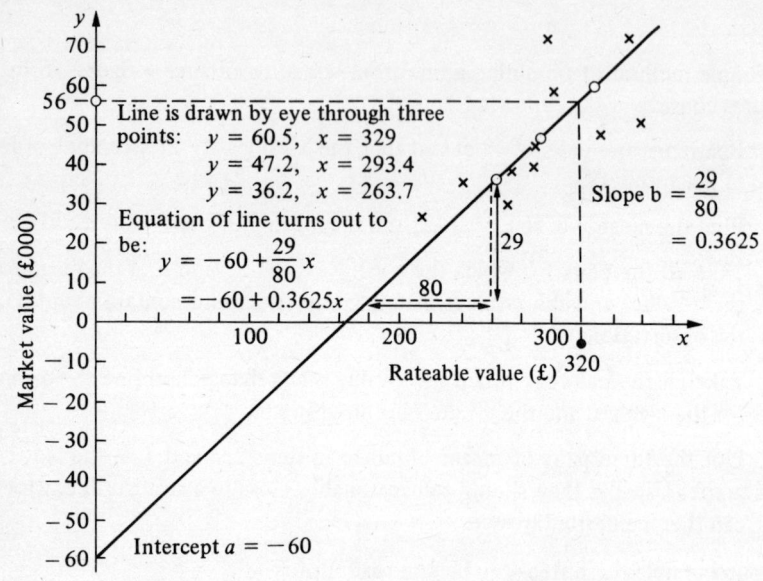

Figure 19.2 *Three-point method regression line*

On this particular scattergraph, the scale has been reduced so that more of the two axes can be shown. This is to illustrate how we can find the equation for the line we have drawn on the graph. The equation of a straight line is:

$y = a + bx$ where y is vertical axis variable (market value)
x is horizontal axis variable (rateable value)
a is the intercept, where the line cuts the y-axis
b is the slope of the line.

In this particular graph the slope turns out to be about three-eighths and the intercept is -60, giving an equation:

$$y = -60 + 0.3625x$$
$$\text{Market value} = -60 + 0.3625 \times \text{rateable value}$$

For prediction we may then use either the line on the graph or the equation. For example, to predict the market value of a house with a rateable value of £320: the graphical method is shown in Fig. 19.2; using the equation:

$$y = -60 + 0.3625x$$
$$= -60 + 0.3625 \times 320 = 56$$

Both methods predict a market value of £56 000.

Note that it is dangerous to make predictions of y-values based on x-values which are well outside the range of those given to us, as the straight-line relationship may be very inaccurate outside the range given. In this example the predicted market value of a detached house with a rateable value of £120 is:
$$y = -60 + 0.3625 \times 120 = -16.5$$
This is obviously a silly result, but the figure of 120 is well outside the range of values given us, the smallest being £215. (If anyone finds a detached house this cheap in Fulwood will they please let me know!)

19.3 Least-squares regression

The three-point method, like fitting 'by eye', relies on visual judgement to position the line, though it simplifies the problem to position it through or near the three points obtained as averages. The other common method of regression has a different approach, however: formulae are given from which we may calculate intercept a and slope b for the regression line $y = a + bx$.

To understand the basis of this method, we note that any fitted regression line gives us predicted values of y (market value) based on the actual values of x (rateable value) for the sample (of eleven houses) which we have obtained. Each of these predictions differs from the actual y-value (actual market value) by a particular amount or 'error'. Fig. 19.3, for example, shows the error made by a particular regression line in predicting the market value of the fourth house.

On what basis should we choose the particular line; in other words, how do we decide values for a and b? Obviously, we want a line which makes the total of these 'errors', in some sense, as small as possible. Then we can hope that it will continue to make the smallest possible errors of prediction in future examples. We need, then, to find formulae for a and b which will make the sum of the errors as small as possible. In fact, for various reasons, we choose formulae for a and b which make the sum of the errors *squared* as small as possible. This is why the method is called *least-squares regression*, and it is the most common method now in use.

Using the mathematical technique of calculus, it can be shown that the values of a and b which minimize the sum of the squares of the errors are:

$$b = \frac{n\Sigma xy - (\Sigma x)(\Sigma y)}{n\Sigma x^2 - (\Sigma x)^2} \quad \text{where } n \text{ is sample size}$$

$$a = \bar{y} - b\bar{x} \quad \text{where } \bar{x} \text{ and } \bar{y} \text{ are the means of the sample values}$$

290–Applied business statistics

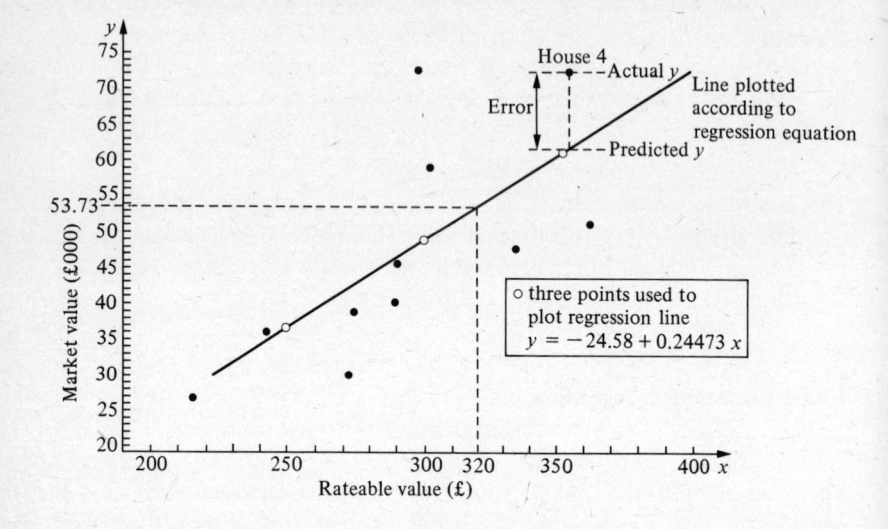

Figure 19.3 *Least-squares regression line*

Three points used to plot regression line:
$y = -24.58 + 0.24473 x$

Table 19.2 *Calculation for least-squares regression*

Market price (y)	Rateable value (x)	x^2	xy
51.0	360	129 600	18 360
59.0	302	91 204	17 818
72.5	298	88 804	21 605
72.5	352	123 904	25 520
45.5	290	84 100	13 195
47.5	333	110 889	15 817.5
38.8	274	75 076	10 631.2
39.5	289	83 521	11 415.5
36.0	242	58 564	8 712
30.0	272	73 984	8 160
27.0	215	46 225	5 805
519.3	3227	965 871	157 039.2

To find these values, it is simplest to set out the calculations in the form of a table like Table 19.2.

$$\text{Slope } b = \frac{n\Sigma xy - (\Sigma x)(\Sigma y)}{n\Sigma x^2 - (\Sigma x)^2}$$

$$= \frac{11 \times 157\,039.2 - 519.3 \times 3227}{11 \times 965\,871 - 3227^2}$$

$$= \frac{51\,650.1}{211\,052} = 0.244\,73$$

Intercept $a = \bar{y} - b\bar{x}$

$$= \frac{519.3}{11} - 0.244\,73 \times \frac{3227}{11}$$

$$= 47.21 - 71.79 = -24.58$$

Thus the regression line $y = a + bx$
$$y = -24.58 + 0.244\,73x$$

To plot this line on our scattergraph we need to find two points on it. Taking three points allows a check on correct plotting. So we take three values for x within the range of x values, and find the corresponding y values on our line:

When $x = 250$ $y = -24.58 + 0.244\,73 \times 250 = 36.60$
When $x = 300$ $y = -24.58 + 0.244\,73 \times 300 = 48.84$
When $x = 350$ $y = -24.58 + 0.244\,73 \times 350 = 61.08$

These three points are plotted on our scattergraph and the regression line is drawn through them. This has been done for the line in Fig. 19.3.

Predictions of market value for any house in the area can now be made on the basis of its rateable value, using either the graph or the regression equation itself. For example, suppose that we hear of a detached house in the area for which the rateable value is £320. We would then estimate its market value as:

$$y = -24.58 + 0.244\,73 \times 320 = 53.73 \text{ (£000)}$$

This is also the value we would read off from the regression line on our graph.

The direction of prediction

The regression line which we have obtained ($y = -24.58 + 0.244\,73x$) makes a particular error in predicting the market value of a given house from its rateable value, but it makes a different error in predicting the rateable value from the market value if we choose to use the line this way around. This is shown for house 4 in Fig. 19.4. It might seem reasonable to expect that the same line would minimize the sum of squared errors in predicting y from x and minimize the sum of squared errors in predicting x from y. But it does not: the line which minimizes squared errors in predicting x from y is also shown in Fig. 19.4. The closer the sample data are to a straight line, the nearer are these two regression lines; they are *identical* only if the sample data lie exactly on a straight line. The formulae and methods given in this book

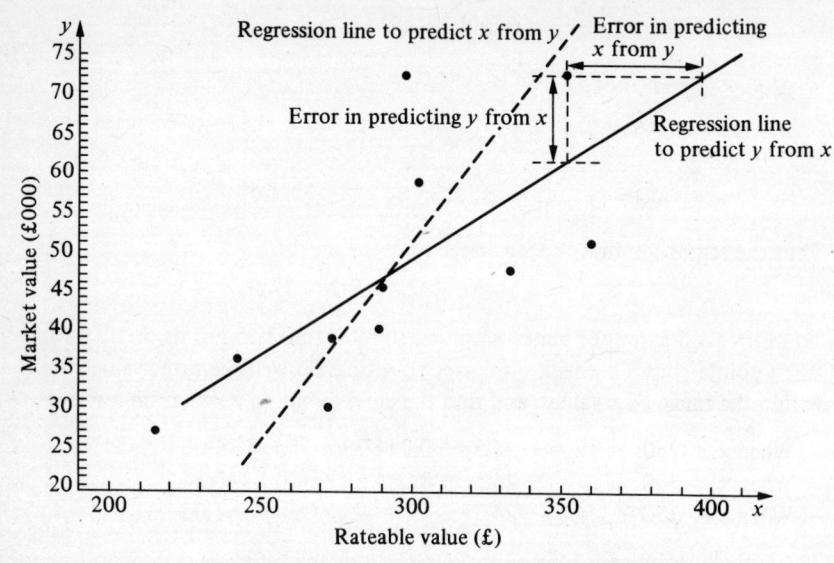

Figure 19.4 *Direction of regression*

assume that we want to obtain what is called *the regression of y on x* - the line to use for predicting y-values from x-values. To obtain the *regression line of x on y* - the line for predicting x-values from y-values - the formulae and procedures should simply be used with x written for y and y written for x. It would, however, be simpler to define as y the variable which is to be predicted; then all the formulae and procedures can be used as they stand, without alteration. Variable y is then sometimes called the *dependent*, and x the *independent* variable.

Handling large numbers

Nowadays, anyone who has many calculations to do will obtain an electronic calculator or computer with a program to do them. On the other hand, access to *any* calculator will make the method shown above adequate to do most occasional regression calculations. Sometimes, however, when we start squaring and multiplying numbers together, we find that the calculator overflows. What do we do then?

In some cases, not only are the numbers themselves large, but there are large differences between them. The obvious approach then is to round off the numbers and then work in, say, thousands or millions, but this does not work in instances where the numbers are large and differences between them small. In such cases rounding the numbers would make them all virtually the same, so we need some other alternative.

The difficulties caused by large numbers will arise in the calculation of slope b. But suppose that we choose any constant, say 25, and subtract it from all the y values to form Y values, and any other constant, say 200, to subtract from all the x values to form X values. The effect of this is illustrated in Fig. 19.5. Subtracting 25 from all the y values moves the x axis up, and subtracting 200 from all the x values moves the y axis across. In both cases

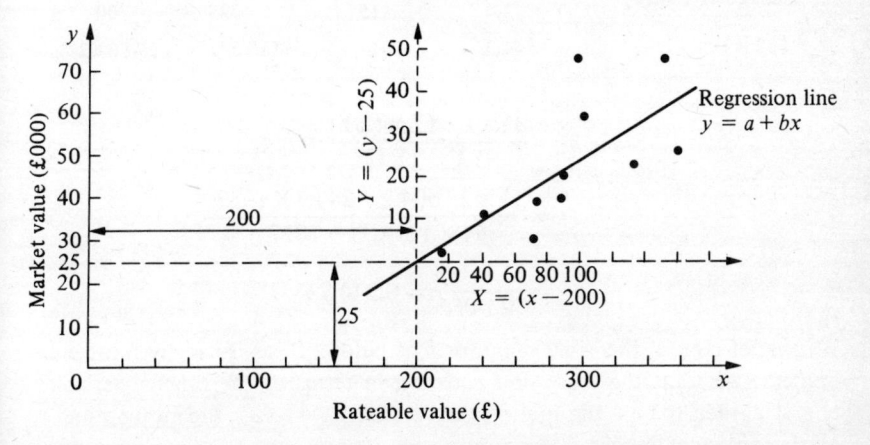

Figure 19.5 *Use of X and Y values*

the *slope* of the regression line is unaltered, for the points remain in the same pattern. This therefore enables us to calculate the slope using the revised X and Y values, which are smaller than the x and y values; the slope, of course, will be unaltered, whatever constants we use. It is convenient to reduce the numbers by subtracting from the x values a round number just below the smallest of them, and from the y values a round number just below their smallest (this avoids cluttering the table up with minus signs); thus here I chose 25 and 200.

In the present example it would probably not be worth while in practice to 'transform' the data in this way by subtracting constants, for the original numbers are reasonably small. However, for demonstration purposes Table 19.3 shows the calculation of the slope b from the X and Y values obtained using the constants 25 and 200.

Table 19.3 *Calculation of least-squares regression for large numbers*

Market price (y)	Rateable value (x)	Y (y−25)	X (x−200)	X^2	XY
51.0	360	26.0	160	25 600	4 160
59.0	302	34.0	102	10 404	3 468
72.5	298	47.5	98	9 604	4 655
72.5	352	47.5	152	23 104	7 220
45.5	290	20.5	90	8 100	1 845
47.5	333	22.5	133	17 689	2 992.5
38.8	274	13.8	74	5 476	1 021.2
39.5	289	14.5	89	7 921	1 290.5
36.0	242	11.0	42	1 764	462
30.0	272	5.0	72	5 184	360
27.0	215	2.0	15	225	30
519.3	3227	244.3	1027	115 071	27 504.2

$$\text{Slope } b = \frac{n\Sigma XY - (\Sigma X)(\Sigma Y)}{n\Sigma X^2 - (\Sigma X)^2}$$

$$= \frac{11 \times 27\,504.2 - 244.3 \times 1027}{11 \times 115\,071 - 1027^2}$$

$$= \frac{51\,650.1}{211\,052} = 0.244\,73$$

This is, of course, the same value for b as before. If, as is usually most convenient, we want the regression equation in terms of the original variables x and y, then to find the intercept a we shall have to use the means $\bar{x}$ and $\bar{y}$ of these original figures:

$$\text{Intercept: } a = \bar{y} - b\bar{x}$$

To find $\bar{x}$ we may use either $\dfrac{\Sigma x}{n}$ or $\left[\dfrac{\Sigma X}{n} + \text{constant}\right]$

$$\bar{x} = \frac{3227}{11} \quad \text{or} \quad \left[\frac{1027}{11} + 200\right] = 293.36$$

To find $\bar{y}$ we may similarly use either $\dfrac{\Sigma y}{n}$ or $\left[\dfrac{\Sigma Y}{n} + \text{constant}\right]$

$$\bar{y} = \frac{519.3}{11} \quad \text{or} \quad \left[\frac{244.3}{11} + 25\right] = 47.21$$

Hence $a = 47.21 - 0.24473 \times 293.36 = -24.58$ (as before).

19.4 Summary

Regression

This is a method of fitting a line (usually a straight line) to a set of pairs of values for two variables thought to be related.

Three-point regression*

This involves fitting a line 'by eye' through three points calculated from the x- and y-values. The first point is the overall mean $\bar{x}$ for the x dimension and the overall mean $\bar{y}$ for the y dimension, that is $(\bar{x}, \bar{y})$. To get the coordinates of the second point we take all the pairs with their x-value less than $\bar{x}$, and find the mean of these x-values and the mean of their corresponding y-values. To get the third point we take all the pairs with their x-value greater than $\bar{x}$, and find the means of the x- and y-values for these pairs.

Least-squares regression

This is a method of regression which gives the straight line that makes the sum of the squared errors in predicting the y-values as small as possible. This line is $y = a + bx$, where:

$$b = \frac{n\Sigma xy - \Sigma x \Sigma y}{n\Sigma x^2 - (\Sigma x)^2}$$

$$a = \bar{y} + b\bar{x}$$

This is a regression of y on x, and is different from the line we would use to predict x from y (the regression of x on y).

Recall and exercise questions

1. In what sense of 'best' is a least-squares regression line the 'line of best fit'?
2. Briefly explain the difference between the 'regression of y on x' and the 'regression of x on y'.

3. A truck manufacturer runs a test on the noise rating of an exhaust system, measuring engine speed (rev/min) against noise (decibels):

Engine speed:	1000	1200	1400	1600	1800	2000
Noise level:	74	75	81	83	84.5	84.5

Find a suitable regression line to predict noise level from engine speed. Use this line (or the associated equation) to predict the noise at:

(a) 1500 rev/min
(b) 700 rev/min

How accurate would you expect these predictions to be?

4. A company's sales of sproggits have risen over recent years as follows:

Sales (£0000)	105	120	130	143	150	164	172
Year	1979	1980	1981	1982	1983	1984	1985

Plot these figures on a graph. Find a regression line using sales as y-variable and year as x-variable, and plot it on your graph.

Use your line to forecast sales in 1986. Comment briefly on your assumptions.

5. A company has been using a test rig to see the effects of brake applications on a set of brake linings. It has found average wear on the top shoe lining (measured at the point of maximum wear) to be as follows:

Wear (inches)	0.056	0.060	0.069	0.122	0.184	0.256	0.280
Test cycles (000 applications)	1	2	3	4	5	6	7

Plot these figures on a scatter diagram.

Find a suitable regression line to forecast how many test cycles would occur before the lining depth of 0.4 inches would be gone and the drum in contact with the rivets.

20
Correlation and Causality

20.1 The correlation coefficient

In the regression methods we looked at in the last chapter we fitted a straight line to some data. Often it will be useful to have a way to measure how accurately a straight-line model really fits the data, or in other words how near the data points are to being in a straight line.

One common measure of this is the *correlation coefficient*, or more fully the *product moment correlation coefficient*, which is usually denoted by the letter r. Some values for r are illustrated in Fig. 20.1. The value of r must always be between -1 and $+1$. If it is $+1$ then the data are perfectly positively correlated: that is, they lie on an exactly straight line with positive slope. If $r = -1$ then the data are perfectly negatively correlated: that is, they lie on an exactly straight line with negative slope. In either case, obviously, prediction of y values from x values would be completely accurate. The nearer r is to $+1$, the nearer the data are to being perfectly positively correlated; the nearer r is to -1, the nearer to being perfectly negatively correlated. A value of r near zero (such as $r = 0.03$) indicates that there is very little correlation between the variables: the value of x gives us very little help in estimating the value of y. The r values of -0.68 and -0.95 both indicate *negative* correlation, but the data with $r = -0.95$ are much closer to a straight line. An r value of $+0.91$ indicates quite a good fit to a *positively* sloped straight line.

Calculation

The formula for the correlation coefficient is:

$$\text{Correlation coefficient } r = \frac{n\Sigma xy - (\Sigma x) \times (\Sigma y)}{\sqrt{[n\Sigma x^2 - (\Sigma x)^2][n\Sigma y^2 - (\Sigma y)^2]}}$$

298–Applied business statistics

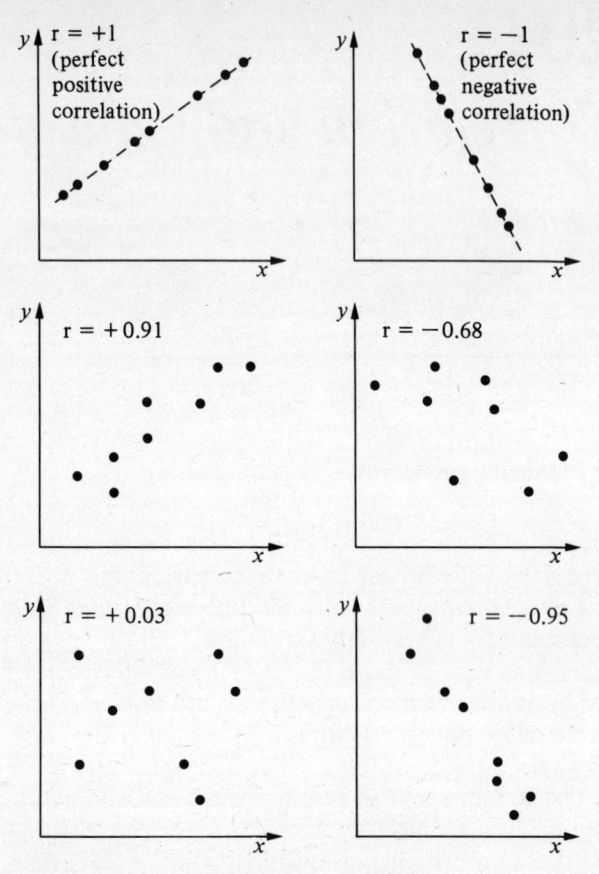

Figure 20.1 *Scattergraphs and correlation coefficients*

Table 20.1 *Calculation of correlation coefficient*

Market price (y)	Rateable value (x)	y^2	x^2	xy
51.0	360	2 601.00	129 600	18 360
59.0	302	3 481.00	91 204	17 818
72.5	298	5 256.25	88 804	21 605
72.5	352	5 256.25	123 904	25 520
45.5	290	2 070.25	84 100	13 195
47.5	333	2 256.25	110.889	15 817.5
38.8	274	1 505.44	75 076	10 631.2
39.5	289	1 560.25	83 521	11 415.5
36.0	242	1 296	58 564	8 712
30.0	272	900	73 984	8 160
27.0	215	729	46 225	5 805
519.3	3227	26 911.69	965 871	157 039.2

The easiest way to find the various summations required is to set out a table similar to that used for regression but with the addition of a y^2 column. This is shown in Table 20.1.

Correlation coefficient

$$r = \frac{n\Sigma xy - (\Sigma x)(\Sigma y)}{\sqrt{[n\Sigma x^2 - (\Sigma x)^2][n\Sigma y^2 - (\Sigma y)^2]}}$$

$$= \frac{11 \times 157\,039.2 - 519.3 \times 3227}{\sqrt{[11 \times 965\,871 - 3227^2][11 \times 26\,911.69 - 519.3^2]}}$$

$$= 0.693$$

Handling large numbers

In section 19.3 we saw that the numbers involved in the calculation of the regression slope b would sometimes become too large for a calculator to deal with. If the differences between the numbers were great, then they could be rounded to work in hundreds or thousands. But if the numbers themselves were large but differences between them were small, this would not do. In such a case we saw that the subtraction of a constant (say 200) from all the x-values and another constant (say 25) from all the y-values merely moved the two axes without altering the slope of the regression line.

All these points apply also to the calculation of the correlation coefficient r. Large numbers with great differences between them may be rounded. For large numbers with small differences between them we may subtract any constant from the x-values and any other constant from the y-values. The effect of this is merely to move the axes without altering the pattern of the points relative to each other (see also Fig. 19.5). Since the correlation coefficient measures only how close this pattern is to being on a straight line, it is unaltered by moving the axes. Thus we may choose constants (like 200 and 25) conveniently below the smallest x- and y-values respectively, transform the data to X-values and Y-values by subtracting 200 from the x-values and 25 from the y-values, and find r from these much smaller X and Y-values. This is shown in Table 20.2.

Correlation coefficient

$$r = \frac{n\Sigma XY - (\Sigma X)(\Sigma Y)}{\sqrt{[n\Sigma X^2 - (\Sigma X)^2][n\Sigma Y^2 - (\Sigma Y)^2]}}$$

$$= \frac{11 \times 27\,504.2 - 244.3 \times 1027}{\sqrt{[11 \times 115\,071 - 1027^2][11 \times 7821.69 - 244.3^2]}}$$

$$= 0.693$$

This is the same value for r as we obtained using the x- and y-values above.

Table 20.2 Calculation of correlation coefficient for large numbers

Market price (y)	Rateable value (x)	(y−25) Y	(x−200) X	Y^2	X^2	XY
51.0	360	26.0	160	676	25 600	4 160
59.0	302	34.0	102	1156	10 404	3 468
72.5	298	47.5	98	2256.25	9 604	4 655
72.5	352	47.5	152	2256.25	23 104	7 220
45.5	290	20.5	90	420.25	8 100	1 845
47.5	333	22.5	133	506.25	17 689	2 992.5
38.8	274	13.8	74	190.44	5 476	1 021.2
39.5	289	14.5	89	210.25	7 921	1 290.5
36.0	242	11.0	42	121	1 764	462
30.0	272	5.0	72	25	5 184	360
27.0	215	2.0	15	4	225	30
		244.3	1027	7821.69	115 071	27 504.2

Interpretation

The value for correlation coefficient r, 0.693, is a positive value and so indicates an upwards sloping line. But just how close to a straight line does a value of 0.693 show the points to be? One way of answering this question is to speak in terms of *explained* and *unexplained variation*. This is illustrated in Fig. 20.2.

If we could *not* use rateable value to estimate market value, we would have to estimate any market value simply as the average for our sample (£47 200). The actual sample values vary about this average, and a reasonable proportion of this variation should be explained by our regression line, if it is any good. For example, the market value of house number 4 is considerably above the average, but part of this difference is 'explained by' the regression line based on rateable value. We could, therefore, conceive of a measure of the *overall* percentage of the variation about the mean figure which is 'explained by' the regression.

In practice, it proves better to deal in terms of errors and distances *squared*; the least-squares regression method gives the line which minimizes the sum of the errors squared (another way of saying the unexplained variations squared). In this case there is a direct measure of the overall percentage of variation 'explained by' the regression line: it is called the *coefficient of determination*. This is the *square* of the correlation coefficient:

Coefficient of determination $r^2 = 0.693^2 = 0.48$ (48 per cent)

In our example, then, nearly 50 per cent of the variation in market value

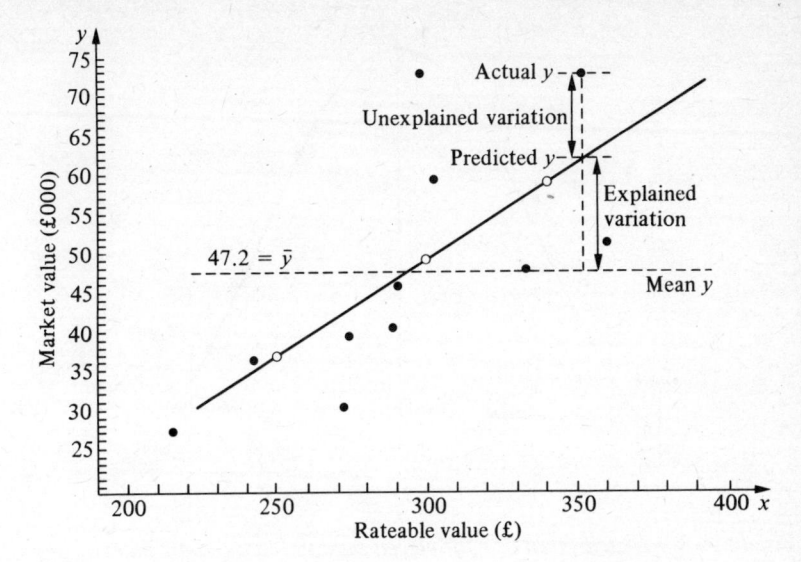

Figure 20.2 *Explained and unexplained variation*

about the mean £47 200 can be 'explained by' a regression line based on rateable value.

20.2 Significance tests on r

The coefficient of determination tells us the percentage of variation explained by the regression. But we might ask a slightly different question: what is the likelihood that the apparent linearity obtained could have arisen by chance, if there is really no repeatable tendency for the points to fall in a straight line? Put another way, suppose that really there were no connection at all between rateable value and market value; how likely would we be to get a correlation coefficient this high simply by accident?

If there were really no connection, then it would be as though we had randomly paired off market prices with rateable values. It is possible to derive a distribution for the correlation coefficient r for eleven randomly paired-off pairs of figures; this distribution looks like Fig. 20.3. The value of r must, of course, lie between -1 and $+1$, but the shape of the distribution looks quite similar to the normal distribution. In such a distribution, 95 per cent of the time a value of r for eleven unconnected pairs would lie between -0.602 and $+0.602$. The value we obtained is 0.693, which is outside this range and

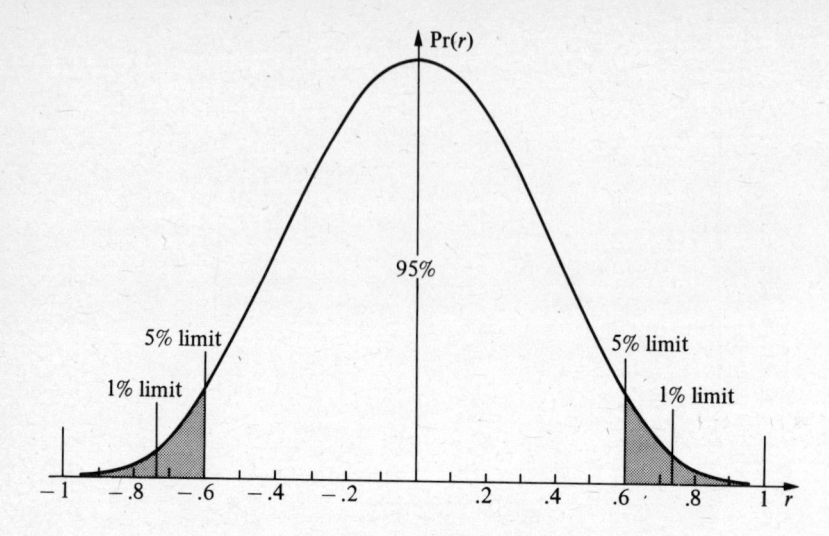

Figure 20.3 *Distribution of r for eleven randomly paired-off pairs of figures*

therefore unlikely to have arisen by chance. It seems more likely that there is a real underlying association between the rateable value and the market price, which would be repeated in further examples. This can be set up formally as a significance test:

Null hypothesis H_0 — There is no repeatable underlying association. In other words, the correlation coefficient for the whole population from which the sample is drawn is zero.

Alternative hypothesis H_1 — There is a repeatable underlying association which is either positive or negative in slope, and the correlation coefficient for the whole population is not zero.

The test result we have obtained is a correlation coefficient $r = 0.693$, and we can look this up in Table 6 (Appendix A), which gives the critical values of r for two-tailed tests (thus accepting positive or negative slopes) at the 5, 1 and 0.1 per cent levels of significance. In this case the number of pairs is eleven, so the critical values are:

5 per cent significance level: 0.6021

1 per cent significance level: 0.7348

0.1 per cent significance level: 0.8470

So our correlation coefficient r is significant at the 5 per cent, but not at the 1 per cent level. There is some, but not strong, evidence to reject H_0 and conclude that the association did not arise by chance but is repeatable.

20.3 Linearity and assumptions

There are a number of points which should be noted. First, however good our linear model may be within the range of sample x-values, it is dangerous to assume that it extends very far outside that range. Other factors may then begin to operate.

Second, both linear regression and correlation fit a *straight line* or linear model to the pairs of values. But consider Figs 20.4 and 20.5. Fig. 20.4 shows fuel consumption in a test of a particular car at constant speeds over a long stretch. The correlation coefficient $r = 0.188$, which means that a straight line would 'explain' less than 4 per cent of the variation. Yet there obviously *is* a relationship between speed and consumption - it is just that it is not linear.

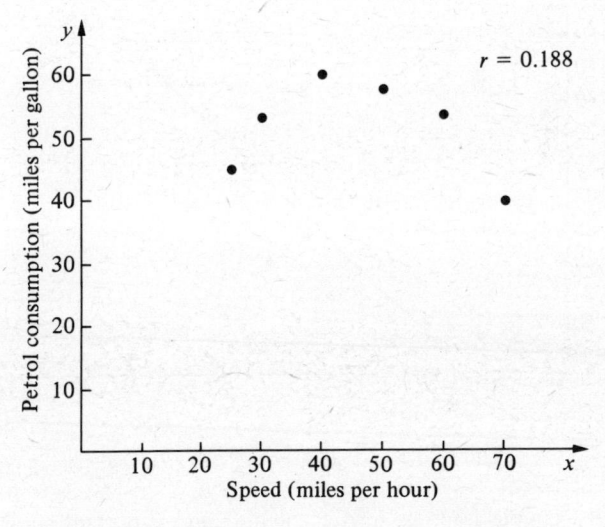

Figure 20.4 *Petrol consumption at constant speed*

Fig. 20.5 shows what is perhaps a more potentially dangerous situation. Here the correlation between the year and the index is high ($r = 0.91$), but a straight line is clearly not very appropriate because the data curves upwards.

Moreover, if we were using our regression line to *predict* future values of the index it would be even more misleading: as the index continues to rise the regression line is likely to be even further out.

One way to deal with this situation is to 'transform' the data using the logarithm or square of the value instead of the value itself. For example:

(a) Log (RPI) = $a + b \times$ year ($r = 0.960$)
or (b) RPI = $a + b \times$ year^2

There are various problems and advantages in this technique, but they are beyond our present scope and most readers will not wish to go into it; something rather similar was covered in section 7.4. A key lesson from these examples is that it is always useful to plot a scatter diagram; this helps us to pick up visually any non-linear pattern which is there.

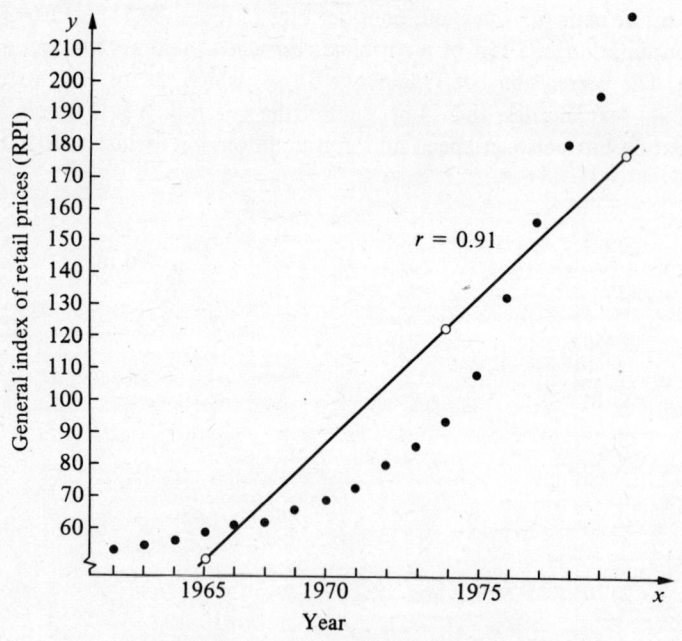

Figure 20.5 *Regression of RPI on year*

The RPI example leads us into a common and important application of regression and correlation, the area of *time series*. The simplest application is where the year (or month or quarter) itself is the x variable, as here. If the past data is near linear, and if there is nothing to suggest any change in trend, then a forecast might be made using regression. But it should be remembered, first, that if any new factors are entering (a general recession, for instance) then a mechanical method like regression may be unsuitable, and, second, it is likely to be a good forecasting tool only for the immediate and not the far future.

20.4 Correlation and causation

I have so far been dealing with correlation and regression as they concern the potential prediction of one variable (y) from another (x). Sometimes, however, we wish to study correlation as an indicator not merely of the possibility of prediction, but of causation.

Let us take as an example smoking and cancer. We might have two kinds of figures showing a correlation:

1. Historical or time-series figures, showing that as the numbers of people smoking rose over time, so did the incidence of lung cancer.
2. Data linking individuals who smoke with the incidence of lung cancer, that is, showing that the incidence of lung cancer is higher in contemporary groups smoking heavily, than in those who smoke less heavily or not at all.

In either case the correlation may be significant: that is, a significance test may effectively rule out any suggestion that the apparent association arose in our sample simply by chance. But would this prove that smoking *caused* lung cancer?

The first situation concerns the correlation of two time-series figures (not with time but with each other) where both are rising over time. Though such series may alert us to possibilities, and may allow prediction, they are unreliable as indicators of causation. They are both likely to be highly correlated with other series also rising over time – the price of potatoes, the total built-up area, or the average time spent in full-time education. Such correlations are very often 'spurious', showing no causal connection, direct or indirect.

The second situation deserves to be treated more seriously, but still does not prove that there is a *direct* causal link. It could be, for example, that more nervous people tend to smoke and that these, *irrespective* of whether or not they smoke, are more likely, because of their nervous disposition, to develop cancer. It might be that urban dwellers tend more often to smoke than those in less urban areas, and that their urban lifestyle, rather than their smoking, is the cause of cancer. As long as the number smoked per day is left to the choice of the smokers themselves, one could always think of some factor which might cause them to choose to smoke; this factor, rather than the smoking itself, could be the cause of the cancer. All that we can do is think of the more obvious possible 'third factors' (nervousness, urban/rural dwelling, and so on) and see if these are correlated with smoking, cancer, or both. It is always conceivable that we might forget or not think of some factor not so far tried, but this gets rather less plausible as the most obvious factors are eliminated. However, the problem would only really be overcome completely if we could actually randomly *assign* people and say how many they must smoke – and that sort of experiment is not allowed in a democracy unless the subjects are dogs.

One last point. Even if the causal link in such instances is indirect through some third common cause rather than direct, we can still validly use regression to *predict y* from *x*.

20.5 Rank correlation*

There are circumstances in which it is useful to find the correlation between two sets of variables where the values are *ranks* rather than actual numbers. For example, if two important customers both rank a set of eight prototypes in order of preference, each prototype would be given a rank order (1 to 8) by each customer:

Prototype	A	B	C	D	E	F	G	H
Customer Whizz Ltd	1	4	6	5	8	7	3	2
Customer Phozz Ltd	2	5	4	7	6	8	1	3

These two sets of ranks can be used in the formula for the correlation coefficient as though they were ordinary numbers. However, there is an alternative formula for this *rank correlation* which gives the same answer:

$$R = 1 - \frac{6 \times \Sigma(\text{difference between ranks})^2}{n(n^2-1)}$$

This may save a little calculation time (not much if using a calculator), but it does require familiarity with yet another formula! The rank correlation for the above set of eight figures works out as 0.762 using either formula – a fairly high positive correlation.

In some examples the variable *can* only be a ranking, not a number on a numerical scale. In this case the rank correlation has to be used, but we may prefer to use it even in some instances where we *are* dealing with numbers on a numerical scale. Data such as that in Fig. 20.5, however extreme its upwards curve, will have a high rank correlation: for Fig. 20.5 it will be + 1. To find the rank correlation in such instances we simply rank both sets of variable (for Fig. 20.5 these are RPI and time) in order of magnitude, and use the ranks to find the correlation coefficient.

20.6 Summary

Correlation coefficient

The product moment correlation coefficient measures the closeness of a set of pairs of values plotted on a scattergraph to being on a straight line.

$$\text{Correlation coefficient} \quad r = \frac{\Sigma xy - \Sigma x \Sigma y}{\sqrt{[n\Sigma x^2 - (\Sigma x)^2][n\Sigma y^2 - (\Sigma y)^2]}}$$

It can be used in a significance test to see if the association is part of a wider pattern.

Coefficient of determination

This is the correlation coefficient squared, and it tells us the percentage of the variation in the y-values which may be explained by the x-values.

Rank correlation*

This may be obtained by ranking the x-values and the y-values in order of magnitude, and using the ranks in the correlation formula instead of the numerical values.

Recall and exercise questions

1. What does the correlation coefficient measure?
2. What is the coefficient of determination and what does it measure?
3. What is 'spurious correlation'?

4. Table 20.3 shows some basic statistics on the car industry in the UK.

Table 20.3 UK car industry data

Year	UK car production (thousands)	UK petroleum consumption (million tonnes)	Import penetration (%)
1971	1742	89.0	19.3
1972	1921	95.4	23.5
1973	1747	96.6	27.4
1974	1534	89.7	27.9
1975	1268	80.3	33.2
1976	1333	78.9	37.9
1977	1328	80.3	45.4
1978	1223	82.0	49.3

(a) Plot the UK car production figures against those for petroleum consumption on a scatter diagram.

(b) Find the correlation coefficient between car production and petrol consumption, and test its significance.

(c) Repeat parts (a) and (b) using the figures for car production against those for import penetration.

(d) Briefly explain why care must be taken in interpreting such results.

5. Table 20.4 shows figures for output per head (1975 = 100) and for total working days lost due to industrial action in the mining and quarrying industry.

Table 20.4 UK mining and quarrying industry

Period		Output per head	Days lost
1978	1st quarter	91.9	67
	2nd quarter	92.9	70
	3rd quarter	95.8	32
	4th quarter	98.0	34
1979	1st quarter	94.0	15
	2nd quarter	96.3	45
	3rd quarter	99.1	37
	4th quarter	98.8	30
1980	1st quarter	99.9	60
	2nd quarter	97.4	40
	3rd quarter	96.8	25
	4th quarter	98.4	32

Find the correlation coefficient between output and days lost, and test it to see whether or not it is statistically significant.

It is frequently asserted that industrial disputes are a major source of loss of output and industrial decline; do your results bear this out?

6. Table 20.5 shows figures for the UK Minimum Lending Rate when this was in operation, over a period during which it fell and then rose again. Figures are also shown for the consumption expenditure as a percentage of total disposable income in the period:

Table 20.5 UK Minimum Lending Rate and consumption expenditure

Period	Consumption expenditure (% disposable income)	MLR
1977 1st quarter	85.6	12.00
2nd quarter	87.3	8.5
3rd quarter	87.1	7.25
4th quarter	84.4	6.00
1978 1st quarter	86.8	6.50
2nd quarter	84.7	8.75
3rd quarter	84.4	10.00
4th quarter	83.5	11.5
1979 1st quarter	83.9	13.0
2nd quarter	86.5	12.5

Do these figures support a view that if interest rates rise the proportion of disposable income spent on consumption should fall?

7. The following are figures (from *Social Trends*) for UK car ownership (cars per 1000 population) and unemployment (%):

Year	1961	1966	1971	1976	1977	1978
Cars	116	181	225	253	260	262
% Unemployed	1.5	1.5	3.5	5.7	6.2	6.1

(a) Find the correlation coefficient between car ownership and unemployment and test its statistical significance.

(b) Does this show that unemployed people tend to buy more cars, or that car ownership causes unemployment?

8. A company's two principal fashion buyers see prototype designs for the new season, and both rank the styles in order of preference. Miss R. E. Bate ranks them: D, H, J, B, F, C, I, E, G, A. Mrs I. O. Yuw ranks the styles: H, D, J, F, E, B, C, I, A, G. How much correspondence would you judge there to be between these two sets of rankings?

9. The figures in Table 20.6 show the total road casualties in Great Britain (in hundreds) and the consumption (in tens of thousands of hectolitres) of alcoholic drinks, excluding spirits, from July 1979 to June 1980:

Table 20.6 Road casualties and alcohol consumption in Great Britain

Year	Month	Road casualties (× 100)	Alcohol consumption (× 10 000 hectolitres)
1979	July	287	648
	August	302	578
	September	299	639
	October	308	541
	November	316	620
	December	322	606
1980	January	249	489
	February	226	496
	March	259	496
	April	241	592
	May	276	443
	June	274	588

(a) We suspect that there is a positive correlation between road casualties and alcohol consumption. Use a significance test on the correlation coefficient r to examine this suspicion, and explain the meaning and implication of your result.

(b) Discuss the assumption that a significant positive correlation would prove that alcohol causes road accidents.

Problems

1. Obtain figures (for example from *Social Trends*) for annual numbers of monochrome and colour TV licences and for cinema admissions. Correlate the cinema admissions with:

    (a) monochrome TV licences

    (b) colour TV licences.

    Do the results signify anything?

2. Table 20.7 relates unemployment to serious criminal offences, by year and by region. It has been suggested that unemployment leads to serious crime, and the correlation between annual unemployment figures and

serious offences has been cited to prove this. Others have denied that the connection is very strong, pointing to the lack of regional correlation.

Write an article analysing the figures in Table 20.7 (including correlation coefficients for both sets of figures), discussing whether or not they support a connection between crime and unemployment.

Table 20.7 *Unemployment and serious crime – by region and by year*

	Regional comparison			Time comparison	
Region	Unemployment (%)	Serious offences per 1000 aged 15–64	Year	UK unemployment (%)	UK serious offences
North	8.6	2.96	1961	1.5	990
Yorks. and Humberside	5.7	1.69	1966	1.5	1479
East Midlands	4.7	2.01	1971	3.5	1878
East Anglia	4.5	3.04	1976	5.7	2441
South East	3.7	0.53	1977	6.2	2984
South West	5.7	1.33	1978	6.1	2884
West Midlands	5.5	1.43	1979	5.8	2875
North West	7.1	1.47			

3. Table 20.8 shows estimates of average rent and rates (£ per square foot) for various urban areas.

Table 20.8 *Average rent and rates for urban areas*

Area	Rates	Rent
Birmingham	1.68	3.81
Bristol	1.46	4.19
Cardiff	1.57	4.87
City of London	13.37	22.85
Edinburgh	3.18	6.47
Glasgow	1.95	6.48
Hull	1.07	2.66
Leeds	1.71	6.49
Liverpool	1.22	5.33
Mayfair	6.46	15.23
Newcastle	2.13	3.81
Southampton	1.58	5.31

(a) Draw a scatter diagram of rates against rent.
(b) Calculate the correlation and rank correlation coefficients.
(c) Discuss whether there is evidence in these figures for a linear relationship generally between rates and rent.

4. The Economist Intelligence Unit (in June 1976) stated that 90 per cent of soft toy sales were for children under five. It would be interesting to know whether soft toy sales have been affected by changing patterns of population.

(a) Obtain from the appropriate *Business Monitor*, *PA494.1*, figures for monetary value of annual soft toy sales.

(b) Use the RPI to convert these sales to values in real terms of 'purchasing power'.

(c) Obtain figures (for example, from *Annual Abstract*) for the UK under-fives population.

(d) Calculate the correlation between expenditure on soft toys in real terms and the UK under-fives population.

21
Time Series

21.1 Time series and basic models

A time series is the result of measuring a value or variable at intervals as time progresses. Examples are monthly sales figures, numbers of absentees in a company, national weekly births, and numbers unemployed.

One frequently finds that in such series two major elements can be distinguished:

(a) a *general trend* which is slowly increasing or decreasing over time;
(b) *seasonal variation* which causes a regular cycle up and down.

Example. Suppose that an ice cream manufacturer supplies a number of colleges with ice cream. His sales may have:

(a) a general upwards trend, due to gradual college expansion; and
(b) a seasonal cycle, for he will obviously sell more in the summer terms than in the spring or autumn terms.

A simple model

What would be the simplest pattern of sales for ice cream we could imagine in a 'statistically idealized world'? It might be one in which the trend was totally static and the seasonal variation was totally regular. For example, we might find that the trend (in £ thousands) was a constant at 11, while the sales (in £ thousands) varied about this trend figure:

> down by 1.7 each spring term;
> up by 2.9 each summer term;
> down by 1.2 each autumn term.

This gives rise to the regular sales pattern shown in Fig. 21.1. This shows the movement of sales over time in our 'idealized world', and also shows the breakdown of the figure for summer 1985 into its component trend (= 11) and seasonal factor (= +2.9).

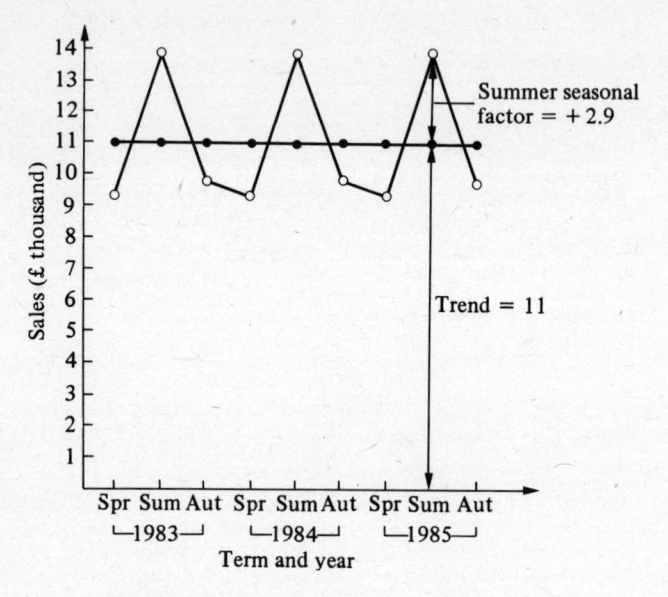

Figure 21.1 *Ice cream sales for simple model*

We should note here that because the seasonal variations or factors are by definition *variations* about the average, over one complete year they should add up to zero:

Spring seasonal factor	+	Summer seasonal factor	+	Autumn seasonal factor	= zero
−1.7	+	+2.9	+	−1.2	= 0

Looked at another way, this means that if we take the average of any complete year's figures (whether calendar year or academic year) the seasonal factors will cancel out (or average to zero) and the result will be the trend or average figure of 11.

Second model

Consider now a second model, still in a fairly idealized world, in which the seasonal factors are still constant but the trend is rising in a straight line. Again:

$$\text{sales} = \text{trend figure} + \text{seasonal factor}$$

But now the trend increases by a constant amount each term. For example, suppose that our ice cream sales are:

    down 1.7 each spring term   (seasonal factor = −1.7)
    up 2.9 each summer term   (seasonal factor = +2.9)
    down 1.2 each autumn term (seasonal factor = −1.2)

But suppose now that the trend began in spring 1983 at 10.0, and rose by a constant amount of 0.2 each term. Table 21.1 shows the overall sales figures with such a trend and seasonal factors. Plotted on a graph this gives us Fig. 21.2.

Table 21.1 *Ice cream sales for second model*

1 Year	2 Season (term)	3 Sales (£ thousands)	4 Trend (£ thousands)	5 Seasonal factor (£ thousands)
1983	Spring	8.3	10.0	−1.7
	Summer	13.1	10.2	+2.9
	Autumn	9.2	10.4	−1.2
1984	Spring	8.9	10.6	−1.7
	Summer	13.7	10.8	+2.9
	Autumn	9.8	11.0	−1.2
1985	Spring	9.5	11.2	−1.7
	Summer	14.3	11.4	+2.9
	Autumn	10.4	11.6	−1.2
1986	Spring	10.1	11.8	−1.7
	Summer	14.9	12.0	+2.9
	Autumn	11.0	12.2	−1.2

21.2 Moving averages

In real life not only are sales not completely regular (of which more presently), but we are usually beginning the other way round. We are not usually given a trend and seasonal factors to use to calculate sales, rather we are given only the sales figures, from which the trend and seasonal factors have to be calculated. How could we have derived the trend and seasonal factors had we been given only the sales for our second simple model?

To see the method, let us look at what happens if we average out any three

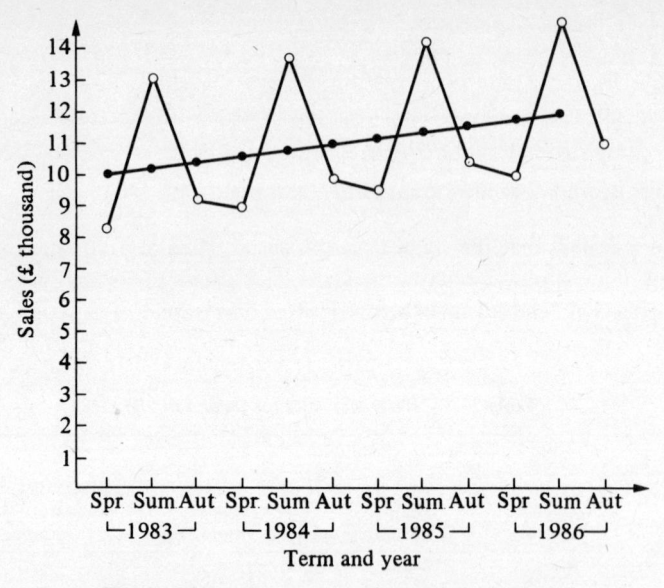

Figure 21.2 *Ice cream sales with rising trend*

successive sales figures. Suppose, for example, we take the average (arithmetic mean) over the academic year 1985-86:

$$\text{Average} = \frac{\text{sales autumn 85} + \text{sales spring 86} + \text{sales summer 86}}{3}$$

But we remember that each sales figure = trend + seasonal factor, so we can substitute:

$$\text{Average} = \frac{\left[\begin{array}{c}\text{trend} \\ \text{autumn 85}\end{array} + \begin{array}{c}\text{autumn} \\ \text{seasonal} \\ \text{factor}\end{array}\right] + \left[\begin{array}{c}\text{trend} \\ \text{spring 86}\end{array} + \begin{array}{c}\text{spring} \\ \text{seasonal} \\ \text{factor}\end{array}\right] + \left[\begin{array}{c}\text{trend} \\ \text{summer 86}\end{array} + \begin{array}{c}\text{summer} \\ \text{seasonal} \\ \text{factor}\end{array}\right]}{3}$$

rearranging:

$$= \frac{\left[\begin{array}{c}\text{trend} \\ \text{autumn 85}\end{array} + \begin{array}{c}\text{trend} \\ \text{spring 86}\end{array} + \begin{array}{c}\text{trend} \\ \text{summer 86}\end{array}\right] + \left[\begin{array}{c}\text{autumn} \\ \text{seasonal} \\ \text{factor}\end{array} + \begin{array}{c}\text{spring} \\ \text{seasonal} \\ \text{factor}\end{array} + \begin{array}{c}\text{summer} \\ \text{seasonal} \\ \text{factor}\end{array}\right]}{3}$$

As we look at this rearranged expression, we may remember that if the seasonal factors are *variations* about the trend, then for one complete year they should add up to zero. Thus we are left with:

$$\text{Average} = \frac{\left[\begin{array}{c}\text{trend} \\ \text{autumn 85}\end{array} + \begin{array}{c}\text{trend} \\ \text{spring 86}\end{array} + \begin{array}{c}\text{trend} \\ \text{summer 86}\end{array}\right]}{3} + \left[\text{zero}\right]$$

But we may go further. Since the trend is assumed to be rising at a constant rate, the trend for spring 1986 is the same amount (= 0.2) above that for autumn 1985 as it is below that for summer 1986. Thus the average of the three trend figures works out to exactly the value of the trend for spring 1986. What we have arrived at, then, is this:

$$\text{Average over three terms of the academic year 1985-86} = \frac{10.4 + 10.1 + 14.9}{3} = \text{trend spring 86} = 11.8$$

If we think about this we should see that the following conclusion can be drawn. If we take the average of *any* three consecutive sales figures, the three seasonal factors will cancel out, and we will be left with the trend figure for the middle period of the three. Therefore we can get all the trend figures except the first and the last by using the method of *moving averages*; that is:

$$\frac{\text{Trend}}{\text{summer 83}} = \text{average of 1st, 2nd and 3rd sales figures}$$

$$\frac{\text{Trend}}{\text{autumn 83}} = \text{average of 2nd, 3rd and 4th sales figures}$$

$$\frac{\text{Trend}}{\text{spring 84}} = \text{average of 3rd, 4th and 5th sales figures}$$

$$\frac{\text{Trend}}{\text{summer 84}} = \text{average of 4th, 5th and 6th sales figures}$$

Thus all except the first and the last trend figures in the fourth column of Table 21.1 could be derived by averaging out the appropriate sets of three consecutive sales figures in the third column. The fifth column could also be found from the previous two columns, since

$$\text{sales} = \text{trend} + \text{seasonal factor}$$

Rearranging this equation:

$$\text{seasonal factor} = \text{sales} - \text{trend}$$

That is, the fifth column entries are those of the third column minus those of the fourth column.

Thus, starting from the sales figures, we can use the moving averages of three consecutive sales figures to derive the trends for all but the first and last terms; and if we subtract these trend figures from the sales in each period we can derive the seasonal factors.

But we must note that the method works only because in adding together sets of *three consecutive* terms' sales we are adding one of each of the three seasonal factors. This enables us to assume that the seasonal factors cancel

318–*Applied business statistics*

out to zero. Were we to use, say, moving averages of two or of four consecutive sales figures with this three-term year the seasonal factors would *not* cancel out, and we would not be able to derive trend figures in this way. We have used what is called a 'three-period moving average' because in this instance there are three seasons (terms) per year.

Non-ideal figures

In real life, of course, sales never follow so regular a pattern as we have assumed. Ice cream sales, for example, will be affected by our unpredictable weather, as well as hundreds of chance occurrences which determine whether individuals do or do not buy ice cream at different times. Moreover, the trend is unlikely to be rising in a perfect straight line. If, therefore, we use the same model and approach, we must expect to find random deviations in the actual figures from whatever model we set up. However, we may find that these deviations are comparatively small if the seasonal factors are fairly regular and the trend does not differ much from a straight line *during any single annual period*.

Let us, then, apply the same approach to some more realistic figures and see what happens. An example of this is given in Table 21.2. The trend figures

Table 21.2   *Calculation of trends and seasonal factors*

1 Year	2 Season (term)	3 Sales (£ thousands)	4a Totals of threes	4b Trend	5 Seasonal factor
1983	Spring Summer Autumn	8.7 12.6 9.5	30.8 30.7	10.27 10.23	+2.33 −0.73
1984	Spring Summer Autumn	8.6 13.5 9.7	31.6 31.8 32.0	10.53 10.60 10.67	−1.93 +2.90 −0.97
1985	Spring Summer Autumn	8.8 13.4 10.2	31.9 32.4 33.0	10.63 10.80 11.0	−1.83 +2.60 −0.80
1986	Spring Summer Autumn	9.4 14.2 10.7	33.8 34.3	11.27 11.43	−1.87 +2.77

in column 4b are derived from the averages of three consecutive sales figures in each case: for example, 10.23 is the average of summer 1983, autumn 1983, and spring 1984. It has been convenient, however, to put in an additional column, 4a, of the totals of the consecutive threes. This is simply a useful intermediate step to finding column 4b, and has no interest in itself.

Column 5, the seasonal factors, is found by subtracting entries in column 4b from those in column 3 as before. But in this instance we see that the model is not totally accurate. The summer seasonal factors, for example, do not all come out to exactly the same figure, as they did before. Rather we have the four slightly different figures: +2.33 (summer 1983), +2.9 (summer 1984), +2.6 (summer 1985) and +2.77 (summer 1986). These, however, are not too different. The most sensible thing to do, therefore, is to average them out:

$$\frac{2.33 + 2.9 + 2.6 + 2.77}{4} = 2.65 \quad \text{best estimate of summer factor}$$

similarly

$$\frac{(-0.73) + (-0.97) + (-0.80)}{3} = -0.83 \quad \text{best estimate of autumn factor}$$

$$\frac{(-1.93) + (-1.83) + (-1.87)}{3} = -1.88 \quad \text{best estimate of spring factor}$$

We can make one more adjustment to these, because they are supposed to add up to zero. In fact if we add them together they come to -0.06. To make them add up to zero, then, we simply add 0.02 to each of them so that they *do* add up to zero. This gives us:

Spring factor	-1.86
Summer factor	$+2.67$
Autumn factor	-0.81

Summary of steps

1. Since there are three seasons (college terms) in each year, we add together successive consecutive threes to give column 4a.
2. Each figure in column 4a is divided by three to give column 4b, the trend.
3. Each trend figure in column 4b is subtracted from sales in column 3 to estimate the seasonal factors.
4. The seasonal factor estimates for each season are averaged to give average seasonal variation.
5. These average seasonal variations for the three seasons are added together. If they do not add up to zero then the same number is added to or subtracted from each, as appropriate, to make the three add up to zero.

Note. The assumption made in the model I have used is that the sales rise or fall by the same amount each time the particular season comes around. This is an *additive model*. It is possible to make the alternative assumption that sales rise or fall by the same *percentage* on the trend. This *multiplicative model* will give different results if the trend is rising or falling markedly. It is similar in technique of application, but has been omitted here for reasons of space.

21.3 Forecasting

One of the most obvious and common uses of trends and seasonal factors is in forecasting future figures. Suppose, for example, that we wish to forecast the ice cream sales for each term of 1987.

The first thing to do is to plot the figures and trend values on a graph, as shown in Fig. 21.3. As shown, the trend on the graph may then be projected forwards. In this example I have assumed a straight-line trend, which seems reasonable if we look at the previous trend movement.

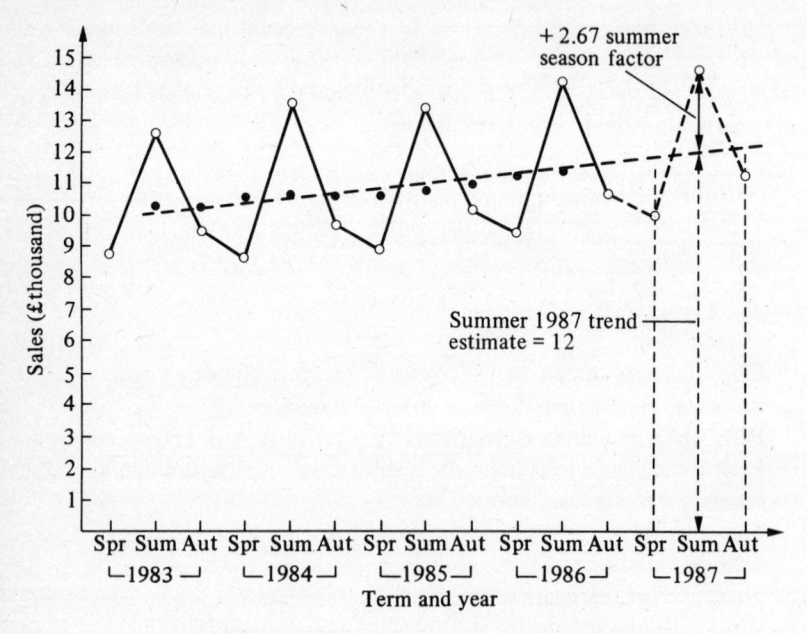

Figure 21.3 *Ice cream sales – forecasting*

For summer 1987 the projected trend gives us a trend forecast of 12. To this trend is added the appropriate seasonal factor, a summer factor of +2.67;

thus for summer 1987 the forecast is: $12+(+2.67) = 14.67$. Similarly, estimates may be made for spring 1987: $11.8+(-1.86) = 10.94$; for autumn 1987: $12.2+(-0.81) = 11.49$.

Generally, forecasting may be done in this way, using trends and seasonal factors. Obviously, however, not all trend lines are as near to straight lines as this one. The four-period model example given in section 21.5 below will illustrate this.

21.4 De-seasonalizing

A second situation in which this kind of time series analysis can be useful is this: when a new figure becomes available, we may wish to know how well or badly we are doing 'for the time of year'. Suppose, for example, that the last figure given (for autumn 1986) has just become available. It is 10.7, whereas the previous term's figure was 14.2 (for summer 1986). This is a drop, but of course in autumn we always sell less ice cream than in summer. How well or badly are we doing 'for the time of year' compared with the previous term? To find this out we *de-seasonalize* the figures.

For summer 1986: note that summer sales are normally up 2.65 on trend, so:

$$\begin{aligned}\text{De-seasonalized figure} &= \text{actual sales} - \text{seasonal factor} \\ &= 14.2 - (+2.65) \\ &= 11.55\end{aligned}$$

For autumn 1986: note that autumn sales are normally down 0.81 on trend, so:

$$\begin{aligned}\text{De-seasonalized figure} &= \text{actual sales} - \text{seasonal factor} \\ &= 10.7 - (-0.81) \\ &= 11.51\end{aligned}$$

21.5 Quarterly figures

In practice, figures where the year is divided into three periods are rare. More commonly, figures are monthly (twelve per year) or quarterly (four per year). The key point in each case is that in order for the seasonal variations to cancel, the number of figures averaged out (three, twelve or four) must be the number

322–*Applied business statistics*

of periods in a year. Thus we take successive twelves or fours, respectively. Where the number of periods is even, we also need to *centre* the figures, as will now be illustrated. I shall illustrate the approach for quarterly figures; that for monthly or bi-monthly is similar.

Suppose, for example, that a footwear manufacturer has the sales figures in Table 21.3. In order for the seasonal factors to cancel out, we have to take

Table 21.3 *Footwear sales figures (× £10 000)*

	1st quarter	2nd quarter	3rd quarter	4th quarter
1983	22	26	26	35
1984	20	24	24	33
1985	22	28	31	39
1986	25			

the averages of successive sets of *four* figures, since there are four periods per year. The calculations are shown in Table 21.4. Column 4a shows the totals

Table 21.4 *De-seasonalizing quarterly figures*

1 Year	2 Season (quarter)	3 Sales	4a Totals of four	4b Trend	4c Centred trend	5 Seasonal factor
1983	1st 2nd 3rd 4th	22 26 26 35	— — 109 107 105	— — 27.25 26.75 26.25	— — 27.00 26.50	— — −1.00 +8.50
1984	1st 2nd 3rd 4th	20 24 24 33	103 101 103 107	25.75 25.25 25.75 26.75	26.00 25.50 25.50 26.25	−6.00 −1.50 −1.50 +6.75
1985	1st 2nd 3rd 4th	22 28 31 39	114 120 123 —	28.50 30.00 30.75 —	27.625 29.25 30.375 —	−5.625 −1.25 +0.625 —
1986	1st	25				

of successive fours of sales figures, and 4b shows the average of these. Column 4b would do as a trend to plot on a graph. In a table, however, it is not convenient because the even number of periods means, for example, that the first average (27.25) falls *between* the 2nd and the 3rd quarters of 1983, the next average (26.75) falls *between* the 3rd and 4th quarters, and so on. It is more convenient, in order to calculate seasonal factors, to obtain *centred trend* figures by averaging out successive pairs of trend figures. This has been done in column 4c, and the seasonal factors are estimated in column 5 by subtracting column 4c from the sales in column 3, as before. For example, the figures of 27.25 and 26.75 in column 4b are averaged to find the centred trend of 27 which therefore comes out opposite to the 3rd quarter of 1983. To find the averaged seasonal variations as before:

1st quarter $\quad \dfrac{(-6.00)+(-5.625)}{2} = -5.813$

2nd quarter $\quad \dfrac{(-1.50)+(-1.25)}{2} = -1.375$

3rd quarter $\quad \dfrac{(-1.00)+(-1.50)+(+0.625)}{3} = -0.625$

4th quarter $\quad \dfrac{(+8.50)+(+6.75)}{2} = +7.625$

We remember that these four seasonal factors should add up to zero. In fact they add up to -0.188, so to make them add up to zero we shall have to add 0.047 to each factor. This gives us adjusted seasonal factors:

1st quarter: $\quad -5.766$
2nd quarter: $\quad -1.328$
3rd quarter: $\quad -0.578$
4th quarter: $\quad +7.672$

To de-seasonalize, remember, we simply deduct the seasonal factor from the actual sales, thus for example:

1985 De-seasonalized 3rd quarter $\quad 31-(-0.578) = 31.578$
1985 De-seasonalized 4th quarter $\quad 39-(+7.672) = 31.328$
1986 De-seasonalized 1st quarter $\quad 25-(-5.766) = 30.766$

To *forecast* we need to draw the graph; it is usually advisable to do this in any case. This is shown in Fig. 21.4, showing footwear sales. Here the trend is not a simple straight line. In 1984 there seems to have been a depression in sales. They picked up again in 1985, but perhaps the expansion showed some signs of slackening in the latter part of 1985. This is confirmed by the fact that the de-seasonalized figures for the last three quarters of the series have been decreasing slightly. The extrapolation of the trend to forecast it in the future is therefore not obvious. It could reasonably be expected to do a number of different things; in the end which particular line is selected will depend on more general factors. What is the market doing? What effect are government policies having? Is this a temporary setback, or the start of a new downturn?

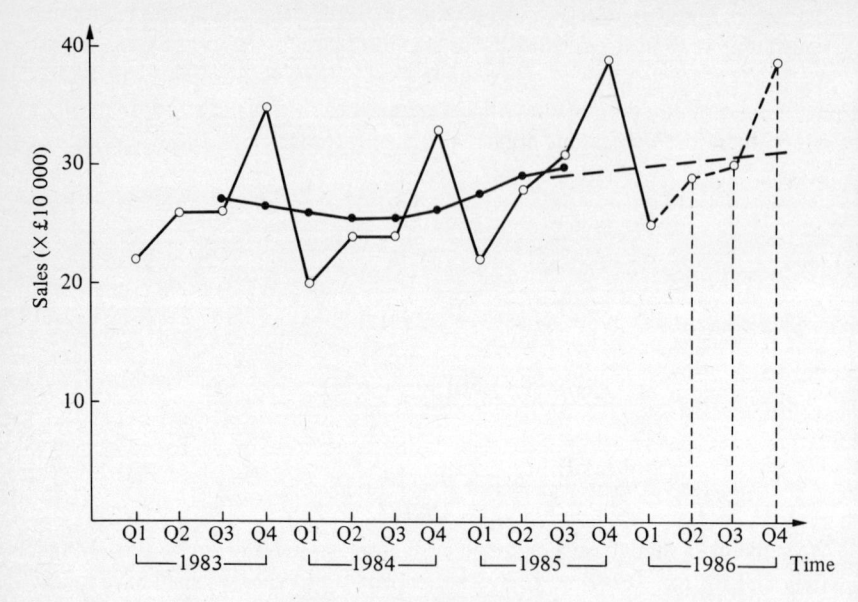

Figure 21.4 *Footwear sales*

The line shown on the graph Fig. 21.4 is a moderately optimistic one. To forecast the sales for the rest of 1986, we take the trend figures and add on the seasonal factors in each case:

1986 2nd quarter $30.4 + (-1.328) \approx 29.1$
1986 3rd quarter $30.8 + (-0.578) \approx 30.2$
1986 4th quarter $31.2 + (+7.672) \approx 38.9$

Note. The figures here have been rounded; indeed, there is little point in keeping three decimal places in the estimated seasonal factors, but I have done so to demonstrate the calculations. Normally, common sense should dictate how many significant figures it is useful to carry.

21.6 Summary

Moving average

This is obtained as follows. The average (arithmetic mean) is taken of the first n values in a series. Then the average is taken of the n values starting with the second and finishing with the $(n+1)$th. Then the average is taken of the third to the $(n+2)$th, and so on, averaging the same number n of variables each time. Such a moving average may be used as a trend.

Seasonal factors

If there is a regular seasonal variation, it is eliminated from the moving average *provided that* the number of successive figures averaged in each case is equal to the number of periods in the year. Seasonal factors can be estimated by subtracting trend from actual figures.

Forecasting

A forecast may be obtained by extending the trend (for example, by eye) and adding on the seasonal variation figure.

De-seasonalizing

Figures are de-seasonalized by subtracting from them the appropriate seasonal variation figure.

Recall and exercise questions

1. What elements might make up a time series figure varying over time?
2. From Table 5.2, take the figures for total UK car production in each quarter:
   (a) plot the figures on a graph;
   (b) find a trend using a four-period centred moving average, and superimpose it on the graph;
   (c) find adjusted seasonal variation figures.
3. From the quarterly UK retail sales data in Table 5.4:
   (a) plot the figures on a graph;
   (b) use moving averages to find and superimpose a trend line;
   (c) find adjusted seasonal variation figures;
   (d) hence forecast the retail sales for the third quarter of 1981;
   (e) de-seasonalize the 1980 figures given.

4. Table 21.5 shows quarterly sales of men's socks.

Table 21.5 Sales of men's socks

Period		Sales (million pairs)
1977	1st quarter	8.83
	2nd quarter	8.29
	3rd quarter	9.50
	4th quarter	11.92
1978	1st quarter	8.46
	2nd quarter	9.60
	3rd quarter	10.22
	4th quarter	12.31
1979	1st quarter	8.25
	2nd quarter	9.40
	3rd quarter	9.73
	4th quarter	11.85
1980	1st quarter	7.22
	2nd quarter	8.30
	3rd quarter	9.47
	4th quarter	11.40

Plot these figures on a graph, and superimpose a trend line.

De-seasonalize the 1980 figures, and forecast the number of pairs that will be sold in the last quarter of 1981.

5. Table 21.6 gives the two-monthly figures for total disposals of maize as animal feed (from the *Monthly Digests*).

Table 21.6 Total disposals maize feedingstuffs

Period	1978	1979	1980	1981
Jan–Feb		365	590	392
Mar–Apr		559	547	
May–Jun		708	452	
Jul–Aug		535	416	
Sep–Oct	594	498	472	
Nov–Dec	541	439	451	

Source: Ministry of Agriculture, Fisheries and Food.

(a) Plot these figures on a graph.
(b) Use a method of moving averages to find and superimpose a trend line.
(c) Calculate adjusted seasonal variation figures.
(d) De-seasonalize the last three figures for 1980 and the first for 1981.
(e) Comment briefly on the accuracy of a seasonal model here.

Problems

Vast numbers of quarterly figures are published in the *Monthly Digest* and in *Business Monitor* PQ series, for example sports (*PQ494.3*), toys (*PQ494.1*), stationery (*PQ495*) and cars (*PQ381*). The *Business Monitor* PM series gives some monthly figures. Any of these could be used to set problems for moving averages and seasonal variations.

1. Table 21.7 (source: *Business Monitors*) shows new fixed-sum credit extended (excluding charges) by finance houses and other consumer credit grantors. These figures are as they appear in the *Business Monitor* SDM 6. A business acquaintance of yours, Mr M. Soak, has recently written to you asking if you can explain how they arrive at the adjusted figures. Write to him with some suggestions.

Table 21.7 *Fixed-sum credit extended by finance houses*

		Credit seasonally adjusted (£ million)	Credit unadjusted (£ million)
1978	3rd quarter	886.8	934.7
	4th quarter	941.4	893.6
1979	1st quarter	985.3	925.7
	2nd quarter	1168.7	1228.2
	3rd quarter	1128.7	1179.1
	4th quarter	1240.2	1189.9
1980	1st quarter	1302.2	1231.4
	2nd quarter	1223.7	1294.7
	3rd quarter	1165.1	1217.3
	4th quarter	1109.7	1057.3
1981	1st quarter	1143.1	1076.8
	2nd quarter	1209.1	1275.8

2. Table 21.8 shows figures for manufacturers' sales of footwear with leather uppers. It is 1981 (a time of economic recession) and your company J. Makaquid Ltd have just acquired small premises suitable for leather shoe production as part of a complex takeover deal. Present a report on the industry based on these figures, showing clearly the general pattern in the industry, seasonal patterns for different lines, and so on.

Table 21.8 *Sales of footwear with leather uppers (million pairs)*

		Total	Men's	Women's	Children's	Sports shoes
1976		17.3	5.5	5.2	5.5	1.1
1977		17.6	5.5	5.2	5.6	1.4
1978		17.9	5.5	5.4	5.6	1.4
1979		18.0	5.3	6.4	5.2	1.2
1980		16.3	4.5	5.8	5.0	0.9
1976	2nd quarter	17.1	5.3	5.0	5.7	1.1
	3rd quarter	16.5	4.9	4.9	5.5	1.2
	4th quarter	16.1	6.0	5.0	4.0	1.1
1977	1st quarter	20.7	6.2	6.0	7.2	1.3
	2nd quarter	17.2	5.4	4.7	5.6	1.5
	3rd quarter	16.6	5.0	5.0	5.3	1.3
	4th quarter	16.4	5.8	5.1	4.2	1.3
1978	1st quarter	19.5	5.8	5.9	6.4	1.4
	2nd quarter	17.7	5.5	5.0	5.6	1.6
	3rd quarter	17.2	5.0	5.2	5.5	1.5
	4th quarter	17.0	5.5	5.7	4.6	1.2
1979	1st quarter	20.0	5.6	6.9	6.2	1.3
	2nd quarter	18.1	5.2	6.1	5.5	1.3
	3rd quarter	17.3	4.9	6.3	4.9	1.2
	4th quarter	16.5	5.2	6.2	4.2	0.9
1980	1st quarter	20.0	5.3	7.5	6.2	1.0
	2nd quarter	15.3	4.1	5.2	5.0	1.0
	3rd quarter	15.6	4.3	5.6	4.7	1.0
	4th quarter	14.2	4.3	5.0	4.2	0.7

Source: Annual Abstract

22
Series and Finance

22.1 Basic concepts

This chapter and the next concern the applications of mathematical series to aspects of finance and business decisions. The subject is central to much numerical work where business decisions involve money paid at different periods.

Arithmetic progressions

An arithmetic progression is a series of numbers in which each is the previous plus some constant, for example:

    7   10   13   16   19   ...   each number is the previous plus 3
    12   10   8   6   4   ...   each number is the previous plus -2

Various formulae for the nth term and the sum of n terms of an arithmetic progression are given in Appendix B, but need not concern us here.

The most obvious use of this series is in the context of *simple interest*, where the same percentage of a capital sum loaned is paid as interest each year. For example a loan of £100 at 15 per cent simple interest per annum is worth:

    £100 + £15   = £115 after one year
    £100 + 2 × £15 = £130 after two years
    £100 + 3 × £15 = £145 after three years, and so on

It is, however, much more common to charge compound interest (see below) than simple interest.

Geometric progressions

In a geometric series each number is the previous one multiplied by a constant factor, for example:

100 110 121 133.1 146.41 ...   each number is the previous multiplied by 1.1

100 90 81 72.9 65.61 ...   each number is the previous multiplied by 0.9

There are three obvious areas of application for geometric progressions:

Compound interest. Suppose that for each £1 left in an account for a year, a bank pays 10p (i.e. 10 per cent) into the account at the end of the year. This interest may then stay in the account and itself collect interest for the next year. What is the effect over several years? Suppose that £100 is put in the account at time zero.

After one year the £100 invested will have earned 10 per cent interest to make:

$$100 + (10 \text{ per cent of } 100) = 100 + 100 \times 0.1 = 100 \times 1.1 = £110$$

After two years the whole of the £110 with which we finished the first year will have earned interest during the second year (at 10 per cent). This will make:

$$110 + (10 \text{ per cent of } 110) = 110 \times 1.1$$
$$= 100 \times 1.1 \times 1.1 = 100 \times 1.1^2 = £121$$

After three years the amount left at the end of the second year will have earned 10 per cent interest, making:

$$100 \times 1.1^3 = £133.10$$

After n years the amount in the account will be:

$$100 \times 1.1^n$$

Generalizing:

Invest a sum P for n years at interest rate r (r is a proportion of 1; for example, for 7 per cent interest $r = 0.07$) and we get: $P \times (1+r)^n$

For example:

£200 invested for five years at 15 per cent interest gives:
$$200 \times 1.15^5 = £402.27$$

Inflation. In Chapter 11 we looked at price indices and purchasing power. Geometric series can help us to work out what prices would be likely to be in *n* years time, given a constant rate of inflation. For example, suppose that prices are assumed to rise at 12 per cent per annum. How much would we need to pay in four years' time to buy the same as we can get for £100 now? Since each year it costs 12 per cent more, after four years the cost will be:

$$100 \times 1.12^4 = £157.35$$

Put another way, in four years' time the *purchasing power* of the pound will be $1/(1.12)^4$ of its present value.

In general, if the rate of inflation is I, where I is a proportion of 1 (for example, for 12 per cent inflation $I = 0.12$), after n years we will have to spend $1 \times (1 + I)^n$ for each pound spent now, to purchase the same goods.

Depreciation. Suppose that for accounting purposes - or in fact - a piece of machinery is assumed to drop in value by 10 per cent each year. If its initial value is £500, then after six years it will be worth:

$$£500 \times 0.9^6 = £265.72$$

Present value

Let us suppose that any available cash can be put into a bank account and earn compound interest (say 12 per cent) at no risk of loss. This gives money a 'time value', as will now be illustrated.

Mr Jerry Bilt is a self-employed jobbing builder employing a few casual labourers. He has set himself the target of saving £60 000 by five years' time to make a bid on the kind of palatial house that he has always dreamed will bring him true happiness. He and his men do some work for his friend Sid Slick, for which he sends Sid a bill for £200. Sid Slick replies with a letter pointing out that he is hard up for cash at present, and since Jerry will not really need it for five years, can he leave paying it until then? Now the point is this. If Jerry had the money *now*, he could get 12 per cent interest so that in five years it would be worth £200 × 1.12⁵ = £352.47, that is:

£200 now is equivalent to £352.47 in five years' time

Now note that this has nothing to do with the *purchasing power* of the pound in five years' time compared with today - even in a politicians' paradise with zero inflation it would still be true. If we wished to ask what sum in five years' time would have the same *purchasing power* as £200 now, then we would be looking at inflation, not at interest rates. For example, if inflation ran at 15 per cent per annum then $200 \times 1.15^5 = £402.27$ would be the sum needed in five years' time to buy what £200 would buy today.

In this case the rate of inflation is not relevant. Even if Jerry had the money now, he could not preserve its purchasing power against inflation (governments may be unwilling to run 'inflation proof' savings schemes when interest rates are below the rate of inflation). Jerry's decision is not whether to purchase now or to purchase later. He does not intend using the money for five years anyway. The decrease in purchasing power will occur whether he has the cash now or later, and is not relevant. What *is* relevant is the interest which Jerry could get if he had the money now. As we have seen, with the interest it could earn, £200 now is equivalent to £352.47 in five years' time (at 12 per cent interest), so we say that £200 is the *present value* (*PV*) of £352.47 in five years' time.

We can, of course, do this the other way around and ask: What is the present value of £200 in five years' time?

If P is the present value:

$$P \times 1.12^5 = 200$$

Dividing both sides by $(1.12)^5$ we get:

$$P = \frac{200}{1.12^5} = £113.49$$

A payment of £200 in five years' time is 'equivalent' (in monetary terms, *not* in purchasing power) to £113.49 now.

Suppose that Sid Slick, noting this argument, offers £100 in one year's time, and £150 in five years' time?

$$\text{Present value} = \frac{100}{(1.12)^1} + \frac{150}{(1.12)^5} = 89.29 + 85.11 = £174.40$$

This is still less than the £200 paid now, and the *net present value* (sometimes written NPV) of the suggested agreement to delay payment is

$$174.40 - 200 = -£25.60 \text{ to Jerry;}$$

an additional payment of £25.60 now would make the agreement fair (excluding any extra payment Jerry may feel appropriate since there is a higher risk of default with Sid Slick than with, say, a bank).

The main point of all this is that, irrespective of inflation or questions of purchasing power, in a capitalist system one cannot simply add together cash payments at different periods. Any cash available earlier could be invested at interest, and so is worth more than later sums.

Tables. Using a calculator, compound interests and present values are easy to manipulate. For those who wish, however, *present value factors* for different interest rates and numbers of years are listed in Table 7 in Appendix A. For example:

for 12 per cent interest the present value factor after one year is
$$0.8929 \left(\frac{1}{(1.12)^1}\right);$$
for 12 per cent interest the present value factor after five years is
$$0.5674 \left(\frac{1}{(1.12)^5}\right).$$

Using present value factors, the present value of proposed repayments of £100 after one year and £150 after five years is:
$$100 \times 0.8929 + 150 \times 0.5674 = 174.40 \quad \text{as above}$$

Present value and interest

The present value of any payment series does, of course, depend on the interest rate assumed. For example, the present value for the proposed advance to Sid Slick of £200, with repayment of £100 in one year's time and £150 after five years, if interest rates were only 5 per cent would be:
$$-200 + 100 \times 0.9524 + 150 \times 0.7835 = £12.76$$
so Jerry Bilt is £12.76 (present value) better off under this arrangement than if he puts his £200 at yearly compounded interest of 5 per cent. We can draw a graph of present value of this loan against interest rate as shown in Fig. 22.1.

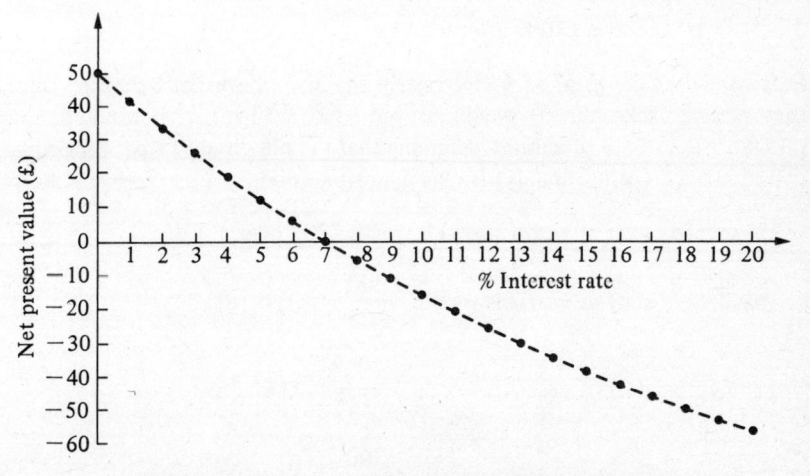

Figure 22.1 *Net present values for Jerry Bilt's loan*

334—Applied business statistics

We can see that for any interest rate above about 7 per cent the net present value is negative, so Jerry Bilt is better off putting his money to earn interest at that rate, while for any interest rate below about 7 per cent the loan is a better proposition. The figure of 7 per cent is called the *internal rate of return* of the loan – it is the interest rate which could just be paid on the capital invested through the financial agreement made.

22.2 Financial agreements

There are a number of situations in which purely financial agreements made between two parties about payments of sums at different times are not affected by any changes in purchasing power of the pound. These include mortgages, personal loans, insurance or assurance (for given amounts, not those linked with profit-sharing schemes), hire-purchase and credit card repayments. In comparing receipts and payments made at different times, the concepts of compound interest and present value must be borne in mind. For example, an insurance salesman tells me that he has a policy with which I just cannot lose (something like this actually happened to me!)

1. I pay a premium of £200 at the beginning of each of the next ten years, making the first payment now.
2. If I die before the ten years is complete then all my premiums are returned to my heirs.
3. If I live, I receive £3000 after ten years.

Is this really a bargain? If I die, clearly my heirs have not benefited since they receive back only the premiums; but what if I live? What would be the present value of my premiums, assuming that I could obtain 11 per cent compound interest (paid annually) from a deposit account?

Payment at start of year 1 (now) is worth £200 present value

Payment at start of year 2 is worth $\dfrac{200}{1.11} = £180.18$

Payment at start of year 3 is worth $\dfrac{200}{1.11^2} = £162.32$

Payment at start of year 4 is worth $\dfrac{200}{1.11^3} = £146.24$ and so on.

Completing this for all ten years, the total present value of payments is £1107.41. What, then, is the present value of receipts?

The present value of £3000 in ten years' time is $\dfrac{3000}{1.11^{10}} = £1056.55$

The *net* present value (NPV) of the policy is

$$£1056.55 - 1107.41 = £-50.86$$

Since the policy's net present value is negative I would do better to put my premiums into a building society or bank account earning 11 per cent.

This kind of calculation of the net present value of a financial agreement – sometimes for comparison with an alternative agreement – is an important aspect of the application of mathematical series to finance. A rather different use of such series involves finding a *true rate of compound interest* on a loan agreement. For example, a loan company offers a loan which they quote as '10 per cent interest'. In fact, a loan of £300 over three years would be repaid in three instalments each of £130. This, they say, is a total of £390 – the extra £90 being 10 per cent interest on £300 over three years. This, however, is misleading, for the whole of the £300 is *not* owing for the whole of the three years. The true rate of interest (r) will be that rate which makes the present value of payments equal to the present value of the amount borrowed. The present value of repayments will be:

$$\dfrac{130}{(1+r)} + \dfrac{130}{(1+r)^2} + \dfrac{130}{(1+r)^3}$$

The value of r which makes this sum equal to the loan of £300 will be the effective rate of compound interest; that is:

$$300 = 130\left(\dfrac{1}{(1+r)} + \dfrac{1}{(1+r)^2} + \dfrac{1}{(1+r)^3}\right)$$

The algebraic solution of this equation for r is beyond the scope of this book. There are, however, two other ways of solving it.

The first is to draw a graph showing the present value of three payments of £130 after one, two and three years respectively (similar to the graph in Fig. 22.1). The value of r which makes the present value of this £300 can then be read off from the graph. (The reader may wish to try this, and show that r comes to about 14.5 per cent.) The second way to solve this problem and find the value of r which makes the net present value of the three annual £130 payments equal to £300 necessitates the introduction of more tables.

Cumulative present value tables

Table 8 (Appendix A) may be used to solve the problem we have been considering. This is a *cumulative present value* table. It gives the *total* present value of payments made at the ends of each of the next n years. In our example, payments of £130 are made at the ends of each of the next three years, so total present value is 130 × (cumulative present value for three years at interest rate r). Since we wish to find the value of r to make this come to 300:

$$\text{(Cumulative present value for three years at interest rate } r) = \frac{300}{130} = 2.308$$

Thus r will be between 0.14 (giving 2.322) and 0.15 (giving 2.283). If we try a guess of 0.145 (14.5 per cent) we can calculate a present value of £299.30, which is near enough to £300. This calculation is useful if we want to know what interest rate we are really paying, as long as the repayments are *equal*. The entries in the table are sometimes called *discount factors*.

Further example

What is the true rate of interest if a loan of £500 is repaid by payments of £160 per annum for four years?

$$500 = 160 \times \text{(cumulative present value for four years at interest } r)$$

Rearranging:

$$\text{Cumulative present value for four years at interest } r = \frac{500}{160} = 3.125$$

The true r lies between 0.10 (giving 3.17) and 0.11 (giving 3.102). If we want a more accurate figure we can use trial and error for r in the expression:

$$160 \left(\frac{1}{(1+r)} + \frac{1}{(1+r)^2} + \frac{1}{(1+r)^3} + \frac{1}{(1+r)^4} \right)$$

to obtain 500.

From the point of view of the lender, the 'true rate of interest' represents a rate of return on his capital laid out, and so it is termed *internal rate of return* (IRR). The net present value, then, is a value based on an assumed rate of interest, while internal rate of return is a kind of interest rate which makes the net present value of a payment series zero. Usually, as illustrated in Fig. 22.1, if IRR on an agreement exceeds the prevailing interest rate, then NPV to the investor will be positive.

22.3 Summary

Simple interest

A fixed percentage of the original capital invested is added on at the end of each year.

Compound interest

The interest given at the end of one year collects interest itself during the next year.

If £P is invested at an interest rate of r (with r as a proportion of one), then after n years compounded annually it is worth $P(1+r)^n$.

Present value

If we assume a prevailing rate of interest r, and a payment X is promised in n years' time, the present value of the payment is the amount which if invested now would amount to X in n years' time. It is given by:

$$\text{Present value} = \frac{X}{(1+r)^n} \quad \text{or use Table 7, Appendix A.}$$

Internal rate of return

This is the rate of interest r which makes the net present value of a series of payments equal to zero.

Recall and exercise questions

1. What is compound interest?
2. What is a present value?
3. What will be the sum available:
    (a) after six years if £500 is invested at 11 per cent simple interest?
    (b) after six years if £500 is invested at 11 per cent annual compound interest?
    (c) after six years if £500 is invested with 5 per cent compound interest added in every six months?
    (d) after five years if £100 is invested at the beginning of each of the five years at 11 per cent annual compound interest?

4. What is the present value of:
   (a) £400 in four years' time if the interest rate is 13 per cent?
   (b) £400 in four years' time if the interest rate is 6 per cent?
   (c) a payment of £200 after one year, £300 after two years, and £500 after four years (assume 11 per cent interest rates)?
   (d) payments of £220 at the end of each of the next five years (at 11 per cent interest)?
   (e) payments of £220 at the beginning of each of the next five years (at 11 per cent interest)?

5. A church recently appealed to members to give seven-year interest-free loans to allow a building extension. If available interest rates are about 12 per cent, what sacrifice in terms of net present value would a loan of £1000 represent?

6. A finance company arranges an advance of £500 now to a person who promises to repay £200 after two years, a further £200 after three years, and a further £300 after four years.
   (a) Construct a table showing the net present value of this arrangement to the finance company for a range of interest values between 0 and 20 per cent.
   (b) Use your table to draw a graph of net present value against r.
   (c) What effective rate of compound interest does this arrangement represent? That is, what value of r makes net present value zero?

7. A loan company advertises that its loans are made at '*only* 10 per cent interest'. In fact, however, their scheme works like this: a loan of, say, £100 is paid back in four end-of-year annual payments of £35 each, making a total of £140, which they say is 40 per cent interest over four years. Use a graphical method (showing net present value against r) to find the true rate of compound interest which such a loan represents.

8. A credit card account begins with a debt of £200. Each month interest of 2 per cent is compounded, but 10 per cent is paid off. How much is owing after 14 months?

9. A twenty-five year mortgage is to be repaid in 25 equal annual instalments at the end of each of the next twenty-five years. The amount of the loan is £20 000, and payments are to be made such that (at the agreed interest rate) the net present value of the total of the payments exactly equals this sum.
   (a) Find the annual payments if the interest rate is 15 per cent.

(b) What would be the effect on payments if the interest rate were 17 per cent instead of 15 per cent?

(c)* Suppose that the agreed interest rate is '15 per cent', but the sums calculated for 'annual' payments are not paid in one sum at the end of each year. Rather, one twelfth of the annual payment is made at the end of each month. Find the true rate of interest (assume that interest is payable on complete months, but is only compounded annually at the end of each year).

Problems

1. If I need to borrow £200 for two years, what will be the difference in cost of doing this by:
   (a) keeping my credit card debt at £200 more during the two years and paying monthly interest of 2 per cent?
   (b) keeping an overdraft of £200 at 17 per cent (compounded annually) during the two years?

2. Most building societies now offer higher interest rates if one is prepared to commit money to investment for longer periods. These often work on a sliding scale of interest rates, depending on length of commitment.

   Visit *two* different societies and obtain leaflets giving details of their schemes. From the two societies work out the best scheme to invest £400:
   (a) with a commitment for five years, money being withdrawn after that time;
   (b) with an initial commitment for two years, but in fact leaving the money in for a further three years before withdrawing it after five years altogether;
   (c) with an initial commitment of only one year, but in fact leaving it invested for five years.

   In each case what sum would be available after five years?

3. (a) For an arithmetic progression with a first term of 1 and a difference of $(-1/12)$ find:
   (i) the 10th term
   (ii) the sum of the first ten terms.

(b) A building society gives 10 per cent interest. The interest is given for all complete months, but is made up only once per year on 1 April. If I invest £100 on the first of each month from April to January, to what will the total sum (with interest) amount when my account is made up the following April?

(c) A local authority allows rates to be paid in two ways:
   (i) in two payments, each of half the sum, paid on 1 April and 1 October;
   (ii) in ten equal payments, made on the first of each month from April to January.

My rates are £200, and I can get interest from the building society mentioned in part (b) above. Which method of payment involves a larger sum in net present value terms?

4. Suppose that a consumer magazine is planning an article on the costs of obtaining a maintained 20-inch colour TV over various periods up to, say, five years. Alternatives might be:

(a) the cheapest available rental you can find, with premiums paid annually in advance;
(b) a discount warehouse (or shop) from which a set can be purchased guaranteed for one or two years, after which a maintenance contract can be taken out indefinitely for the annual payment in advance of a certain sum of money;
(c) a discount warehouse (or shop) offering either a five-year guarantee included in the price of its sets, or an extension to a five-year guarantee period on payment at time of purchase of a single insurance premium.

Obtain details of such alternative schemes available in your area, and use them to write a suitable article.

23
Investment Appraisal

23.1 Business objectives in agreements

What exactly is the meaning of net present value (NPV) and internal rate of return (IRR) in the context of agreements to loan or of investments where returns are known and fixed?

The NPV depends on the prevailing rate of interest, and could be thought of in the following way. Suppose that a particular project has an initial outlay, but promises an ultimate return which brings a positive NPV. An investor could borrow at the prevailing rate of interest a total sum T where:

$$T = (\text{outlay on project} + \text{NPV of project})$$

He would invest the outlay in the project, but could spend the rest of the borrowed sum T on what he liked. The ensuing returns on the project would then exactly pay off the total debt T including the interest which that debt incurred during its duration.

The NPV is a kind of sum of money. The IRR, on the other hand, is a kind of rate of interest. It was introduced in the previous chapter as the rate of interest which would make the NPV on a project zero. But this is equivalent to saying that it is the rate of interest which can be offered on any capital invested in the project. This, in itself, says nothing about whether the sums involved are large or small. If, however, a company always invests in projects with a high IRR then, assuming it can find enough of these to take up all its capital, it will receive a high return on that capital. For a given amount of capital to invest, maximizing its IRR will also maximize the return in NPV terms.

There are two other common criteria for choosing between projects besides NPV and IRR, and it will be useful to compare all four in relation to four alternative financial agreements. The projects are shown in the first part of Table 23.1.

Table 23.1 Four financial agreements appraised

		Project W	Project X	Project Y	Project Z
Initial outlay (£ thousand)		100	50	100	100
Receipts at end of year (£ thousand)	1	19.2	10.3	0	55.3
	2	19.2	10.3	0	55.3
	3	19.2	10.3	0	
	4	19.2	10.3	0	
	5	19.2	10.3	0	
	6	19.2	10.3	200	
	7	19.2	10.3		
	8	19.2	10.3		
	9	19.2	10.3		
	10	19.2	10.3		
NPV (at 11 per cent) (£ thousand)		13.1	10.7	6.9	−5.3
Internal rate of return		14%	16%	12.2%	7%
Payback period (years)		6	5	6	2
Average rate of return		9.2%	10.6%	16.7%	5.3%

Net present value

The prevailing rate of interest is presumed to be 11 per cent, which allows NPV to be calculated using tables or formulae. For example, using formulae:

$$\text{NPV of W} = -100 + \frac{19.2}{(1.11)} + \frac{19.2}{(1.1)^2} + \frac{19.2}{(1.1)^3} + \ldots + \frac{19.2}{(1.11)^{10}} = 13.069$$

This can easily be calculated; alternatively, the cumulative PV table (Appendix A, Table 8) figure for ten years' annual payments at 11 per cent is 5.889, so present value of £19.2 at the end of each of the next ten years is

$$19.2 \times 5.889 = 113.069$$

Subtracting the initial outlay of 100 gives the NPV of 13.069.

Internal rate of return

This represents the effective rate of compound interest which receipts represent as a return on outlay. Suppose that we have a series of 'returns' commencing now with amount A, then B at the end of year one, C at the end of year two, D at the end of year three, and so on. Some of these will, of course, be negative (payments into the project) and some positive (receipts from the project). This will give:

$$\text{NPV} = A + \frac{B}{(1+r)} + \frac{C}{(1+r)^2} + \frac{D}{(1+r)^3} + \ldots \text{etc.}$$

The IRR is the value of r which makes the NPV zero. In general this may be difficult to obtain algebraically, and a graphical method of plotting various NPV results for various r values may have to be used. But there are two special cases where algebra or tables can be used. The first is where, as in projects W, X and Z, a single initial payment is followed by a number of years of equal returns. The tables in Appendix A, Table 8 give us the cumulative present value of £1 at various rates of interest, at the end of each of the next n years. Thus for project W, where we have £19.2 return at the end of each of the next 10 years:

NPV of W = $-100 + 19.2 \times$ (cumulative PV at interest rate r for ten years)

Setting this to zero and rearranging we obtain:

$$\text{Cumulative PV at interest rate } r \text{ for ten years} = \frac{100}{19.2} = 5.21$$

In Appendix A, Table 8 we find that the nearest value to 5.21 for ten years' equal payments is the figure of 5.216 which is given for a 14 per cent interest rate. The value of the IRR is therefore near to 14 per cent. The IRR values for projects X and Z can be found similarly.

The IRR for project Y is an even simpler special case. For the rate of r which makes its NPV zero, the £100 initial outlay must be exactly balanced by the present value of the £200 after six years. Thus r may be found from:

$100 = 200 \times$ (present value at rate r of £1 after six years)

The value of interest rate which gives us the nearest to (100/200) in the present value tables for £1 after six years will be 12 per cent, which gives 0.5066. In fact we can find a more precise figure if we can take the sixth root of a number since $1 + r = (200/100)$. This will give us the more precise figure of 12.2 per cent for the IRR of project Y.

Payback criterion

This simply looks for the project which (without considerations of present value or interest) repays the initial outlay the quickest. For example, for project W after five years £96 of the £100 has been repaid, and after six years the whole £100 has been recovered with £15.20 extra. The *payback period* is therefore six years; it can be found easily for the other projects as shown.

Average rate of return

This method, like payback, also ignores present value and compound interest. Its rationale is as follows. The net return on the project, ignoring present value, is total receipts − initial outlay. It might make sense to look at this as a percentage of the initial outlay:

$$\frac{\text{total receipts} - \text{initial outlay}}{\text{initial outlay}} \times 100$$

As it stands, however, this would take no account of the time for which money was tied up in the project. It is therefore more useful to take the percentage return on initial outlay *per year of the project duration*:

Average rate of return
$$= \frac{\text{total receipts} - \text{initial outlay}}{\text{initial outlay}} \div \text{number of years of project} \times 100$$

For Project W:
$$\text{Average rate of return} = \frac{(10 \times 19.2) - 100}{100} \div 10 \times 100 = 9.2 \text{ per cent}$$

As may be seen from the table, each of these four criteria suggests a different choice of project, so what does it all mean? Having explained the basic meanings of the four methods, I prefer not to list abstract 'advantages and disadvantages' for each, but rather to try to consider more generally what might be relevant business objectives and how to relate these to them.

Let us for the present suppose that the four projects are loan agreements and the investor is a bank. Which would it choose? One question we must ask is whether the choice is between a *type* of project, or whether only one may be done. That is, could we, for example, choose to spend our £100 000 either on one project W or two like project X? If we have such a choice, 14 per cent internal rate of return on one £100 000 loan will bring in less return (in present value terms) than a 16 per cent internal rate of return for *two* £50 000 projects - at least for any normal projects which have high initial outlay and returns over a period of time. The IRR would then be a sensible criterion. But if we are limited to only one project, then the higher NPV may be the better criterion.

There are, however, other considerations. One thing that a bank must maintain is *liquidity* - the ability to turn an investment into cash if necessary. We note that project Y ties up all the money in the agreement until the sixth year, W and B return some earlier and some later, while Z returns everything more quickly. Obviously, if a bank could borrow unlimited amounts at, say, 11 per cent interest to make its investments liquidity would never be a problem. But this is not normally the case, and the bank might wish to choose a project with a lower IRR but which did not tie up money in non-liquid assets for so long. The payback criterion applied blindly would be very silly, but it can be looked at as one consideration.

The average rate of return (ARR) is widely used, but seems to be an irrelevant figure in a society where compound interest is always available on cash investment. The only defence of it I have seen is that it is simple to apply – an argument which is rendered obsolete by the development of easy-to-use package programs in desk-top computers if by nothing else.

23.2 General business objectives

We have been considering financial agreements and arrangements where nothing but straight and agreed money payments are involved. Now let us consider a different kind of example.

Mr Jack Kupp is deciding whether or not to invest his pools winnings in a garage for car repairs. The sums involved are as follows:

Outlay £400 000 for buying land and building garage;
£25 000 for hydraulic ramp;
£15 000 for other ancillary equipment.

Returns Estimated at a net £20 000 per year in initial period, after payment of rates, wages etc.

There are a number of things to note about this:

1. As well as the annual income, the outlay is purchasing various assets. The building itself is likely to appreciate in value as values of buildings rise. The rate of appreciation is *not* the same as the rate of inflation measured by retail price index, but it will relate in some way to inflation. Valuation of the machinery is more complex. Secondhand machinery is worth less the older it gets and this rate of *depreciation* must be allowed for. On the other hand, inflation means that the price of new machinery is continually rising, so secondhand prices and values will also rise. In money terms, therefore, the machinery could increase or decrease in value, depending on whether inflation or depreciation has most effect.

2. The annual returns will be linked in some way to prices, and so will rise with inflation. It would be quite unrealistic to project £20 000 as the net income (or *residual income*) for the next ten years. In general, for those whose income rises with inflation but who have an outstanding capital *debt*, inflation makes things easier by increasing the ratio of income to interest payments. The assumption that inflation is always wholly bad for everyone is naive.

3. It is possible to calculate an NPV for this project only if we assume that it has a finite length *and* if we assume selling prices for assets at the end of that time. This may not be realistic.

There is a possibility that one could become so involved in complex NPV and IRR calculations as almost to lose sight of these basic problems in investment appraisal. A detailed analysis of these problems is certainly beyond the scope of this present text, but readers should at least be aware that in this kind of investment appraisal, various considerations beyond simple criteria need to be presented.

Recall and exercise questions

1. Briefly explain the meaning of the following four criteria for business investment decision:
   (a) payback;
   (b) average rate of return;
   (c) internal rate of return;
   (d) net present value.

2. Apply each of the four criteria named in Question 1 to the following financial agreements:
   (a) an advance of £1500 to be made now, repaid by five payments of £500 at the end of each of the next five years;
   (b) an advance of £1500 to be made now, to be repaid by seven payments of £400 at the end of each of the next seven years.
   (c) An advance of £1500 to be made now, to be repaid in a lump sum of £2600 at the end of four years.

   Do the NPV calculations for an interest rate of 12 per cent and for one of 14 per cent.

Problem

1. Mr and Mrs N. Terprize are considering purchase of a retail business for £60 000. Of this, £30 000 would be paid out of savings and the sale of their present property. Their bank will advance them a mortgage to cover the rest of the outlay, charging 14 per cent interest. In the first year they hope to be able to pay off £3200 (assume that this is paid at the end of

the year for simplicity). With inflation, however, their residual income is likely to rise, and they hope to be able to increase the amount they pay back by 10 per cent each year: £3520 at the end of the second year, £3872 at the end of third, and so on. How long will it be before they have paid back all the loan on this basis? What complications (other than unpredictability) might arise in practice?

Appendix A
Statistical tables

1. Binomial tables
2. Poisson tables
3. Normal tables
4. *t*-tables
5. Chi-square tables
6. Correlation significance tables
7. Present values
8. Cumulative present values

My thanks are given to my colleagues Mr Paul Cockram, Mr Eddie Taylor and Mr Brian Wignall for permission to use the tables in this Appendix.

350–*Applied business statistics*

Table 1 Cumulative binomial probabilities

The table gives $Pr(r \leq x)$ for various values of n and p. $Pr(r \leq x) = \sum_{\substack{\text{sum from} \\ i = 0}}^{\text{to } i = x} \dfrac{n!}{i!(n-i)!} p^i(1-p)^{n-i}$

$n = 2$

p =	0.01	0.02	0.03	0.04	0.05	0.06	0.08	0.10	0.15	0.20	0.25	0.30	0.40	0.50
x														
0	0.9801	0.9604	0.9409	0.9216	0.9025	0.8836	0.8464	0.8100	0.7225	0.6400	0.5625	0.4900	0.3600	0.2500
1	0.9999	0.9996	0.9991	0.9984	0.9975	0.9964	0.9936	0.9900	0.9775	0.9600	0.9375	0.9100	0.8400	0.7500
2	1.0000	1.0000	1.0000	1.0000	1.0000	1.0000	1.0000	1.0000	1.0000	1.0000	1.0000	1.0000	1.0000	1.0000

$n = 3$

p =	0.01	0.02	0.03	0.04	0.05	0.06	0.08	0.10	0.15	0.20	0.25	0.30	0.40	0.50
x														
0	0.9703	0.9412	0.9127	0.8847	0.8574	0.8306	0.7787	0.7290	0.6141	0.5120	0.4219	0.3430	0.2160	0.1250
1	0.9997	0.9988	0.9974	0.9953	0.9927	0.9896	0.9818	0.9720	0.9392	0.8960	0.8438	0.7840	0.6480	0.5000
2	1.0000	1.0000	1.0000	0.9999	0.9999	0.9998	0.9995	0.9990	0.9966	0.9920	0.9844	0.9730	0.9360	0.8750
3	1.0000	1.0000	1.0000	1.0000	1.0000	1.0000	1.0000	1.0000	1.0000	1.0000	1.0000	1.0000	1.0000	1.0000

$n = 4$

p =	0.01	0.02	0.03	0.04	0.05	0.06	0.08	0.10	0.15	0.20	0.25	0.30	0.40	0.50
x														
0	0.9606	0.9224	0.8853	0.8493	0.8145	0.7807	0.7164	0.6561	0.5220	0.4096	0.3164	0.2401	0.1296	0.0625
1	0.9994	0.9977	0.9948	0.9909	0.9860	0.9801	0.9656	0.9477	0.8905	0.8192	0.7383	0.6517	0.4752	0.3125
2	1.0000	1.0000	0.9999	0.9998	0.9995	0.9992	0.9981	0.9963	0.9880	0.9728	0.9492	0.9163	0.8208	0.6875
3	1.0000	1.0000	1.0000	1.0000	1.0000	1.0000	1.0000	0.9999	0.9995	0.9984	0.9961	0.9919	0.9744	0.9375
4	1.0000	1.0000	1.0000	1.0000	1.0000	1.0000	1.0000	1.0000	1.0000	1.0000	1.0000	1.0000	1.0000	1.0000

$n = 5$

$p =$	0.01	0.02	0.03	0.04	0.05	0.06	0.08	0.10	0.15	0.20	0.25	0.30	0.40	0.50
x														
0	0.9510	0.9039	0.8587	0.8154	0.7738	0.7339	0.6591	0.5905	0.4437	0.3277	0.2373	0.1681	0.0778	0.0313
1	0.9990	0.9962	0.9915	0.9852	0.9774	0.9681	0.9456	0.9185	0.8352	0.7373	0.6328	0.5282	0.3370	0.1875
2	1.0000	0.9999	0.9997	0.9994	0.9988	0.9980	0.9955	0.9914	0.9734	0.9421	0.8965	0.8369	0.6826	0.5000
3		1.0000	1.0000	1.0000	1.0000	0.9999	0.9998	0.9995	0.9978	0.9933	0.9844	0.9692	0.9130	0.8125
4						1.0000	1.0000	1.0000	0.9999	0.9997	0.9990	0.9976	0.9898	0.9688
5									1.0000	1.0000	1.0000	1.0000	1.0000	1.0000

$n = 10$

$p =$	0.01	0.02	0.03	0.04	0.05	0.06	0.08	0.10	0.15	0.20	0.25	0.30	0.40	0.50
x														
0	0.9044	0.8171	0.7374	0.6648	0.5987	0.5386	0.4344	0.3487	0.1969	0.1074	0.0563	0.0282	0.0060	0.0010
1	0.9957	0.9838	0.9655	0.9418	0.9139	0.8824	0.8121	0.7361	0.5443	0.3758	0.2440	0.1493	0.0464	0.0107
2	0.9999	0.9991	0.9972	0.9938	0.9885	0.9812	0.9599	0.9298	0.8202	0.6778	0.5256	0.3828	0.1673	0.0547
3	1.0000	1.0000	0.9999	0.9996	0.9990	0.9980	0.9942	0.9872	0.9500	0.8791	0.7759	0.6496	0.3823	0.1719
4			1.0000	1.0000	0.9999	0.9998	0.9994	0.9984	0.9901	0.9672	0.9219	0.8497	0.6331	0.3770
5					1.0000	1.0000	1.0000	0.9999	0.9986	0.9936	0.9803	0.9527	0.8338	0.6230
6								1.0000	0.9999	0.9991	0.9965	0.9894	0.9452	0.8281
7									1.0000	0.9999	0.9996	0.9984	0.9877	0.9453
8										1.0000	1.0000	0.9999	0.9983	0.9893
9												1.0000	0.9999	0.9990
10													1.0000	1.0000

Table 1 (cont.)

Table 1 Cumulative binomial probabilities

The table gives $Pr(r \leqslant x)$ for various values of n and p. $Pr(r \leqslant x) = \sum_{\substack{\text{sum from} \\ i=0}}^{\text{to } i = x} \dfrac{n!}{i!(n-i)!} p^i (1-p)^{n-i}$

$n = 15$

$p =$	0.01	0.02	0.03	0.04	0.05	0.06	0.08	0.10	0.15	0.20	0.25	0.30	0.40	0.50
x														
0	0.8601	0.7386	0.6333	0.5421	0.4633	0.3953	0.2863	0.2059	0.0874	0.0352	0.0134	0.0047	0.0005	0.0000
1	0.9904	0.9647	0.9270	0.8809	0.8290	0.7738	0.6597	0.5490	0.3186	0.1671	0.0802	0.0353	0.0052	0.0005
2	0.9996	0.9970	0.9906	0.9797	0.9638	0.9429	0.8870	0.8159	0.6042	0.3980	0.2361	0.1268	0.0271	0.0037
3	1.0000	0.9998	0.9992	0.9976	0.9945	0.9896	0.9727	0.9444	0.8227	0.6482	0.4613	0.2969	0.0905	0.0176
4		1.0000	0.9999	0.9998	0.9994	0.9986	0.9950	0.9873	0.9383	0.8358	0.6865	0.5155	0.2173	0.0592
5			1.0000	1.0000	0.9999	0.9999	0.9993	0.9978	0.9832	0.9389	0.8516	0.7216	0.4032	0.1509
6					1.0000	1.0000	0.9999	0.9997	0.9964	0.9819	0.9434	0.8689	0.6098	0.3036
7							1.0000	1.0000	0.9994	0.9958	0.9827	0.9500	0.7869	0.5000
8									0.9999	0.9992	0.9958	0.9848	0.9050	0.6964
9									1.0000	0.9999	0.9992	0.9963	0.9662	0.8491
10										1.0000	0.9999	0.9993	0.9907	0.9408
11											1.0000	0.9999	0.9981	0.9824
12												1.0000	0.9997	0.9963
13													1.0000	0.9995
14														1.0000
15														

Statistical tables

$n = 20$

$p =$	0.01	0.02	0.03	0.04	0.05	0.06	0.08	0.10	0.15	0.20	0.25	0.30	0.40	0.50
x														
0	0.8179	0.6676	0.5438	0.4420	0.3585	0.2901	0.1887	0.1216	0.0388	0.0115	0.0032	0.0008	0.0000	0.0000
1	0.9831	0.9401	0.8802	0.8103	0.7358	0.6605	0.5169	0.3917	0.1756	0.0692	0.0243	0.0076	0.0005	0.0000
2	0.9990	0.9929	0.9790	0.9561	0.9245	0.8850	0.7879	0.6769	0.4049	0.2061	0.0913	0.0355	0.0036	0.0002
3	1.0000	0.9994	0.9973	0.9926	0.9841	0.9710	0.9294	0.8670	0.6477	0.4114	0.2252	0.1071	0.0160	0.0013
4		1.0000	0.9997	0.9990	0.9974	0.9944	0.9817	0.9568	0.8298	0.6296	0.4148	0.2375	0.0510	0.0059
5			1.0000	0.9999	0.9997	0.9991	0.9962	0.9887	0.9327	0.8042	0.6172	0.4164	0.1256	0.0207
6				1.0000	1.0000	0.9999	0.9994	0.9976	0.9781	0.9133	0.7858	0.6080	0.2500	0.0577
7						1.0000	0.9999	0.9996	0.9941	0.9679	0.8982	0.7723	0.4159	0.1316
8							1.0000	0.9999	0.9987	0.9900	0.9591	0.8867	0.5956	0.2517
9								0.9999	0.9998	0.9974	0.9861	0.9520	0.7553	0.4119
10								1.0000	1.0000	0.9994	0.9961	0.9829	0.8725	0.5881
11										0.9999	0.9991	0.9949	0.9435	0.7483
12										1.0000	0.9998	0.9987	0.9790	0.8684
13											1.0000	0.9997	0.9935	0.9423
14												1.0000	0.9984	0.9793
15													0.9997	0.9941
16													1.0000	0.9987
17														0.9998
18														1.0000

If n is large (say more than 20), then for values not covered in this table one can approximate the binomial distribution by:

(a) A normal distribution with mean $= np$ and standard deviation $\sqrt{np(1-p)}$ if p is between 0.1 and 0.9

(b) A Poisson with mean $m = np$ if p is less than 0.1 or greater than 0.9

Table 2 Cumulative Poisson probabilities

The table gives $Pr(R \leq x)$ for various values of m. $Pr(R \leq x) = \sum_{i=0}^{i=x} \dfrac{e^{-m} m^i}{i!}$

$m =$	0.10	0.20	0.30	0.40	0.50	0.60	0.70	0.80	0.90	1.00
x										
0	0.9048	0.8187	0.7408	0.6703	0.6065	0.5488	0.4966	0.4493	0.4066	0.3679
1	0.9953	0.9825	0.9631	0.9384	0.9098	0.8781	0.8442	0.8088	0.7725	0.7358
2	0.9998	0.9989	0.9964	0.9921	0.9856	0.9769	0.9659	0.9526	0.9371	0.9197
3	1.0000	0.9999	0.9997	0.9992	0.9982	0.9966	0.9942	0.9909	0.9865	0.9810
4		1.0000	1.0000	0.9999	0.9998	0.9996	0.9992	0.9986	0.9977	0.9963
5				1.0000	1.0000	1.0000	0.9999	0.9998	0.9997	0.9994
6							1.0000	1.0000	1.0000	0.9999

$m =$	1.10	1.20	1.30	1.40	1.50	1.60	1.70	1.80	1.90	2.00
x										
0	0.3329	0.3012	0.2725	0.2466	0.2231	0.2019	0.1827	0.1653	0.1496	0.1353
1	0.6990	0.6626	0.6268	0.5918	0.5578	0.5249	0.4932	0.4628	0.4337	0.4060
2	0.9004	0.8795	0.8571	0.8335	0.8088	0.7834	0.7572	0.7306	0.7037	0.6767
3	0.9743	0.9662	0.9569	0.9463	0.9344	0.9212	0.9068	0.8913	0.8747	0.8571
4	0.9946	0.9923	0.9893	0.9857	0.9814	0.9763	0.9704	0.9636	0.9559	0.9473
5	0.9990	0.9985	0.9978	0.9968	0.9955	0.9940	0.9920	0.9896	0.9868	0.9834
6	0.9999	0.9997	0.9996	0.9994	0.9991	0.9987	0.9981	0.9974	0.9966	0.9955
7	1.0000	1.0000	0.9999	0.9999	0.9998	0.9997	0.9996	0.9994	0.9992	0.9989
8			1.0000	1.0000	1.0000	1.0000	0.9999	0.9999	0.9998	0.9998
9							1.0000	1.0000	1.0000	1.0000

$m =$	2.10	2.20	2.30	2.40	2.50	2.60	2.70	2.80	2.90	3.00
x										
0	0.1225	0.1108	0.1003	0.0907	0.0821	0.0743	0.0672	0.0608	0.0550	0.0498
1	0.3796	0.3546	0.3309	0.3084	0.2873	0.2674	0.2487	0.2311	0.2146	0.1991
2	0.6496	0.6227	0.5960	0.5697	0.5438	0.5184	0.4936	0.4695	0.4460	0.4232
3	0.8386	0.8194	0.7993	0.7787	0.7576	0.7360	0.7141	0.6919	0.6696	0.6472

Statistical tables

x	3.50	4.00	4.50	5.00	5.50	6.00	7.00	8.00	9.00	10.00
4	0.9379	0.9275	0.9162	0.9041	0.8912	0.8774	0.8629	0.8477	0.8318	0.8153
5	0.9796	0.9751	0.9700	0.9643	0.9580	0.9510	0.9433	0.9349	0.9258	0.9161
6	0.9941	0.9925	0.9906	0.9884	0.9858	0.9828	0.9794	0.9756	0.9713	0.9665
7	0.9985	0.9980	0.9974	0.9967	0.9958	0.9947	0.9934	0.9919	0.9901	0.9881
8	0.9997	0.9995	0.9994	0.9991	0.9989	0.9985	0.9981	0.9976	0.9969	0.9962
9	0.9999	0.9999	0.9999	0.9998	0.9997	0.9996	0.9995	0.9993	0.9991	0.9989
10	1.0000	1.0000	1.0000	1.0000	0.9999	0.9999	0.9999	0.9998	0.9998	0.9997
11					1.0000	1.0000	1.0000	1.0000	0.9999	0.9999

$m =$

x	3.50	4.00	4.50	5.00	5.50	6.00	7.00	8.00	9.00	10.00
0	0.0302	0.0183	0.0111	0.0067	0.0041	0.0025	0.0009	0.0003	0.0001	0.0000
1	0.1359	0.0916	0.0611	0.0404	0.0266	0.0174	0.0073	0.0030	0.0012	0.0005
2	0.3208	0.2381	0.1736	0.1247	0.0884	0.0620	0.0296	0.0138	0.0062	0.0028
3	0.5366	0.4335	0.3423	0.2650	0.2017	0.1512	0.0818	0.0424	0.0212	0.0103
4	0.7254	0.6288	0.5321	0.4405	0.3575	0.2851	0.1730	0.0996	0.0550	0.0293
5	0.8576	0.7851	0.7029	0.6160	0.5289	0.4457	0.3007	0.1912	0.1157	0.0671
6	0.9347	0.8893	0.8311	0.7622	0.6860	0.6063	0.4497	0.3134	0.2068	0.1301
7	0.9733	0.9489	0.9134	0.8666	0.8095	0.7440	0.5987	0.4530	0.3239	0.2202
8	0.9901	0.9786	0.9597	0.9319	0.8944	0.8472	0.7291	0.5925	0.4557	0.3328
9	0.9967	0.9919	0.9829	0.9682	0.9462	0.9161	0.8305	0.7166	0.5874	0.4579
10	0.9990	0.9972	0.9933	0.9863	0.9747	0.9574	0.9015	0.8159	0.7060	0.5830
11	0.9997	0.9991	0.9976	0.9945	0.9890	0.9799	0.9467	0.8881	0.8030	0.6968
12	0.9999	0.9997	0.9992	0.9980	0.9955	0.9912	0.9730	0.9362	0.8758	0.7916
13	1.0000	0.9999	0.9997	0.9993	0.9983	0.9964	0.9872	0.9658	0.9281	0.8645
14			0.9999	0.9998	0.9994	0.9986	0.9943	0.9827	0.9585	0.9165
15			1.0000	0.9999	0.9998	0.9995	0.9976	0.9918	0.9780	0.9513
16				1.0000	0.9999	0.9998	0.9990	0.9963	0.9889	0.9730
17					1.0000	0.9999	0.9996	0.9984	0.9947	0.9857
18						1.0000	0.9999	0.9993	0.9978	0.9928
19							0.9999	0.9997	0.9989	0.9965
20							1.0000	0.9999	0.9996	0.9984
21								1.0000	0.9998	0.9993
22									0.9999	0.9997
23									1.0000	0.9999
24										1.0000

Table 3 Normal distribution table

The table shows values of z against the size of the smaller of the two areas into which a line at plus or minus z standard deviations divides the total area.

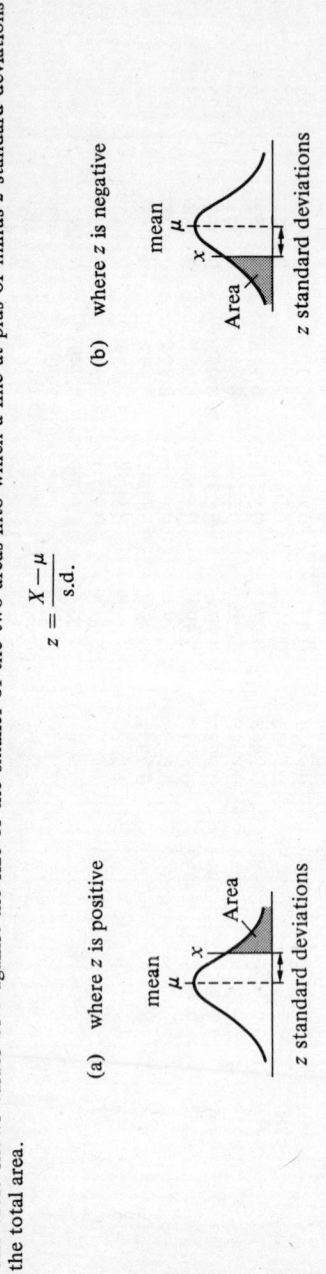

(a) where z is positive

(b) where z is negative

$$z = \frac{X - \mu}{\text{s.d.}}$$

z	.00	.01	.02	.03	.04	.05	.06	.07	.08	.09
0.0	0.50000	0.49601	0.49202	0.48803	0.48405	0.48006	0.47608	0.47210	0.46812	0.46414
0.1	0.46017	0.45620	0.45224	0.44828	0.44433	0.44038	0.43644	0.43251	0.42858	0.42465
0.2	0.42074	0.41683	0.41294	0.40905	0.40517	0.40129	0.39743	0.39358	0.38974	0.38591
0.3	0.38209	0.37828	0.37448	0.37070	0.36693	0.36317	0.35942	0.35569	0.35197	0.34827
0.4	0.34458	0.34090	0.33724	0.33360	0.32997	0.32636	0.32276	0.31918	0.31561	0.31207
0.5	0.30854	0.30503	0.30153	0.29806	0.29460	0.29116	0.28774	0.28434	0.28096	0.27760
0.6	0.27425	0.27093	0.26763	0.26435	0.26109	0.25785	0.25463	0.25143	0.24825	0.24510
0.7	0.24196	0.23885	0.23576	0.23270	0.22965	0.22663	0.22363	0.22065	0.21770	0.21476
0.8	0.21186	0.20897	0.20611	0.20327	0.20045	0.19766	0.19489	0.19215	0.18943	0.18673
0.9	0.18406	0.18141	0.17879	0.17619	0.17361	0.17106	0.16853	0.16602	0.16354	0.16109
1.0	0.15866	0.15625	0.15386	0.15151	0.14917	0.14686	0.14457	0.14231	0.14007	0.13786
1.1	0.13567	0.13350	0.13136	0.12924	0.12714	0.12507	0.12302	0.12100	0.11900	0.11702
1.2	0.11507	0.11314	0.11123	0.10935	0.10749	0.10565	0.10383	0.10204	0.10027	0.09853
1.3	0.09680	0.09510	0.09342	0.09176	0.09012	0.08851	0.08691	0.08534	0.08379	0.08226
1.4	0.08076	0.07927	0.07780	0.07636	0.07493	0.07353	0.07215	0.07078	0.06944	0.06811

Statistical tables—357

1.5	0.06681	0.06552	0.06426	0.06301	0.06178	0.06057	0.05938	0.05821	0.05705	0.05592
1.6	0.05480	0.05370	0.05262	0.05155	0.05050	0.04947	0.04846	0.04746	0.04648	0.04551
1.7	0.04457	0.04363	0.04272	0.04182	0.04093	0.04006	0.03920	0.03836	0.03754	0.03673
1.8	0.03593	0.03515	0.03438	0.03362	0.03288	0.03216	0.03144	0.03074	0.03005	0.02938
1.9	0.02872	0.02807	0.02743	0.02680	0.02619	0.02559	0.02500	0.02442	0.02385	0.02330
2.0	0.02275	0.02222	0.02169	0.02118	0.02068	0.02018	0.01970	0.01923	0.01876	0.01831
2.1	0.01786	0.01743	0.01700	0.01659	0.01618	0.01578	0.01539	0.01500	0.01463	0.01426
2.2	0.01390	0.01355	0.01321	0.01287	0.01255	0.01222	0.01191	0.01160	0.01130	0.01101
2.3	0.01072	0.01044	0.01017	0.00990	0.00964	0.00939	0.00914	0.00889	0.00866	0.00842
2.4	0.00820	0.00798	0.00776	0.00755	0.00734	0.00714	0.00695	0.00676	0.00657	0.00639
2.5	0.00621	0.00604	0.00587	0.00570	0.00554	0.00539	0.00523	0.00508	0.00494	0.00480
2.6	0.00466	0.00453	0.00440	0.00427	0.00415	0.00402	0.00391	0.00379	0.00368	0.00357
2.7	0.00347	0.00336	0.00326	0.00317	0.00307	0.00298	0.00289	0.00280	0.00272	0.00264
2.8	0.00256	0.00248	0.00240	0.00233	0.00226	0.00219	0.00212	0.00205	0.00199	0.00193
2.9	0.00187	0.00181	0.00175	0.00169	0.00164	0.00159	0.00154	0.00149	0.00144	0.00139
3.0	0.00135	0.00131	0.00126	0.00122	0.00118	0.00114	0.00111	0.00107	0.00104	0.00100
3.1	0.00097	0.00094	0.00090	0.00087	0.00084	0.00082	0.00079	0.00076	0.00074	0.00071
3.2	0.00069	0.00066	0.00064	0.00062	0.00060	0.00058	0.00056	0.00054	0.00052	0.00050
3.3	0.00048	0.00047	0.00045	0.00043	0.00042	0.00040	0.00039	0.00038	0.00036	0.00035
3.4	0.00034	0.00032	0.00031	0.00030	0.00029	0.00028	0.00027	0.00026	0.00025	0.00024
3.5	0.00023	0.00022	0.00022	0.00021	0.00020	0.00019	0.00019	0.00018	0.00017	0.00017
3.6	0.00016	0.00015	0.00015	0.00014	0.00014	0.00013	0.00013	0.00012	0.00012	0.00011
3.7	0.00011	0.00010	0.00010	0.00010	0.00009	0.00009	0.00008	0.00008	0.00008	0.00008
3.8	0.00007	0.00007	0.00007	0.00006	0.00006	0.00006	0.00006	0.00005	0.00005	0.00005
3.9	0.00005	0.00005	0.00004	0.00004	0.00004	0.00004	0.00004	0.00004	0.00003	0.00003
4.0	0.00003	0.00003	0.00003	0.00003	0.00003	0.00003	0.00002	0.00002	0.00002	0.00002
4.1	0.00002	0.00002	0.00002	0.00002	0.00002	0.00002	0.00002	0.00002	0.00002	0.00001
4.2	0.00001	0.00001	0.00001	0.00001	0.00001	0.00001	0.00001	0.00001	0.00001	0.00001
4.3	0.00001	0.00001	0.00001	0.00001	0.00001	0.00001	0.00001	0.00001	0.00001	0.00001
4.4	0.00001	0.00001	0.00001	0.00001	0.00001	0.00001	0.00001	0.00001	0.00000	0.00000
4.5	0.00000	0.00000	0.00000	0.00000	0.00000	0.00000	0.00000	0.00000	0.00000	0.00000
4.6	0.00000	0.00000	0.00000	0.00000	0.00000	0.00000	0.00000	0.00000	0.00000	0.00000
4.7	0.00000	0.00000	0.00000	0.00000	0.00000	0.00000	0.00000	0.00000	0.00000	0.00000
4.8	0.00000	0.00000	0.00000	0.00000	0.00000	0.00000	0.00000	0.00000	0.00000	0.00000
4.9	0.00000	0.00000	0.00000	0.00000	0.00000	0.00000	0.00000	0.00000	0.00000	0.00000

Table 4 *Critical values for student's t-test*

Degrees of freedom	Sample size	One-sided significance tests			Two-sided significance tests		
		5% level	1% level	0.1% level	5% level	1% level	0.1% level
1	2	6.31	31.82	318.31	12.71	63.66	636.62
2	3	2.92	6.96	22.33	4.30	9.92	31.60
3	4	2.35	4.54	10.21	3.18	5.84	12.92
4	5	2.13	3.75	7.17	2.78	4.60	8.61
5	6	2.02	3.36	5.89	2.57	4.03	6.87
6	7	1.94	3.14	5.21	2.45	3.71	5.96
7	8	1.89	3.00	4.79	2.36	3.50	5.41
8	9	1.86	2.90	4.50	2.31	3.36	5.04
9	10	1.83	2.82	4.30	2.26	3.25	4.78
10	11	1.81	2.76	4.14	2.23	3.17	4.59
11	12	1.80	2.72	4.02	2.20	3.11	4.44
12	13	1.78	2.68	3.93	2.18	3.05	4.32
13	14	1.77	2.65	3.85	2.16	3.01	4.22
14	15	1.76	2.62	3.79	2.14	2.98	4.14
15	16	1.75	2.60	3.73	2.13	2.95	4.07
16	17	1.75	2.58	3.69	2.12	2.92	4.01
17	18	1.74	2.57	3.65	2.11	2.90	3.97
18	19	1.73	2.55	3.61	2.10	2.88	3.92
19	20	1.73	2.54	3.58	2.09	2.86	3.88
20	21	1.72	2.53	3.55	2.09	2.85	3.85
21	22	1.72	2.52	3.53	2.08	2.83	3.82
22	23	1.72	2.51	3.50	2.07	2.82	3.79
23	24	1.71	2.50	3.48	2.07	2.81	3.77
24	25	1.71	2.49	3.47	2.06	2.80	3.75
25	26	1.71	2.49	3.45	2.06	2.79	3.73
26	27	1.71	2.48	3.43	2.06	2.78	3.71
27	28	1.70	2.47	3.42	2.05	2.77	3.69
28	29	1.70	2.47	3.41	2.05	2.76	3.67
29	30	1.70	2.46	3.40	2.05	2.76	3.66
39	40	1.68	2.42	3.31	2.02	2.70	3.55
49	50	1.67	2.40	3.26	2.01	2.68	3.50
99	100	1.66	2.37	3.17	1.99	1.63	3.39
∞	∞	1.64	2.33	3.09	1.96	2.58	3.29

Table 5 Chi-square tables

The table shows critical values for chi-square statistics with various degrees of freedom.

Degrees of freedom	Critical values 5% significance level	1% significance level	0.1% significance level
1	3.84	6.63	10.83
2	5.99	9.21	13.82
3	7.81	11.34	16.27
4	9.49	13.28	18.47
5	11.07	15.09	20.52
6	12.59	16.81	22.46
7	14.07	18.48	24.32
8	15.51	20.09	26.12
9	16.92	21.67	27.88
10	18.31	23.21	29.59
11	19.68	24.72	31.26
12	21.03	26.22	32.91
13	22.36	27.69	34.53
14	23.68	29.14	36.12
15	25.00	30.58	37.70
16	26.30	32.00	39.25
17	27.59	33.41	40.79
18	28.87	34.81	42.31
19	30.14	36.19	43.82
20	31.41	37.57	45.31
21	32.67	38.93	46.80
22	33.92	40.29	48.27
23	35.17	41.64	49.73
24	36.42	42.98	51.18
25	37.65	44.31	52.62
26	38.89	45.64	54.05
27	40.11	46.96	55.48
28	41.34	48.28	56.89
29	42.56	49.59	58.30
30	43.77	50.89	59.70
31	44.99	52.19	61.10
32	46.19	53.49	62.49
33	47.40	54.78	63.87
34	48.60	56.06	65.25
35	49.80	57.34	66.62
36	51.00	58.62	67.99
37	52.19	59.89	69.35
38	53.38	61.16	70.70
39	54.57	62.43	72.05
40	55.76	63.69	73.40
41	56.94	64.95	74.74
42	58.12	66.21	76.08
43	59.30	67.46	77.42
44	60.48	68.71	78.75
45	61.66	69.96	80.08
46	62.83	71.20	81.40
47	64.00	72.44	82.72
48	65.17	73.68	84.04
49	66.34	74.92	85.35
50	67.50	76.15	86.66

Table 6 Correlation coefficient

The table shows critical values for the correlation coefficient, for various numbers of observations (where by 'observations' we mean a pair of figures, one x-value and one y-value).

Number of observations	One-sided significance tests			Two-sided significance tests		
	5% level	1% level	0.1% level	5% level	1% level	0.1% level
3	0.9877	0.9995	1.0000	0.9969	0.9999	1.0000
4	0.9000	0.9800	0.9980	0.9500	0.9900	0.9990
5	0.8054	0.9343	0.9859	0.8783	0.9587	0.9911
6	0.7293	0.8822	0.9633	0.8114	0.9172	0.9741
7	0.6694	0.8329	0.9350	0.7545	0.8745	0.9509
8	0.6215	0.7887	0.9049	0.7067	0.8343	0.9249
9	0.5822	0.7498	0.8751	0.6664	0.7977	0.8983
10	0.5494	0.7155	0.8467	0.6319	0.7646	0.8721
11	0.5214	0.6851	0.8199	0.6021	0.7348	0.8470
12	0.4973	0.6581	0.7950	0.5760	0.7079	0.8233
13	0.4762	0.6339	0.7717	0.5529	0.6835	0.8010
14	0.4575	0.6120	0.7501	0.5324	0.6614	0.7800
15	0.4409	0.5923	0.7301	0.5140	0.6411	0.7604
16	0.4259	0.5742	0.7114	0.4973	0.6226	0.7419
17	0.4124	0.5577	0.6940	0.4821	0.6055	0.7247
18	0.4000	0.5425	0.6777	0.4683	0.5897	0.7084
19	0.3887	0.5285	0.6624	0.4555	0.5751	0.6932
20	0.3783	0.5155	0.6481	0.4438	0.5614	0.6788
21	0.3687	0.5034	0.6346	0.4329	0.5487	0.6652
22	0.3598	0.4921	0.6219	0.4227	0.5368	0.6524
23	0.3515	0.4815	0.6099	0.4132	0.5256	0.6402
24	0.3438	0.4716	0.5986	0.4044	0.5151	0.6287
25	0.3365	0.4622	0.5878	0.3961	0.5052	0.6178
26	0.3297	0.4534	0.5776	0.3882	0.4958	0.6074
27	0.3233	0.4451	0.5679	0.3809	0.4869	0.5974
28	0.3172	0.4372	0.5587	0.3739	0.4785	0.5880
29	0.3115	0.4297	0.5499	0.3673	0.4705	0.5790
30	0.3061	0.4226	0.5415	0.3610	0.4629	0.5703
40	0.2638	0.3665	0.4741	0.3120	0.4026	0.5007
50	0.2353	0.3281	0.4267	0.2787	0.3610	0.4514
60	0.2144	0.2997	0.3911	0.2542	0.3301	0.4143
70	0.1982	0.2776	0.3632	0.2352	0.3060	0.3850
80	0.1852	0.2597	0.3405	0.2199	0.2864	0.3611
90	0.1745	0.2449	0.3215	0.2072	0.2702	0.3412
100	0.1654	0.2324	0.3054	0.1966	0.2565	0.3242
200	0.1166	0.1644	0.2173	0.1388	0.1818	0.2310
300	0.0951	0.1343	0.1777	0.1133	0.1485	0.1891
400	0.0824	0.1163	0.1541	0.0981	0.1287	0.1639
500	0.0736	0.1040	0.1379	0.0877	0.1151	0.1467

Table 7 Present value of £1 after n years at various interest rates

$$\text{Present value} = \frac{1}{(1+r)^n} \quad \text{where } r \text{ is interest rate as a proportion of 1}$$

Year (n)	5%	6%	7%	8%	9%	10%	11%	12%	13%	14%	15%	16%	17%	18%	19%	20%
1	.9524	.9434	.9346	.9259	.9174	.9091	.9009	.8929	.8850	.8772	.8696	.8621	.8547	.8475	.8403	.8333
2	.9070	.8900	.8734	.8573	.8417	.8264	.8116	.7972	.7831	.7695	.7561	.7432	.7305	.7182	.7062	.6944
3	.8638	.8396	.8163	.7938	.7722	.7513	.7312	.7118	.6931	.6750	.6575	.6407	.6244	.6086	.5934	.5787
4	.8227	.7921	.7629	.7350	.7084	.6830	.6587	.6355	.6133	.5921	.5718	.5523	.5337	.5158	.4987	.4823
5	.7835	.7473	.7130	.6806	.6499	.6209	.5935	.5674	.5428	.5194	.4972	.4761	.4561	.4371	.4190	.4019
6	.7462	.7050	.6663	.6302	.5963	.5645	.5346	.5066	.4803	.4556	.4323	.4104	.3898	.3704	.3521	.3349
7	.7107	.6651	.6227	.5835	.5470	.5132	.4817	.4523	.4251	.3996	.3759	.3538	.3332	.3139	.2959	.2791
8	.6768	.6274	.5820	.5403	.5019	.4665	.4339	.4039	.3762	.3506	.3269	.3050	.2848	.2660	.2487	.2326
9	.6446	.5919	.5439	.5002	.4604	.4241	.3909	.3606	.3329	.3075	.2843	.2630	.2434	.2255	.2090	.1938
10	.6139	.5584	.5083	.4632	.4224	.3855	.3522	.3220	.2946	.2697	.2472	.2267	.2080	.1911	.1756	.1615
11	.5847	.5268	.4751	.4289	.3875	.3505	.3173	.2875	.2607	.2366	.2149	.1954	.1778	.1619	.1476	.1346
12	.5568	.4970	.4440	.3971	.3555	.3186	.2858	.2567	.2307	.2076	.1869	.1685	.1520	.1372	.1240	.1122
13	.5303	.4688	.4150	.3677	.3262	.2897	.2575	.2292	.2042	.1821	.1625	.1452	.1299	.1163	.1042	.0935
14	.5051	.4423	.3878	.3405	.2992	.2633	.2320	.2046	.1807	.1597	.1413	.1252	.1110	.0985	.0876	.0779
15	.4810	.4173	.3624	.3152	.2745	.2394	.2090	.1827	.1599	.1401	.1229	.1079	.0949	.0835	.0736	.0649
16	.4581	.3936	.3387	.2919	.2519	.2176	.1883	.1631	.1415	.1229	.1069	.0930	.0811	.0708	.0618	.0541
17	.4363	.3714	.3166	.2703	.2311	.1978	.1696	.1456	.1252	.1078	.0929	.0802	.0693	.0600	.0520	.0451
18	.4155	.3503	.2959	.2502	.2120	.1799	.1528	.1300	.1108	.0946	.0808	.0691	.0592	.0508	.0437	.0376
19	.3957	.3305	.2765	.2317	.1945	.1635	.1377	.1161	.0981	.0829	.0703	.0596	.0506	.0431	.0367	.0313
20	.3769	.3118	.2584	.2145	.1784	.1486	.1240	.1037	.0868	.0728	.0611	.0514	.0433	.0365	.0308	.0261
21	.3589	.2942	.2415	.1987	.1637	.1351	.1117	.0926	.0768	.0638	.0531	.0443	.0370	.0309	.0259	.0217
22	.3418	.2775	.2257	.1839	.1502	.1228	.1007	.0826	.0680	.0560	.0462	.0382	.0316	.0262	.0218	.0181
23	.3256	.2618	.2109	.1703	.1378	.1117	.0907	.0738	.0601	.0491	.0402	.0329	.0270	.0222	.0183	.0151
24	.3101	.2470	.1971	.1577	.1264	.1015	.0817	.0659	.0532	.0431	.0349	.0284	.0231	.0188	.0154	.0126
25	.2953	.2330	.1842	.1460	.1160	.0923	.0736	.0588	.0471	.0378	.0304	.0245	.0197	.0160	.0129	.0105
26	.2812	.2198	.1722	.1352	.1064	.0839	.0663	.0525	.0417	.0331	.0264	.0211	.0169	.0135	.0109	.0087
27	.2678	.2074	.1609	.1252	.0976	.0763	.0597	.0469	.0369	.0291	.0230	.0182	.0144	.0115	.0091	.0073
28	.2551	.1956	.1504	.1159	.0895	.0693	.0538	.0419	.0326	.0255	.0200	.0157	.0123	.0097	.0077	.0061
29	.2429	.1846	.1406	.1073	.0822	.0630	.0485	.0374	.0289	.0224	.0174	.0135	.0105	.0082	.0064	.0051
30	.2314	.1741	.1314	.0994	.0754	.0573	.0437	.0334	.0256	.0196	.0151	.0116	.0090	.0070	.0054	.0042

Table 8 Cumulative present values: present value of £1 at each of the next n years at various interest rates

$$\text{Cumulative present value} = \frac{P}{r}\left(1 - \frac{1}{(1+r)^n}\right) \text{ where } r \text{ is interest rate as a proportion of 1}$$

Years (n)	1%	2%	3%	4%	5%	6%	7%	8%	9%	10%
1	0.9901	0.9804	0.9709	0.9615	0.9524	0.9434	0.9346	0.9259	0.9174	0.9091
2	1.9704	1.9416	1.9135	1.8861	1.8594	1.8334	1.8080	1.7833	1.7591	1.7355
3	2.9410	2.8839	2.8286	2.7751	2.7232	2.6730	2.6243	2.5771	2.5313	2.4869
4	3.9020	3.8077	3.7171	3.6299	3.5460	3.4651	3.3872	3.3121	3.2397	3.1699
5	4.8534	4.7135	4.5797	4.4518	4.3295	4.2124	4.1002	3.9927	3.8897	3.7908
6	5.7955	5.6014	5.4172	5.2421	5.0757	4.9173	4.7665	4.6229	4.4859	4.3553
7	6.7282	6.4720	6.2303	6.0021	5.7864	5.5824	5.3893	5.2064	5.0330	4.8684
8	7.6517	7.3255	7.0197	6.7327	6.4632	6.2098	5.9713	5.7466	5.5348	5.3349
9	8.5660	8.1622	7.7861	7.4353	7.1078	6.8017	6.5152	6.2469	5.9952	5.7590
10	9.4713	8.9826	8.5302	8.1109	7.7217	7.3601	7.0236	6.7101	6.4177	6.1446
11	10.3676	9.7868	9.2526	8.7605	8.3064	7.8869	7.4987	7.1390	6.8052	6.4951
12	11.2551	10.5753	9.9540	9.3851	8.8633	8.3838	7.9427	7.5361	7.1607	6.8137
13	12.1337	11.3484	10.6350	9.9856	9.3936	8.8527	8.3577	7.9038	7.4869	7.1034
14	13.0037	12.1062	11.2961	10.5631	9.8986	9.2950	8.7455	8.2442	7.7862	7.3667
15	13.8651	12.8493	11.9379	11.1184	10.3797	9.7122	9.1079	8.5595	8.0607	7.6061
16	14.7179	13.5777	12.5611	11.6523	10.8378	10.1059	9.4466	8.8514	8.3126	7.8237
17	15.5623	14.2919	13.1661	12.1657	11.2741	10.4773	9.7632	9.1216	8.5436	8.0216
18	16.3983	14.9920	13.7535	12.6593	11.6896	10.8276	10.0591	9.3719	8.7556	8.2014
19	17.2260	15.6785	14.3238	13.1339	12.0853	11.1581	10.3356	9.6036	8.9501	8.3649
20	18.0456	16.3514	14.8775	13.5903	12.4622	11.4699	10.5940	9.8181	9.1285	8.5136
21	18.8570	17.0112	15.4150	14.0292	12.8212	11.7641	10.8355	10.0168	9.2922	8.6487
22	19.6604	17.6580	15.9369	14.4511	13.1630	12.0416	11.0612	10.2007	9.4424	8.7715
23	20.4558	18.2922	16.4436	14.8568	13.4886	12.3034	11.2722	10.3711	9.5802	8.8832
24	21.2434	18.9139	16.9355	15.2470	13.7986	12.5504	11.4693	10.5288	9.7066	8.9847
25	22.0232	19.5235	17.4131	15.6221	14.0939	12.7834	11.6536	10.6748	9.8226	9.0770
26	22.7952	20.1210	17.8768	15.9828	14.3752	13.0032	11.8258	10.8100	9.9290	9.1609
27	23.5596	20.7069	18.3270	16.3296	14.6430	13.2105	11.9867	10.9352	10.0266	9.2372
28	24.3164	21.2813	18.7641	16.6631	14.8981	13.4062	12.1371	11.0511	10.1161	9.3066
29	25.0658	21.8444	19.1885	16.9837	15.1411	13.5907	12.2777	11.1584	10.1983	9.3696
30	25.8077	22.3965	19.6004	17.2920	15.3725	13.7648	12.4090	11.2578	10.2737	9.4269

Statistical tables

Years (n)	Interest rate										
	11%	12%	13%	14%	15%	16%	17%	18%	19%	20%
1	0.9009	0.8929	0.8850	0.8772	0.8696	0.8621	0.8547	0.8475	0.8403	0.8333
2	1.7125	1.6901	1.6681	1.6467	1.6257	1.6052	1.5852	1.5656	1.5465	1.5278
3	2.4437	2.4018	2.3612	2.3216	2.2832	2.2459	2.2096	2.1743	2.1399	2.1065
4	3.1024	3.0373	2.9745	2.9137	2.8550	2.7982	2.7432	2.6901	2.6386	2.5887
5	3.6959	3.6048	3.5172	3.4331	3.3522	3.2743	3.1993	3.1272	3.0576	2.9906
6	4.2305	4.1114	3.9975	3.8887	3.7845	3.6847	3.5892	3.4976	3.4098	3.3255
7	4.7122	4.5638	4.4226	4.2883	4.1604	4.0386	3.9224	3.8115	3.7057	3.6046
8	5.1461	4.9676	4.7988	4.6389	4.4873	4.3436	4.2072	4.0776	3.9544	3.8372
9	5.5370	5.3282	5.1317	4.9464	4.7716	4.6065	4.4506	4.3030	4.1633	4.0310
10	5.8892	5.6502	5.4262	5.2161	5.0188	4.8332	4.6586	4.4941	4.3389	4.1925
11	6.2065	5.9377	5.6869	5.4527	5.2337	5.0286	4.8364	4.6560	4.4865	4.3271
12	6.4924	6.1944	5.9176	5.6603	5.4206	5.1971	4.9884	4.7932	4.6105	4.4392
13	6.7499	6.4235	6.1218	5.8424	5.5831	5.3423	5.1183	4.9095	4.7147	4.5327
14	6.9819	6.6282	6.3025	6.0021	5.7245	5.4675	5.2293	5.0081	4.8023	4.6106
15	7.1909	6.8109	6.4624	6.1422	5.8474	5.5755	5.3242	5.0916	4.8759	4.6755
16	7.3792	6.9740	6.6039	6.2651	5.9542	5.6685	5.4053	5.1624	4.9377	4.7296
17	7.5488	7.1196	6.7291	6.3729	6.0472	5.7487	5.4746	5.2223	4.9897	4.7746
18	7.7016	7.2497	6.8399	6.4674	6.1280	5.8178	5.5339	5.2732	5.0333	4.8122
19	7.8393	7.3658	6.9380	6.5504	6.1982	5.8775	5.5845	5.3162	5.0700	4.8435
20	7.9633	7.4694	7.0248	6.6231	6.2593	5.9288	5.6278	5.3527	5.1009	4.8696
21	8.0751	7.5620	7.1016	6.6870	6.3125	5.9731	5.6648	5.3837	5.1268	4.8913
22	8.1757	7.6446	7.1695	6.7429	6.3587	6.0113	5.6964	5.4099	5.1486	4.9094
23	8.2664	7.7184	7.2297	6.7921	6.3988	6.0442	5.7234	5.4321	5.1668	4.9245
24	8.3481	7.7843	7.2829	6.8351	6.4338	6.0726	5.7465	5.4509	5.1822	4.9371
25	8.4217	7.8431	7.3300	6.8729	6.4641	6.0971	5.7662	5.4669	5.1951	4.9476
26	8.4881	7.8957	7.3717	6.9061	6.4906	6.1182	5.7831	5.4804	5.2060	4.9563
27	8.5478	7.9426	7.4086	6.9352	6.5135	6.1364	5.7975	5.4919	5.2151	4.9636
28	8.6016	7.9844	7.4412	6.9607	6.5335	6.1520	5.8099	5.5016	5.2228	4.9697
29	8.6501	8.0218	7.4701	6.9830	6.5509	6.1656	5.8204	5.5098	5.2292	4.9747
30	8.6938	8.0552	7.4957	7.0027	6.5660	6.1772	5.8294	5.5168	5.2347	4.9789

Appendix B
List of Formulae

* means may be omitted without loss of continuity
** means not actually included in this book, but sometimes used in the context indicated.

Errors (Chapter 2)

Absolute error$(A+B)$ = Absolute error(A) + Absolute error(B)
Absolute error$(A-B)$ = Absolute error(A) − Absolute error(B)
Relative error$(A \times B)$ ≏ Relative error(A) + Relative error(B)
Relative error$(A \div B)$ ≏ Relative error(A) − Relative error(B)

Averages and dispersion (Chapters 9 and 10)

Arithmetic mean = $\dfrac{\text{sum of all the figures in the group}}{\text{number of figures in the group}} = \dfrac{\Sigma x}{n}$

or $\quad \dfrac{\Sigma fx}{n}$ for ungrouped frequencies

or $\quad \dfrac{\Sigma fm}{n}$ for grouped frequencies

* or $\quad A + \dfrac{1}{n}\Sigma f(m-A)$ for large numbers

Median = $\dfrac{n-1}{2}$ observation

= $\begin{pmatrix}\text{lower boundary}\\ \text{of median class}\end{pmatrix} + \left[\dfrac{\left(\frac{n}{2}\right) - \begin{pmatrix}\text{cumulative frequency of}\\ \text{class before median class}\end{pmatrix}}{(\text{frequency } f \text{ in median class})} \times \begin{pmatrix}\text{class interval}\\ \text{of median class}\end{pmatrix}\right]$

Estimate of mode

$$= \begin{pmatrix}\text{lower boundary}\\ \text{of modal class}\end{pmatrix} + \left[\frac{\begin{pmatrix}(f/CI)\text{ for}\\ \text{modal class}\end{pmatrix} - \begin{pmatrix}(f/CI)\text{ for}\\ \text{preceding class}\end{pmatrix}}{2 \times \begin{pmatrix}(f/CI)\text{ for}\\ \text{modal class}\end{pmatrix} - \begin{pmatrix}\text{sum of }(f/CI)\text{s for}\\ \text{classes either side}\end{pmatrix}} \times \begin{pmatrix}CI\text{ of}\\ \text{modal class}\end{pmatrix}\right]$$

* Geometric mean of n numbers = $\sqrt[n]{\text{all numbers multiplied together}}$

* Harmonic mean of n numbers = $\dfrac{1}{\dfrac{1}{n}\Sigma\dfrac{1}{x}}$

* Pearson's coefficient of skewness = $\dfrac{3\,(\text{mean} - \text{median})}{\text{standard deviation}}$

Mean deviation = average over all values of deviation from the mean

** $= \dfrac{1}{n}\Sigma\,|x - \bar{x}|$ where | | means that each value is taken as positive irrespective of its sign

** $= \dfrac{1}{n}\Sigma f|x - \bar{x}|$ or $\dfrac{1}{n}\Sigma f|m - \bar{x}|$ for ungrouped and grouped frequency tables

Variance = Average over all values of square of deviation of value from the mean

$= \dfrac{1}{n}\Sigma(x - \bar{x})^2$ (for individual observations)

Standard deviation = $\sqrt{\text{variance}}$

$= \sqrt{\dfrac{1}{n}\Sigma(x - \bar{x})^2}$ or $\sqrt{\dfrac{1}{n}\Sigma x^2 - \bar{x}^2}$ for individuals

or $\sqrt{\Sigma f(x - \bar{x})^2}$ or $\sqrt{\Sigma f x^2 - \bar{x}^2}$ for ungrouped frequency tables

or $\sqrt{\Sigma f(m - \bar{x})^2}$ or $\sqrt{\Sigma f m^2 - \bar{x}^2}$ for grouped frequency tables

*or $\sqrt{\Sigma f d^2 - \left(\dfrac{1}{n}\Sigma f d\right)^2}$ for large numbers

Coefficient of dispersion = $\dfrac{\text{standard deviation}}{\text{arithmetic mean}} \times 100$ per cent

$$\text{Quartile deviation} = \frac{Q_3 - Q_1}{2}$$

$$\text{Quartile coefficient of dispersion} = \frac{Q_3 - Q_1}{Q_3 + Q_1}$$

Index numbers (Chapter 11)

Laspeyres price index

$$= \frac{\text{total for all items of} \left[\text{price} \binom{\text{in year for which}}{\text{index is required}} \times \text{quantity (in base year)} \right]}{\text{total for all items of} \left[\text{price (in base year)} \times \text{quantity (in base year)} \right]} \times 100$$

Paasche price index

$$= \frac{\text{total for all items of} \left[\text{price} \binom{\text{in year for which}}{\text{index is required}} \times \text{quantity} \binom{\text{in year for which}}{\text{index is required}} \right]}{\text{total for all items of} \left[\text{price (in base year)} \times \text{quantity} \binom{\text{in year for which}}{\text{index is required}} \right]} \times 100$$

Price index using weights and price-relatives

$$(a) = \frac{\text{total for all items of} \left[\text{price-relative} \binom{\text{in year for which}}{\text{index is required}} \times \text{weight for item} \right]}{\text{total for all items of} \left[\text{weight for item} \right]}$$

or $(b) = \dfrac{\Sigma R_i W}{\Sigma W}$ where R_i is price-relative in year for which index is required
W is weight for each item

Laspeyres quantity index

$$= \frac{\text{total for all items of} \left[\text{price (in base year)} \times \text{quantity} \binom{\text{in year for which}}{\text{index is required}} \right]}{\text{total for all items of} \left[\text{price (in base year)} \times \text{quantity (in base year)} \right]} \times 100$$

Paasche quantity index

$$= \frac{\text{total for all items of} \left[\text{price} \binom{\text{in year for which}}{\text{index is required}} \times \text{quantity} \binom{\text{in year for which}}{\text{index is required}} \right]}{\text{total for all items of} \left[\text{price} \binom{\text{in year for which}}{\text{index is required}} \times \text{quantity (in base year)} \right]} \times 100$$

Probability and distributions (Chapters 12, 13, 14 and 15)

Complementary events

$$Pr(A \text{ occurs}) + Pr(A \text{ does not occur}) = 1$$

Conditional probability

$Pr(B \text{ given } A)$ is the probability that B will occur given that A has occurred

List of formulae

* **Addition law**

$$Pr(A \text{ or } B \text{ or both}) = Pr(A) + Pr(B) - Pr(\text{both})$$

For *mutually exclusive* events this simplifies and extends to:

$$Pr(A \text{ or } B \text{ or } C) = Pr(A) + Pr(B) + Pr(C)$$

* **Multiplication law**

$$Pr(A \text{ and } B) = Pr(A) \times Pr(B \text{ given } A)$$

For *independent events* this simplifies and extends to:

$$Pr(A \text{ and } B \text{ and } C) = Pr(A) \times Pr(B) \times Pr(C)$$

* Binomial probability $= Pr(r = x) = \dfrac{n!}{r!(n-r)!} p^r (1-p)^{n-r}$

* Poisson probability $= Pr(R = x) = \dfrac{e^{-m} m^x}{x!}$

Normal distribution z-value $= \dfrac{X - \text{mean}}{\text{standard deviation}}$

$$= \dfrac{X-\mu}{\text{s.d.}(x)} \quad \text{or} \quad \dfrac{X-\mu}{\sigma} \quad \text{for distributions of individual observations}$$

$$= \dfrac{X-\mu}{\text{s.d.}(\bar{x})} = \dfrac{X-\mu}{\text{s.e.}(\bar{x})}$$

$$= \dfrac{X-\mu}{(\text{s.d.}(x)/\sqrt{n})} \quad \text{or} \quad \dfrac{X-\mu}{(\sigma/\sqrt{n})} \Bigg\} \text{ for distributions of sampling means}$$

Standard error of the mean $= \text{s.e.}(\bar{x}) = \text{s.d.}(\bar{x}) = \dfrac{\text{s.d.}(x)}{\sqrt{n}}$

Significance tests and confidence intervals (Chapters 16, 17, 18)

Confidence limits for a population mean are given by $\bar{x} \pm z \left(\dfrac{\text{s.d.}(x)}{\sqrt{n}} \right)$

Confidence limits for a population proportion are given by $p \pm z \sqrt{\dfrac{p(1-p)}{n}}$

t-test value $\quad t = \dfrac{\bar{x}-\mu}{\sqrt{\dfrac{\dfrac{1}{n-1}\Sigma(x-\bar{x})^2}{n}}} \quad$ or $\quad \dfrac{\bar{x}-\mu}{(s/\sqrt{n})}$

Difference of two means test statistic

$$z = \dfrac{(\bar{x}\text{ for sample 1})-(\bar{x}\text{ for sample 2})}{\sqrt{\dfrac{\text{s.d.(sample 1)}^2}{(\text{size of sample 1})} + \dfrac{\text{s.d.(sample 2)}^2}{(\text{size of sample 2})}}} \quad \text{or} \quad \dfrac{\bar{x}_1-\bar{x}_2}{\sqrt{\dfrac{s_1^2}{n_1}+\dfrac{s_2^2}{n_2}}}$$

Proportions test statistic

$$z = \dfrac{P-P}{\sqrt{\dfrac{P(1-P)}{n}}}$$

Difference between two proportions test statistic

$$z = \dfrac{p_1-p_2}{\sqrt{\left(\dfrac{n_1p_1+n_2p_2}{n_1+n_2}\right)\left(1-\dfrac{n_1p_1+n_2p_2}{n_1+n_2}\right)\left(\dfrac{1}{n_1}+\dfrac{1}{n_2}\right)}}$$

which is sometimes approximated by

$$\dfrac{p_1-p_2}{\sqrt{\dfrac{p_1(1-p_1)}{n_1}+\dfrac{p_2(1-p_2)}{n_2}}}$$

Chi-squared test statistic

$$\chi^2 = \Sigma \dfrac{(O-E)^2}{E}$$

Regression and correlation (Chapters 19 and 20)

Regression of y on x

$$\text{Slope } b = \dfrac{n\Sigma xy - \Sigma x \Sigma y}{n\Sigma x^2 - (\Sigma x)^2} \quad \text{or} \quad \dfrac{n\Sigma XY - \Sigma X \Sigma Y}{n\Sigma X^2 - (\Sigma X)^2}$$

$$\text{Intercept } a = \bar{y} - b\bar{x}$$

Correlation coefficient

$$r = \frac{n\Sigma xy - \Sigma x \Sigma y}{\sqrt{[n\Sigma x^2 - (\Sigma x)^2][n\Sigma y^2 - (\Sigma y)^2]}}$$

$$\text{or} \quad = \frac{n\Sigma XY - \Sigma X \Sigma Y}{\sqrt{[n\Sigma X^2 - (\Sigma X)^2][n\Sigma Y^2 - (\Sigma Y)^2]}}$$

Coefficient of determination $= r^2$

** Rank correlation coefficient

This may be found using the rankings instead of actual numerical values in the ordinary formula. Alternatively:

$$\text{Coefficient} = 1 - \frac{6\Sigma D^2}{n(n^2-1)} \quad \text{where } D \text{ is the difference between each pair of ranks}$$

Series and interest (Chapters 22 and 23)

Compound interest after n years $= P(1+r)^n$

$$\text{Present value} = \frac{X}{(1+r)^n}$$

** Arithmetic progression

kth term $= a + (n-1)d \quad$ where $a =$ starting value
$\qquad\qquad\qquad\qquad\qquad\qquad d =$ difference

Sum of n terms $= \dfrac{n}{2}(2a + (n-1)d)$

** Geometric progression

kth term $= a \times R^{n-1} \quad$ where $a =$ starting value
$\qquad\qquad\qquad\qquad\qquad R =$ ratio of successive values

** Sum after n periods with P set aside at the end of each at interest r

$$= \frac{P}{r}[(1+r)^n - 1]$$

** Sum after n periods with P set aside at the beginning of each

$$= \frac{P}{r}[(1+r)^n - 1] \times (1+r)$$

** Present value of series of payments each of P at the end of each of the next n years (discounted at interest r)

$$= -\frac{P}{r}\left[1 - \frac{1}{(1+r)^n}\right]$$

Index

Absolute error, 9
Accuracy, 9, 12–15, 53
Alternative hypotheses, 251, 272–273
Approximation, *see* Errors
Arithmetic mean, 111–116
   grouped frequency tables, 114–115
   for individual data, 111–112
   large numbers, 115–116
   ungrouped frequency tables, 113
Arithmetic progression, 329
Assumed mean, 143
Attributes, 6
Attribute sampling, 267
Average rate of return, 343
Averages, 111–124
   arithmetic mean, 111–116
   choosing which average, 127–129
   geometric mean, 123
   harmonic mean, 123–124
   median, 116–120
   mode, 120–123

Band curve charts, 83–84
Bar charts, 68–71
   component, 70–71
   multiple, 70
   percentage component, 71
Base year, 151
Bias, 28–29, 222–265
   sources of, 35–43
Bimodal, 125–126
Binomial distribution, 197–203
   definition of, 197
   formula for, 198
   use of tables, 201–203
Breaks in axes, 79

Calculators, 4–6
Causality and correlation, 305–306
Central limit theorem, 236
Centred moving averages, 322
Chain based indices, 157–159
Charts, 64–75
   band curve, 83–84
   bar, 68–71
   pictograms, 66–68
   pie, 71–74
   Z-charts, 90–92
Chi-squared tests, 280–282
Class –
   choice of boundaries, 57–60
   class intervals, 56
   class limits, 56
Coefficients –
   correlation (product moment), 297–301
   determination, 300
   quartile coefficient of dispersion, 145
   rank correlation, 306–307
   skewness, 135
   variation, 140–142
Collection of data, 27–44
Compensating errors, 12
Component bar charts, 70–71
Compound interest, 330–331
Conditional probability, 182–183
Confidence intervals, 265–268
   for means, 265–267
   for proportions, 267–268
Contingency tables, 280–282
Continuous numbers, 7
Continuous probability distributions, 208–211
Continuous variables, 7, 59–60

372–Index

Correlation coefficient, product moment, 297–301
  and causality, 305–306
  interpretation, 300
  and linearity, 303–305
  negative, 297–298
  positive, 297–298
  significance tests on, 301–303
Correlation coefficient, rank, 306–307
Cost of living, 164
Counts, 6
Critical values, 256, 272–280
Cumulative errors, 12
Cumulative frequency curves, 104–106
Cumulative present values, 335–336

Data –
  collection of, 27–44
  primary and secondary, 18–20, 27
Degrees of freedom, 276, 282
De-seasonalizing, 52, 321
Diagrams –
  bar charts, 68–71
  pictograms, 66–68
  pie charts, 71–74
Discrete numbers 7, 56–59
Dispersion, 134–145
  mean deviation, 135–136
  quartile coefficient of dispersion, 144
  range, 134–135
  semi-interquartile range, 144
  standard deviation, 137–144
  variance, 137
Distributions –
  binomial, 197–203
  continuous, 208–211
  frequency, *see* Frequency tables
  normal, 212–222
  probability, 188–190
  of r (correlation coefficient), 301–303
  of $\bar{x}$ (sampling mean), 232–241
  of χ^2 (chi-square), 280–282

Errors, 9–15, 28–29, 262–265
  absolute, 9
  and bias, 28–29, 262–263
  compensating, 12
  cumulative, 12
  laws of, 10–12
  and regression, 289
  relative, 9
  sources of, 35–43
  standard error, 236–239
Estimation, 262–269

Events –
  definition, 179
  independent, 187
  mutually exclusive, 184–185
Expected values, 188–189

Factorial numbers, 200
Family Expenditure Survey, 162–163
Financial agreements, 334–336
Forecasting –
  using moving averages, 320–321
  by regression, 291–292
Frequency –
  definition, 6, 56
  density, 97–98
  polygons, 103–104
  tables, 54

Gallup Poll, 18, 35
General Index of Retail Prices, 160–165
Geometric –
  mean, 123
  progression, 330
Gini coefficients, 107
Graphs –
  band curve charts, 83–84
  layer, 83
  ratio scale graphs, 84–90
  scattergraphs, 77–79, 285–286
  simple time graphs, 80–83

Harmonic mean, 123–124
Histograms, 94–103
  with unequal class intervals, 96–98
Independent events, 187
Index numbers, 148–173
  General Index of Retail Prices, 160–165
  index of average earnings, 164–165
  Index of Industrial Production, 169–170
  interpreting index numbers, 155–157
  Laspeyres price index, 151–152
  Laspeyres quantity index, 166
  linking and chaining, 157–158
  Paasche price index, 152–153
  Paasche quantity index, 166–167
  weighting price relatives, 154–155
Inflation, 52, 331
Interest, 320–336
Internal rate of return, 336, 342–343
Interviewers, 42–43
Investment appraisal criteria, 341–346
  average rate of return, 343–344
  internal rate of return, 342–343
  present value, 341–342
  payback, 343

Index—373

Kinds of numbers —
  attributes, 6
  counts, 6
  discrete, 7, 56–59
  continuous, 7
  measurements, 6

Laspeyres price index, 151–152
Laspeyres quantity index, 166
Layer graphs, 83
Least squares regression, 289–294
Linking, 157–159
Lorenz curves, 106–107

Mean —
  arithmetic, 111–116
  deviation, 135–136
  geometric, 123
  harmonic, 123–124
Measurements, 6
Median, 116–120
  grouped data, 118–120
  ungrouped data, 116–117
Mode, 120–123
Moving averages, 315–320
  centred, 322
Multiple bar charts, 70
Multiplication law of probability, 185–188
Mutually exclusive events, 184–185

Net present value, 332, 341–342
Non-response, 36–38, 42
Normal distribution, 212–222, 236
  application to sampling means, 236
  tables, 220–222
Null hypotheses, 251, 272–273
Numbers, see Kinds of numbers
Open-ended class intervals, 59, 98–99, 115

Paasche price index, 152–153
Paasche quantity index, 166–167
Panels, 31
Payback criterion, 343
Percentage component bar charts, 71
Percentage frequency densities, 97–100
Pictograms, 66–68
Pie charts, 71–74
Pilot survey, 43
Poisson distribution, 203–205
  formula for, 203
  tables, 204–205
Precision, 28–29, 262–265
Prediction —
  using moving averages, 320
  using regression, 291–292
Present values, 332–334
Price index —
  Laspeyres, 151–152
  Paasche, 152–153
Price relatives, 149
  weighting of, 154
Primary data, 18, 27
Probability —
  addition law of, 183–185
  conditional probability, 182
  continuous probability distributions, 208–211
  definitions, 178–180
  distributions, 188–190
  multiplication law of, 185–188
  probability density, 211
Product moment correlation coefficient, 297–301
Published sources, see Sources
Purchasing power, 156–157

Quantity indices —
  Laspeyres, 166
  Paasche, 166–167
Quartiles, 119–120, 144–145
  coefficient of quartile dispersion, 145
  semi-interquartile range, 144
Questionnaire design, 38–41
Quota sampling, 32–36, 268

Random sampling, 30–31
  multistage, 31
  simple, 31
  stratified, 31
Range, 134–135
Rank correlation, 306–307
Ratio-scale graph, 84–90
Regression, 285–294
  direction of prediction, 291–292
  by eye, 286
  least squares, 289–294
  three-point method, 287–289
Relative errors, 9
Reliability of estimates, 27–28
Rounding, 7–9

Sampling —
  attribute, 267
  fraction, 30
  frame, 30, 36
  mean, distribution of, 232–241
  mean, standard error of, 236–238
  quota, 32–34
  random, 30–31

374–Index

Scattergraphs, 77–79, 285–286
Seasonally adjusted series, *see* Deseasonalizing
Seasonal variations, 313
Secondary data, 18–20
Semi-interquartile range, 144
Semi-logarithmic graph, *see* Ratio-scale graph
Series, 329–330
  arithmetic, 329
  geometric, 330
Significance tests –
  chi-squared, 280–282
  concepts of, 244–248, 272–273
  correlation coefficient, 301–302
  difference of two means, 276–277
  difference of two proportions, 278–279
  one and two-tailed, 253–256
  proportions, 277–278
  sampling mean where s.d. known, 244–256
  sampling mean where s.d. unknown, 273–276
  t-tests, 275–276
Simple time graphs, 80–83
Skewness, 124–125
  Pearson's coefficient of, 125
Social class, 32
Sources of published figures –
  Annual Abstract, 24
  Annual Census of Production, 25, 169
  British Business, 24, 169, 170
  British Labour Statistics Yearbook, 25
  Business Monitors, 21, 22, 24, 25, 168, 169
  Census of Population, 22
  Department of Employment Gazette, 22, 24, 170
  Economic Progress Report, 24
  Economic Trends, 22, 24, 169, 170
  Economist, 25
  Employment Gazette, 22, 24, 120
  Family Expenditure Survey, 21, 25, 162–163
  Financial Statistics, 22, 25
  Financial Times, 25, 165
  General Household Survey, 25
  Household Expenditure Survey, 162
  Monthly Digest of Statistics, 24, 169
  National Food Survey, 21
  National Income and Expenditure (Blue Book), 24
  New Earnings Survey, 24
  Price Indexes for Current Cost Accounting, 24
  Regional Trends, 22, 24
  Social Trends, 22, 24
  Statistical News, 170
  Time Rates of Wages and Hours of Work, 25
Spearman's coefficient of correlation, *see* Rank correlation
Standard deviation, 137–144
  for grouped data, 139–144
  using large numbers, 141–144
  of sampling mean, 236–238
  for ungrouped data, 137–139
Standard error of the mean, 236–238

Tables, 60–62
Tabulation, 49–62
  measurable variables, 54–56
  time series, 52–53
Tasks of statistics, 3
Tests of significance, *see* Significance tests
Time series, 52–53, 313–324
  forecasts, 320–321
  moving averages on, 315–320
  quarterly figures, 321–324
Trend, using moving averages, 315–318
Truncation, 7–9

Uniform distribution, 126
Unimodal distribution, 124

Validity, 27–28
Variable, 6
  continuous, 7, 59–60
  discrete, 7, 56–59
Variance, 135–137

Weighting index numbers, 154–155, 162–163

Z-charts, 90–92
z-values, 215